Meta-Photonic Devices for Polarization Manipulation

Polarized light is vital in advanced manufacturing, medical imaging, communication, and fundamental physics, making polarization manipulation central to photonics. This book examines the theory and design of meta-photonic devices, presenting innovative methods for controlling polarization with precision and efficiency.

Meta-photonics manipulates light at the subwavelength scale, offering multifunctionality, high integration, and scalable fabrication beyond conventional optical devices. This book explores groundbreaking advancements in polarizing optics, including polarization conversion, generation, chiral photonics, vector light fields, and 3D polarization control. It also addresses emerging physics, such as bound states in the continuum (BICs) and non-Hermitian systems, as well as on-chip polarization control and AI-driven inverse design techniques.

Targeted at researchers and engineers in photonics and electromagnetic wave studies, the book provides universal principles applicable across the electromagnetic spectrum, delivering practical insights into the transformative potential of meta-photonic polarization devices.

Jitao Li (李继涛) (PhD) is working at Southwest Petroleum University, China. He is an IEEE Senior Member and has been recognized in Stanford University's World's Top 2% Scientists list since 2023. He is interested in Optoelectronics, Microelectronics, and Microwave-Terahertz technologies.

Meta-Photonic Devices for Polarization Manipulation

Disruption and Revolution

Jitao Li

CRC Press
Taylor & Francis Group
Boca Raton London New York

CRC Press is an imprint of the
Taylor & Francis Group, an **informa** business

This book was supported by the National Natural Science Foundation of China (Grant No. 12504504), and the Sichuan Science and Technology Program (Grant No. 2025ZNSFSC0847).

First edition published 2027
by CRC Press
2385 NW Executive Center Drive, Suite 320, Boca Raton FL 33431

and by CRC Press
4 Park Square, Milton Park, Abingdon, Oxon, OX14 4RN

CRC Press is an imprint of Taylor & Francis Group, LLC

ISBN: 978-1-041-32788-2 (hbk)
ISBN: 978-1-041-32845-2 (pbk)
ISBN: 978-1-003-78482-1 (ebk)

DOI: 10.1201/9781003784821

Typeset in Minion
by KnowledgeWorks Global Ltd.

Contents

Preface

Meta-devices can manipulate various physical properties of light, such as amplitude, phase, polarization, frequency, orbital angular momentum, and spin angular momentum, at the subwavelength scale. I have been researching optoelectronic devices since 2016, most of which are related to polarization manipulation, which is an important topic in the field of meta-photonics. However, the current scientific community lacks a specialized book on polarization manipulation in meta-photonic devices, which hinders many people who want to work in this field. Therefore, this book focuses on meta-photonic polarization devices to help researchers and engineers systematically understand this field in terms of knowledge and technology.

This book has accumulated my years of research achievements. I believe it will serve as a valuable guide for researchers and engineers in the field of meta-photonic polarization devices. This confidence comes from the following points. (i) This book has a rigorous logical arrangement in its chapter content, which can help readers gradually delve into complex research from simple principles. (ii) This book holds both academic and engineering value, as it includes rigorous reasoning processes that help readers understand devices' operating principles. Moreover, it encompasses device design processes to help readers master device design techniques. (iii) This book covers numerous cutting-edge topics, presenting both fundamental physical principles and novel physical phenomena, including polarization generation, chirality control, vector light field, 3D polarization, bound states in the continuum, and non-Hermitian systems. Moreover, it introduces both spatial polarization devices and on-chip polarization devices, as well as conventional forward design methods and emerging inverse design approaches based on artificial intelligence.

I would like to express my gratitude to some collaborators and graduate students for their indirect contributions to this book. Section 6.2 of this book refers to the research findings of my collaborator Prof. Zhen Yue (岳震), who is currently working at Jiangsu University, China. Section 7.2.1 of this book refers to the research findings of my collaborator Prof. Jie Li (李杰), who is currently working at Chengdu University of Information Technology, China. My three graduate students, Can Pan (潘灿), Lvyi Zhong (钟绿怡), and Maoqing Guo (郭茂清), assisted me in completing Chapters 9, 10, and 11, respectively.

Jitao Li (李继涛)

Written on 1 March 2026 at Southwest Petroleum University, China

CHAPTER 1

Introduction

Abstract

Traditional polarization devices are often limited to a single function and suffer from drawbacks such as bulky size (e.g. birefringent crystals) and complex fabrication processes (e.g. thin-film polarizers). In contrast, metadevices can manipulate the polarization state of light at the subwavelength scale, endowing them with capabilities unattainable by traditional devices, including multifunctionality, high integration, and ease of batch fabrication. This chapter will introduce the fundamental principles of metadevices, including the generalized Snell's law and the advanced phase principle. The generalized Snell's law and advanced phase responses are the two cornerstones underpinning the extraordinary light-manipulation capabilities of metadevices. Finally, the overall framework of this book will be introduced to help readers quickly understand the content structure of the book.

1.1 POLARIZATION OF LIGHT

Polarization is one of the transverse wave properties of light, referring to the specific orientation of the electric field vibration direction in space. This characteristic is widely applied in science, technology, and daily life. For example, in imaging and display fields, polarizers are core components of liquid crystal displays, which control the polarization direction to adjust the brightness of each pixel and enable image display. In photography, polarizing filters eliminate reflections from non-metallic surfaces such as water and glass, enhancing colour saturation, and are widely used in landscape photography and scientific research documentation. In the field of optical inspection and material analysis, polarized light microscopes analyse the effects of samples on polarized light to observe the microstructure of mineral crystals, biological tissues, and polymer materials. In industry, polarization detection is employed to analyse the internal stress distribution of transparent materials (such as glass and plastic products), ensuring product quality. In communication and information technology, polarization modulation in fibre-optic communication enhances

DOI: 10.1201/9781003784821-1

channel capacity, while quantum communication relies on photonic polarization states to encode information, achieving highly secure transmission. In astronomy and remote sensing, by analysing the polarization information of celestial radiation, astronomers can infer the direction of interstellar magnetic fields and the properties of circumstellar dust. Remote sensing technology also utilizes polarization characteristics to monitor environmental parameters such as atmospheric aerosols and ocean surface conditions. In the biological and medical fields, certain biological structures (such as the cornea and connective tissues) exhibit birefringence, and polarization imaging can assist in diagnosing lesions. Polarization-sensitive techniques are also applied in medical devices such as endoscopes to enhance tissue contrast. In daily life and consumer electronics, 3D movies use orthogonal polarized light to separate left and right eye images, creating a stereoscopic visual experience with polarized glasses. Car anti-glare rearview mirrors and sunglasses often incorporate polarizing filters to reduce interference from scattered reflections. By manipulating the vibration direction of light, polarization technology not only expands humanity's ability to perceive the material world but also plays an indispensable role in precision science, industrial production, and daily life.

Light waves are electromagnetic waves, and the description of polarization applies to all electromagnetic waves without loss of generality. In a plane perpendicular to the direction of propagation, the electric field of an electromagnetic wave oscillates periodically over time. The form of this oscillation defines the polarization state of the electromagnetic wave, which is one of its inherent properties. By depicting the trajectory of the electric vector over a complete period and fully projecting it onto the *x-y* plane (with the propagation direction along the *z*-axis), the polarization state of the electromagnetic wave can be mathematically expressed. Consider a plane wave propagating in vacuum at the speed of light *c*, with linear frequency *f* and angular frequency $\omega = 2\pi f$. Its electric vector can be expressed as the composition of two components along the *x* and *y* directions:

$$E(z,t) = E_x\hat{x} + E_y\hat{y}, \tag{1.1}$$

where $\hat{x}, \hat{y}$ are the basis vectors along the *x* and *y* directions, with electromagnetic travelling wave E_x and E_y given by:

$$\begin{cases} E_x = a_x \cos[\omega(t - z/c) + \varphi_x] = a_x \cos[\omega t - k_x z + \varphi_x] \\ E_y = a_y \cos[\omega(t - z/c) + \varphi_y] = a_y \cos[\omega t - k_y z + \varphi_y] \end{cases}. \tag{1.2}$$

This equation indicates that both components are periodic functions of $t - z/c$, with initial phases φ_x and φ_y. By projecting the motion of the electric vector entirely onto the *x-y* plane (i.e. ignoring the *z/c* term related to propagation) and using a complex representation based on Euler's formula:

$$\begin{cases} E_x = a_x e^{i\omega t} e^{i\varphi_x} \\ E_y = a_y e^{i\omega t} e^{i\varphi_y} \end{cases}. \tag{1.3}$$

This can be represented by a 2 × 1 matrix, known as the Jones matrix. Let $\varphi = \varphi_y - \varphi_x$, and the Jones matrix of the normalized electric field of an electromagnetic wave can be written as: $E = \frac{1}{\sqrt{a_x^2 + a_y^2}} e^{i\omega t} e^{i\varphi_x} \begin{pmatrix} a_x \\ a_y e^{i\varphi} \end{pmatrix}$. This shows that once the maximum amplitudes of the x and y components and their phase difference are known, the polarization state can be expressed in Jones matrix form. The normalized vectors of several typical linearly polarized waves and circularly polarized waves can be expressed as:

$$H = \begin{pmatrix} 1 \\ 0 \end{pmatrix}, V = \begin{pmatrix} 0 \\ 1 \end{pmatrix}, D = \frac{1}{\sqrt{2}} \begin{pmatrix} 1 \\ 1 \end{pmatrix}, A = \frac{1}{\sqrt{2}} \begin{pmatrix} 1 \\ -1 \end{pmatrix}, R = \frac{1}{\sqrt{2}} \begin{pmatrix} 1 \\ i \end{pmatrix}, L = \frac{1}{\sqrt{2}} \begin{pmatrix} 1 \\ -i \end{pmatrix}. \tag{1.4}$$

Among them, H and V represent x- and y-polarized field vectors, respectively; D and A represent linear polarization vectors with x-axis angles of $\pi/4$ and $-\pi/4$, respectively; and R and L represent right-handed and left-handed circular polarization vectors, respectively. It is easy to obtain the following relationships:

$$R = \frac{H + iV}{\sqrt{2}}, L = \frac{H - iV}{\sqrt{2}}, D = \frac{H + V}{\sqrt{2}}, A = \frac{H - V}{\sqrt{2}}. \tag{1.5}$$

Further, from Eq. (1.2), we get the trajectory equation of polarized light:

$$\frac{E_x^2}{a_x^2} + \frac{E_y^2}{a_y^2} - 2\cos\varphi \frac{E_x E_y}{a_x a_y} = \sin^2\varphi. \tag{1.6}$$

This equation describes an ellipse in the x-y plane. Specifically, when $\varphi = 0$, it represents a linearly polarized state; when $a_x = a_y$ and $\varphi = \pi/2$, it represents a circularly polarized state.

Equation (1.6) cannot describe the orientation of elliptical polarization. As shown in Fig. 1.1(a), an orientation angle Φ and an ellipticity χ can be introduced to represent any polarization state in the x-y coordinate system. The orientation angle Φ is the angle between the x-axis and the major axis of the ellipse, while the ellipticity χ is determined

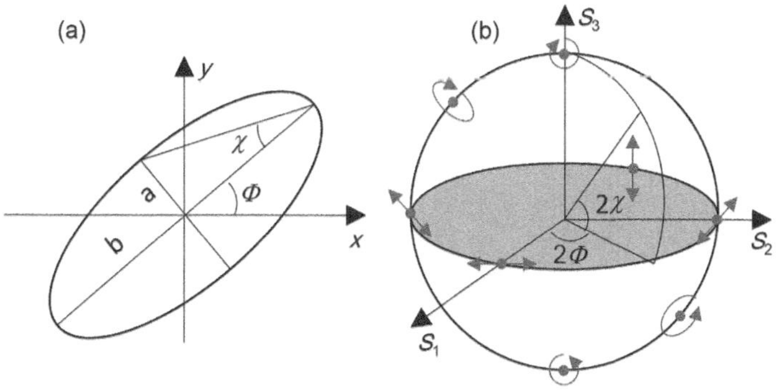

FIGURE 1.1 (a) Representation of the polarization state of an electromagnetic wave in the x-y coordinate system and (b) its representation on the Poincaré sphere.

by the ratio of the minor axis to the major axis: a/b. It is not difficult to see that rotating an elliptical polarization $\begin{pmatrix} \cos\chi \\ \pm i\sin\chi \end{pmatrix}$ that coincides with the *x-y* coordinate by Φ yields an elliptical polarization with an orientation angle Φ, which can be represented by the Jones matrix as:

$$\begin{aligned} \begin{pmatrix} E_x \\ E_x \end{pmatrix} &= R(\Phi)\begin{pmatrix} \cos\chi \\ \pm i\sin\chi \end{pmatrix} = \begin{pmatrix} \cos\Phi & -\sin\Phi \\ \sin\Phi & \cos\Phi \end{pmatrix}\begin{pmatrix} \cos\chi \\ \pm i\sin\chi \end{pmatrix} \\ &= \begin{pmatrix} \cos\Phi\cos\chi \mp i\sin\Phi\sin\chi \\ \sin\Phi\cos\chi \pm i\cos\Phi\sin\chi \end{pmatrix} \end{aligned}, \tag{1.7}$$

where the symbol "±" in the formula is used to distinguish between right-handed and left-handed elliptical polarization states. $R(\Phi)$ is a rotation matrix, and its mathematical form is directly written here. The specific derivation will be introduced in Chapter 3.

However, this representation based on orientation angle and ellipticity does not include intensity information. To address this, the following four parameters can be introduced:

$$\begin{cases} S_0 = a_x^2 + a_y^2 \\ S_1 = S_0\cos 2\chi\cos 2\Phi \\ S_2 = S_0\cos 2\chi\sin 2\Phi \\ S_3 = S_0\sin 2\chi \end{cases}. \tag{1.8}$$

These four parameters S_0, S_1, S_2, and S_3 are known as the Stokes parameters. The three parameters S_1, S_2, and S_3 can be mapped to a Cartesian coordinate system to represent any polarization state including intensity. Such a coordinate system is called the Poincaré sphere. As shown in Fig. 1.1(b), each point on the surface of the Poincaré sphere represents a polarization state. For example, when $\chi = 0$, there is $S_3 = 0$, representing points on the equator of the Poincaré sphere, which correspond to different linearly polarized states. When $2\Phi = 0$ and $2\Phi = \pi$, they represent *x* and *y* polarization, respectively. At the poles of the Poincaré sphere, there are $S_1 = S_2 = 0$ and $2\chi = \pm\pi/2$, representing right-handed circularly polarized (RCP) and left-handed circularly polarized (LCP) waves, respectively. Other positions on the sphere represent elliptically polarized waves. Alternatively, the four Stokes parameters can also be expressed in terms of the amplitudes of the *x* and *y* components and their phase difference φ. Equation (1.8) is equivalent to:

$$\begin{cases} S_0 = I_H + I_V = a_x^2 + a_y^2 \\ S_1 = I_H - I_V = a_x^2 - a_y^2 \\ S_2 = I_D - I_A = 2a_x a_y\cos\varphi \\ S_3 = I_R - I_L = 2a_x a_y\sin\varphi \end{cases}. \tag{1.9}$$

where the subscripts "*H, V, D, A, R*, and *L*" have the same meanings as described in Eq. (1.4), and *I* represents the corresponding light intensity of the subscript.

1.2 GENERALIZED SNELL'S LAW AND PHASE RESPONSES IN META-PHOTONICS

The human visual system cannot detect the polarization information of light; it can only be perceived using polarizing devices. Polarizing devices operate based on the polarization characteristics of light, making this information observable. Traditional polarizing devices primarily utilize the birefringence effect of natural crystals or multilayer dielectric thin films that separate polarized light near the Brewster angle. The inherently small separation angle of natural birefringent crystals dictates that crystal polarizers require significant thickness to adequately separate the two polarization states, resulting in bulky devices that are difficult to miniaturize and integrate. Multilayer film structures, as alternatives, leverage birefringence properties several times greater than those of natural crystals, which substantially reduces thickness. However, fabricating such multilayer structures involves numerous film layers, making the manufacturing process relatively complex. Moreover, the polarization selectivity of these multilayer films typically achieves high extinction ratios only within a limited wavelength and angular ranges. Therefore, there are significant constraints on the application of both types of polarizers.

Given the stringent demands of optical systems for component quality and size, reducing the volume of polarizing devices while maintaining optical performance is an important research direction. Meta-photonic devices, with their unique light-field modulation mechanisms, offer advantages such as compactness, thin profiles, and high functional integration, making them a promising solution for miniaturizing and reducing the weight of polarizers. The fundamental units of meta-photonic devices are periodic structures with dimensions smaller than the operating wavelength – known as subwavelength periodic structures (including photonic crystals and metasurfaces/metamaterials), which enable sophisticated manipulation of light polarization states, such as polarization conversion, polarization multiplexing, and vector beam generation. With the advancement of new principles and technologies in meta-photonics, polarization control is evolving towards higher efficiency, miniaturization, and multifunctionality.

A key characteristic of meta-photonic devices is that their scattering behaviour follows the generalized Snell's law, which challenges traditional understanding of light propagation. The generalized Snell's law was first proposed by Capasso's team [1] in 2011 and serves as a significant extension to the conventional Snell's law.

Snell's Law. As shown in Fig. 1.2(a), for an infinite, flat, and uniform interface, the entire system exhibits translational symmetry along the direction parallel to the interface (i.e. the physical laws remain unchanged when arbitrarily shifting within the interface plane). According to Noether's theorem, such symmetry necessarily implies the existence of a conserved quantity in that direction, which is the tangential momentum. The conservation of tangential momentum is a fundamental premise of Snell's law. In a medium, the momentum of a photon is determined by its material wavelength (i.e. de Broglie wavelength). The wavelength of light in a medium is given by $\lambda_n = \lambda_0/n$, where λ_0 is the vacuum wavelength and n is

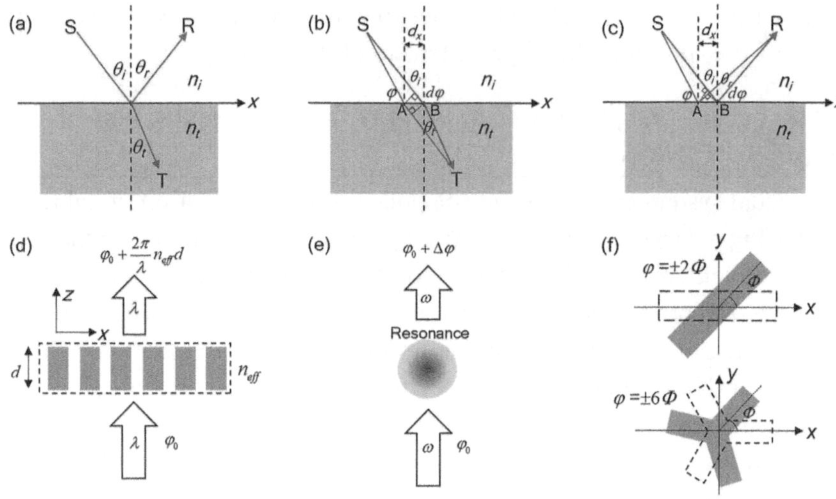

FIGURE 1.2 Schematic diagram of the generalized Snell's law and phase response in metasurface devices. (a) Reflection and refraction laws for an infinite, flat, and uniform interface. (b, c) Refraction and reflection laws for an interface with broken translational symmetry. (d) Acquisition of propagation phase via optical path accumulation. (e) Resonance phase induced by light-excited electromagnetic resonance. (f) Acquisition of geometric phase through structural rotation.

the refractive index of the medium. Thus, the momentum of a photon in the medium is $p = h/\lambda_n = np_0$, where h is Planck's constant and $p_0 = h/\lambda_0$ is the photonic momentum in vacuum. Accordingly, the tangential momentum of the incident photon is $p_i\sin\theta_i = (n_ip_0)\sin\theta_i$, and that of the refracted photon is $p_t\sin\theta_t = (n_tp_0)\sin\theta_t$, where the subscripts "$i$" and "$t$" denote the incident and transmitted media, respectively. Based on the conservation of tangential momentum, there is $p_i\sin\theta_i = p_t\sin\theta_t$, and thus Snell's law is derived as:

$$n_i \sin\theta_i = n_t \sin\theta_t. \tag{1.10}$$

For reflection, it can be treated as a special case where $n_i = n_t$. Substituting this into Eq. (1.10) yields the reflection angle equal to the incident angle, i.e. $\theta_i = \theta_t$.

Generalized Snell's Law. Snell's law applies to flat, infinite, and uniform interfaces. When a periodic, subwavelength structure is introduced on the interface, this translational symmetry is broken. In such cases, the momentum acquired by light at the interface is no longer solely determined by the material refractive index but is also influenced by the geometric structure. This leads to the generalized Snell's law. Consider a carefully designed metasurface capable of introducing a spatially varying and abrupt phase shift $\varphi(x, y)$ to the wavefront as light passes through. Considering a one-dimensional scenario, assume the interface exhibits a varying phase $d\varphi$ along the dx direction, with light propagating in the x-z plane; and considering two infinitely close light paths (e.g. the S→A→T and S→B→T-paths in Fig. 1.2(b)), which irradiate points A(x) and B($x + dx$) on the interface, respectively. When two light paths are extremely close, they are approximately perpendicular to their equiphase planes. It is straightforward to calculate that the phase accumulated along dx by the incident wave (S→B→T path) at the interface is $k_0n_i\sin\theta_i \cdot dx$, the phase discontinuity at point B on the interface is $d\varphi$, and the phase accumulated along dx by the transmitted wave (S→A→T path) at the interface is $k_0 n_t \sin\theta_t \cdot dx$. The total phase of the incident wave, the interface itself, and the transmitted/reflected wave must match at the interface. This matching condition can be derived from Fermat's principle: light always propagates along the path that extremizes the optical path length between two points. This implies that if two infinitely close actual light paths originate from the same point S and converge at the same point T, their phase difference must be zero. To satisfy Fermat's principle, the total phase change of these two light paths after passing through the interface must be equal, leading to:

$$\begin{aligned} & k_0 n_i \sin\theta_i \cdot dx + d\varphi = k_0 n_t \sin\theta_t \cdot dx \\ & \Rightarrow k_0 n_i \sin\theta_i + \frac{d\varphi}{dx} = k_0 n_t \sin\theta_t \\ & \Rightarrow n_t \sin\theta_t - n_i \sin\theta_i = \frac{\lambda_0}{2\pi}\frac{d\varphi}{dx}. \end{aligned} \tag{1.11}$$

This equation is the refraction behaviour described by the generalized Snell's law. Similarly, as shown in Fig. 1.2(c), the reflection behaviour described by the generalized Snell's law can be derived as:

$$\sin\theta_r - \sin\theta_i = \frac{\lambda_0}{2\pi n_i}\frac{d\varphi}{dx}. \tag{1.12}$$

Equations (1.11) and (1.12) indicate that for a given incident angle, if a constant discontinuous phase gradient is introduced along the interface, the refracted and reflected beams can be directed arbitrarily. By utilizing an array of optical resonant structures with subwavelength spacing and varying phase shifts, discontinuous phase changes along the interface can be achieved, enabling control over anomalous refraction and reflection. Furthermore, if $d\varphi/dx = 0$, i.e. no phase discontinuity along dx, the generalized Snell's law transits to the conventional Snell's law. Thus, the conventional Snell's law can be viewed as a special case of the generalized Snell's law.

Rich phase responses are the key factor enabling meta-photonic devices to achieve powerful polarization control capabilities. These phase responses can be classified into three main types: propagation phase, resonance phase, and geometric phase.

Propagation Phase (Dynamic Phase). Similar to traditional optical devices that rely on phase accumulation through optical path length, propagation-phase metadevices achieve phase modulation via the optical path difference generated during electromagnetic wave propagation. As shown in Fig. 1.2(d), the propagation phase is typically related to the incident light wavelength, the effective refractive index of the material, and its thickness. When light of wavelength λ is normally incident on a medium surface with an effective refractive index n_{eff} and thickness d, the accumulated phase delay can be expressed as:

$$\varphi = \frac{2\pi}{\lambda} n_{eff} d. \tag{1.13}$$

Propagation-phase metadevices can utilize high-refractive-index media or modify the size of subwavelength structural units to adjust the effective refractive index of the material, thereby ensuring sufficiently small device thickness. The propagation phase is a very common type of phase response in nature, accumulating phase through the propagation distance of light. As such, it is insensitive to frequency and polarization. Furthermore, since the propagation of light is inherently a dynamic process related to time, the propagation phase is also referred to as the dynamic phase in the field of meta-photonics.

Resonance Phase. Resonance phase emerged early in the field of metasurface optics and was commonly used in early metal/plasmonic metamaterials. When Capasso's team [1] proposed the generalized Snell's law in 2011, the resonance phase served as an example. They designed a V-shaped gold nanoantenna with dual resonance characteristics, enabling complete control over the optical wavefront. These subwavelength metal antenna structures can be regarded as optical resonators. When the wavelength of incident light matches the resonance wavelength of the structure, displacement currents are induced within the resonant structure, generating strongly scattered electromagnetic fields. By altering the opening angle of the V-shaped antenna to detune it from the resonance peak, the phase shift is correspondingly modified. Thus, phase control can be achieved by adjusting the geometry and size of the resonant structural units within the nanostructure array, thereby altering the resonance wavelength.

As shown in Fig. 1.2(e), when light illuminates metallic or dielectric microstructures, electromagnetic resonance can be excited, accompanied by significant phase variations. In metallic microstructures, localized surface plasmon resonance is primarily excited, while in dielectric microstructures, Mie resonance is dominant. The resonance phase requires a specific matching condition between the light wavelength and the microstructure size, making it wavelength-sensitive and limiting its operational bandwidth. By modifying parameters such as structural size and shape, both the resonance frequency and phase can be tuned.

Geometric Phase. The geometric phase, first proposed by the British physicist Berry in 1984 [2], refers to the accumulated change in the phase of a quantum system's state during cyclic evolution in parameter space. It arises from the interaction between the photonic spin angular momentum (SAM) and orbital angular momentum (OAM). Berry considered a quantum system whose Hamiltonian depends on certain parameters. When these parameters are slowly and continuously varied and eventually returned to their initial values, the system's wavefunction acquires an additional phase. The geometric phase is a topological phase whose value depends solely on the geometric path of the system's evolution in parameter space. Suppose a unit cell of a metasurface is rotated by an angle Φ relative to the x-axis, as shown in Fig. 1.2(f). In this case, the Jones matrix $\boldsymbol{T}$ of the unit cell can be expressed as:

$$\begin{aligned}
\boldsymbol{T} &= \mathbf{R}(\Phi)\boldsymbol{T}_0\mathbf{R}^{-1}(\Phi) \\
&= \begin{pmatrix} \cos\Phi & -\sin\Phi \\ \sin\Phi & \cos\Phi \end{pmatrix} \begin{pmatrix} t_{xx} & 0 \\ 0 & t_{yy} \end{pmatrix} \begin{pmatrix} \cos\Phi & \sin\Phi \\ -\sin\Phi & \cos\Phi \end{pmatrix}, \\
&= \begin{pmatrix} \cos^2\Phi\cdot t_{xx} + \sin^2\Phi\cdot t_{yy} & \cos\Phi\sin\Phi\cdot(t_{xx}-t_{yy}) \\ \cos\Phi\sin\Phi\cdot(t_{xx}-t_{yy}) & \sin^2\Phi\cdot t_{xx} + \cos^2\Phi\cdot t_{yy} \end{pmatrix}
\end{aligned} \tag{1.14}$$

where $\boldsymbol{R}(\Phi)$ is the rotation matrix, $\boldsymbol{T}_0$ is the Jones matrix before rotation, and t_{xx} and t_{yy} represent the co-polarized transmission coefficients for the *x*- and *y*-polarized components, respectively. The mathematical derivation of $\boldsymbol{R}(\Phi)$ and $\boldsymbol{T}$ is briefly presented here, with a more detailed discussion provided in Chapter 3. Thus, when the incident light field is circularly polarized, represented as $[1, -i\sigma]^T$, the output light field can be expressed as:

$$\boldsymbol{E}_o = \boldsymbol{T}\begin{pmatrix} 1 \\ -i\sigma \end{pmatrix} = \frac{t_{xx}+t_{yy}}{2}\begin{pmatrix} 1 \\ -i\sigma \end{pmatrix} + \frac{t_{xx}-t_{yy}}{2}e^{-2i\sigma\Phi}\begin{pmatrix} 1 \\ i\sigma \end{pmatrix}, \tag{1.15}$$

where σ takes the value −1 or 1, corresponding to RCP and LCP waves, respectively. The output light field contains both a co-polarized circularly polarized component and a cross-polarized circularly polarized component carrying an additional geometric phase of $-2\sigma\Phi$. The geometric phase is twice the rotation angle Φ of the principal axis. Consequently, the geometric phase depends only on the polarization of the incident light and the rotation of the structure, independent of the incident wavelength and material properties. By varying the rotation angle of the unit structures, phase modulation can be achieved. This type of phase is also known as the Pancharatnam-Berry (PB) phase. The classical PB phase shift has a uniquely determined relationship with the rotation angle of the unit cell. The PB phase is symmetric for orthogonal circular polarizations, meaning that the geometric phases for LCP and RCP have the same magnitude but opposite signs. In recent years, the PB phase has been further extended to high-order PB phases. The high-order geometric phase considers the rotational symmetry and lattice coupling effects of the structure, allowing the phase shift to not be limited to a double relationship with the structural rotation angle [3]. More specifically, Eq. (1.15) only applies to the C_2 symmetry case in a tetragonal lattice. If other symmetry cases in a tetragonal lattice are considered, the more general form of geometric phase expression is:

$$\varphi = \begin{cases} \pm 2n\Phi, n \text{ is odd} \\ \pm n\Phi, n \text{ is even} \end{cases}, n \neq 4,8,12,\cdots, \tag{1.16}$$

where n means that the structure has n-fold rotational symmetry. For example, for a structure with C_3 symmetry in a tetragonal lattice, the geometric phase shift is $\pm 6\Phi$, as shown in Fig. 1.2(f). For hexagonal lattices, the general form of the geometric phase is written as:

$$\varphi = \begin{cases} \dfrac{2n}{n-6m}\Phi \\ \dfrac{-4n}{n-6m}\Phi \end{cases}, m = \left[\frac{n}{6}\right], n \neq 3,6,9,\cdots, \tag{1.17}$$

where "$m=[\cdot]$" denotes the nearest integer function. Furthermore, the above discussion is limited to the linear range [4]. Geometric phase can be further extended to nonlinear PB phases that are primarily applied in the study of nonlinear optical metasurfaces.

1.3 THE CHAPTER STRUCTURE OF THIS BOOK

This book consists of 11 chapters. In terms of chapter arrangement, this book presents a clear progressive structure, allowing readers to advance from basic concepts to in-depth knowledge. An overall framework is shown in Fig. 1.3. Chapter 2 provides a comprehensive overview of the development trends in meta-photonics polarizers, which helps readers quickly understand this field. Chapter 3 is based on birefringence-like effects and resonance effects, introducing fundamental polarization conversion to help readers understand the most fundamental principles and design methods of polarization conversion in detail.

After understanding birefringence-like effects, introducing multiple meta-atom interference effects can further help us understand how to obtain polarized waves from unpolarized waves, which is the content of Chapter 4. Chapter 5 further delves into chiral photonic manipulation based on resonance effects. Chapters 3–5 all study scalar light fields, while Chapter 6 begins to explore the principles of vector light fields and device design, where spin-decoupling will be introduced.

Chapters 3–6 focus on polarization manipulation in a 2D plane, without paying attention to polarization changes along the propagation path. Therefore, Chapter 7 investigates longitudinal light field manipulation, i.e. polarization manipulation in 3D space, which requires readers to understand the multiple meta-atom interference techniques proposed in Chapter 4 and the spin-decoupling technology proposed in Chapter 6.

Chapters 3–7 can be summarized as polarization manipulation under the framework of fundamental physics principles, while Chapters 8 and 9 will introduce polarization manipulation enabled by novel physical phenomena, including bound states in the continuum

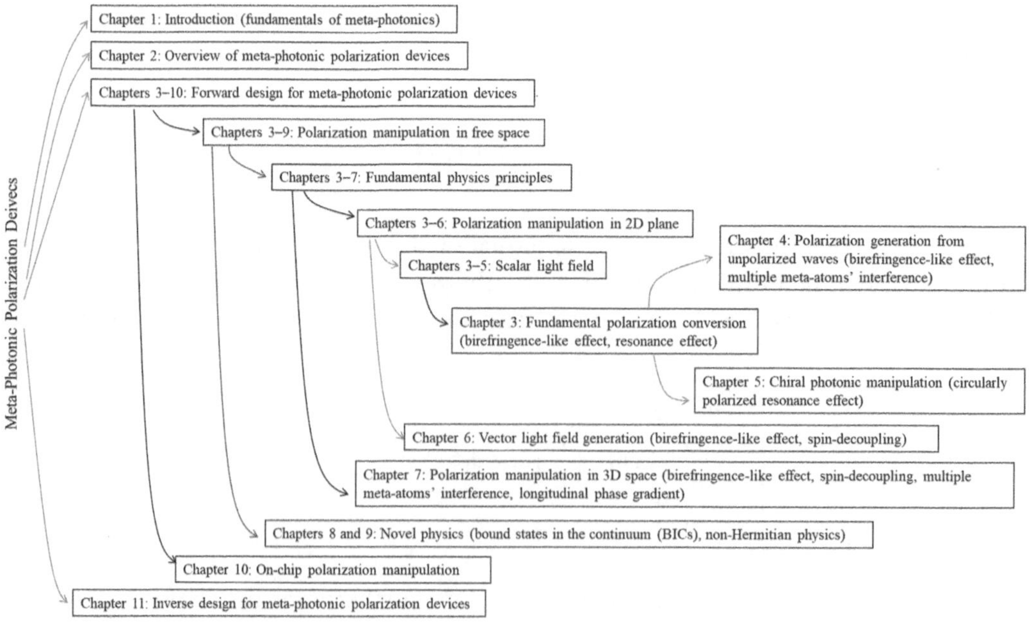

FIGURE 1.3 The overall framework of this book.

(BICs) and non-Hermitian physics that have been introduced in the field of meta-photonics in recent years.

Polarization manipulation is mainly aimed at free space and thus extensively and systematically introduced in Chapters 2–9. In addition to free space, polarized light is also attempting to connect with on-chip systems, which will be further introduced in Chapter 10. Chapters 2–10 all focus on the conventional forward design process from principles to devices, while with the development of artificial intelligence, emerging inverse design techniques have been reported, which will be presented in Chapter 11.

1.4 CHAPTER SUMMARY

Meta-photonics is an emerging science that manipulates light at subwavelength scales. Compared to traditional optics, it elevates the control over light polarization to an unprecedented level. This chapter begins by introducing the fundamental knowledge of polarization optics, elaborating on the role of polarization and its mathematical representation. It then presents the basic principles of meta-photonics. Compared to traditional optical devices, meta-photonic devices adhere to the generalized Snell's law and incorporate rich phase responses, which form the foundation for polarization manipulation. Finally, this chapter explains the overall framework of the book to help readers understand the content arrangement.

REFERENCES

1. Yu, N.F., Genevet, P., Kats, M.A., et al. (2011) 'Light propagation with phase discontinuities: generalized laws of reflection and refraction', Science, 334(6054), 333–337.
2. Berry, M.V. (1984) 'Quantal phase factors accompanying adiabatic changes', Proceedings of the Royal Society A: Mathematical, Physical and Engineering Sciences, 392(1802), 45–57.
3. Xie, X., Pu, M.B., Jin, J.J., et al. (2021) 'Generalized Pancharatnam-Berry phase in rotationally symmetric meta-atoms', Physical Review Letters, 126(18), 183902.
4. Tal, M., Haim, D.B. and Ellenbogen, T. (2022) 'Geometric phase opens new frontiers in nonlinear frequency conversion of light', Frontiers of Physics, 17(1), 12302.

CHAPTER 2

Overview of Meta-Photonic Polarization Devices

Abstract

Polarization is a fundamental property of light described based on free space. The research on spatial polarization state manipulation in the field of meta-photonics is very rich. Herein, an overview of the development of meta-photonic devices for spatial light polarization manipulation will be provided. The polarization manipulation achieved by meta-photonic devices has evolved from scalar light fields to vector light fields, from two-dimensional planes to three-dimensional spaces, and has further developed connections with novel physical phenomena, such as bound states in the continuum and non-Hermitian physics.

Meta-photonic devices primarily refer to photonic crystals and electromagnetic metasurfaces/metamaterials, which are artificially designed periodic array devices composed of subwavelength structural units. These devices interact with electromagnetic waves at subwavelength scales, enabling the manipulation of inherent properties such as polarization, phase, amplitude, and frequency. They exhibit novel characteristics that traditional electromagnetic devices cannot achieve. The history of using meta-photonic devices to manipulate the polarization state of electromagnetic waves dates back to the 1990s. In 1998, researchers, including Ebbesen from the NEC Research Institute in Princeton, USA, reported on their exploration of the optical properties of cylindrical cavities in submicron metal films [1]. They discovered that arrays of such apertures displayed highly unusual zero-order transmission spectra at wavelengths larger than the array period constant. Notably, sharp transmission peaks were observed at wavelengths approximately ten times the diameter of the cylinders, where the transmission efficiency exceeded predictions from standard aperture theory by an order of magnitude. To clarify the origin of this phenomenon, they tested the dependence of the transmission spectra on all possible variables, including aperture size, periodicity, thickness, and metal type. The results revealed that

DOI: 10.1201/9781003784821-2

the spectral evolution was related to surface plasmons (SPs). SPs are oscillations of surface charges at metal interfaces, excited when their momentum matches that of incident photons and the grating. The momentum matching condition is given by $k_{sp} = k_x \pm nG_x \pm mG_y$, where k_{sp} denotes the SP wavevector of the metal, k_x represents the wavevector component of incident photons in the grating plane, and G_x and G_y are the grating momentum wavevectors of the array. Further studies showed that when isotropic aperture structures were replaced with anisotropic grating structures [2], the array exhibited anomalous responses only to incident p-polarized plane waves, while no anomalous response was observed for s-polarized plane waves. This demonstrated that the interaction between incident waves and SPs is sensitive to the excitation polarization, indicating its potential for polarization manipulation.

Building on this theoretical foundation, Beruete et al. [3] proposed a prism-like device in 2010. This device is based on a periodic arrangement of corrugated slit structures, with slits distributed along the y-axis and exhibiting a wavy pattern along the y-direction. Photonic band structure analysis revealed that the structure supports related modes for s-polarization (y-polarization) and p-polarization (x-polarization) within specific frequency ranges. Notably, these modes exhibit a normal positive slope under p-polarization but a negative slope under s-polarization, which is characteristic of materials with an effective negative refractive index. This birefringent property, where the refractive index is positive for one polarization and negative for the other, causes s-polarized and p-polarized waves to propagate in different directions, enabling independent propagation control for different polarization states. With the advancement of metasurface theory and technology, the manipulation of electromagnetic wave polarization states has become increasingly sophisticated. Especially with the introduction of phase distribution design, the polarization control capabilities have grown more intriguing [4–6]. For example, it is common to use metallic rectangular bars or split-ring resonators to achieve electromagnetic wave polarization conversion [7,8]. Here, resonance phases are obtained by adjusting the size of rectangular bars or split-ring resonators, geometric phases are achieved through structural rotation, and specific phase distribution arrangements of unit cells enable simultaneous wavefront manipulation. The material systems have also expanded from metals to all-dielectric materials. All-dielectric materials, with their advantages of high refractive index and low loss, facilitate more efficient polarization control. For instance, high-efficiency conversion among multiple polarization states can be achieved using all-dielectric birefringent rectangular bars, and multiple polarization conversion modules can be integrated into a single device to create an efficient all-dielectric metasurface with diverse polarization manipulation capabilities [9]. Additionally, materials with tunable physical properties, such as graphene whose Fermi level can be adjusted, have been incorporated into polarization metasurfaces. By reversibly altering the physical properties of these materials, dynamic modulation of polarization states becomes possible [10,11]. Today, spatial polarization manipulation in meta-photonics has evolved from simple to complex functionalities, from two-dimensional planes to three-dimensional spaces, and has further developed connections with novel physical phenomena, such as bound states in the continuum (BICs) and non-Hermitian physics [12].

2.1 POLARIZATION MANIPULATION IN 2D PLANE

To accommodate complex functional scenarios, most polarization-manipulating metasurfaces consider polarization conversion. Although the meta-atoms may vary widely, one commonly used design principle is to mimic birefringence: creating a fast axis and a slow axis for electromagnetic wave transmission within the meta-atom. The polarization state of the input light can be decomposed into two polarization components aligned with the fast and slow axes. After passing through the meta-atom, these components are subjected to different phase delays, and when they recombine, various polarization changes occur, such as linear polarization rotation, handedness flipping of circular polarization, linear-to-circular conversion, and linear-to-elliptical conversion [9]. Chiral meta-atoms are another approach to achieving polarization conversion [13–15], often used for circular polarization chirality control and linear-to-circular polarization conversion. Chiral meta-atoms exhibit distinct geometric features and can be categorized into intrinsic and extrinsic chirality. Intrinsically chiral meta-atoms are geometrically incapable of coinciding with their mirror images through any in-plane rotation, much like how a left hand cannot match a right hand regardless of in-plane rotation. Structures with such geometry can generate transmitted waves with flipped handedness for normally incident circularly polarized light or reflected waves with preserved handedness (unlike mirror reflections, which flip the handedness of circular polarization). In contrast, extrinsically chiral meta-atoms lack inherent chiral geometric features [16] and cannot manipulate the handedness of normally incident waves. However, they can produce transmitted waves with flipped handedness or reflected waves with preserved handedness for obliquely incident circularly polarized light.

Typically, polarization-converting metasurfaces establish a unique functional relationship between the input and output polarization states, meaning the output polarization is constrained by the input polarization and changes accordingly when the input polarization varies. However, in certain special cases, this constraint can be broken. Wang et al. decomposed arbitrary polarization states into two orthogonal elliptical polarization components [17] and derived a Jones matrix that produces a stable output polarization state independent of the incident polarization, successfully disrupting the fixed functional relationship between input and output polarizations. This Jones matrix comprises two parts, realized by two birefringent rectangular bars with a $\pi/4$ rotation angle separation. One bar applies phase delays of 0 and π to light passing through its fast and slow axes, respectively, while the other applies delays of -2χ and 2χ, where χ determines the ellipticity of the final output polarization. The complete unit cell formed by these two birefringent rectangular bars essentially performs the function of transmitting only the handedness-flipped state of one polarization component of arbitrarily polarized incident light, while completely blocking the transmission of the orthogonal component. Furthermore, by controlling χ, the ellipticity can be adjusted to achieve outputs of linear, elliptical, or circular polarization. By rotating the entire two-bar structure while maintaining the $\pi/4$ rotation angle separation, the azimuth angle of linear or elliptical polarization outputs can be controlled. Thus, this technique can generate polarized beams at any position on the Poincaré sphere, regardless of the input polarization. In practice, it enables direct extraction of arbitrary

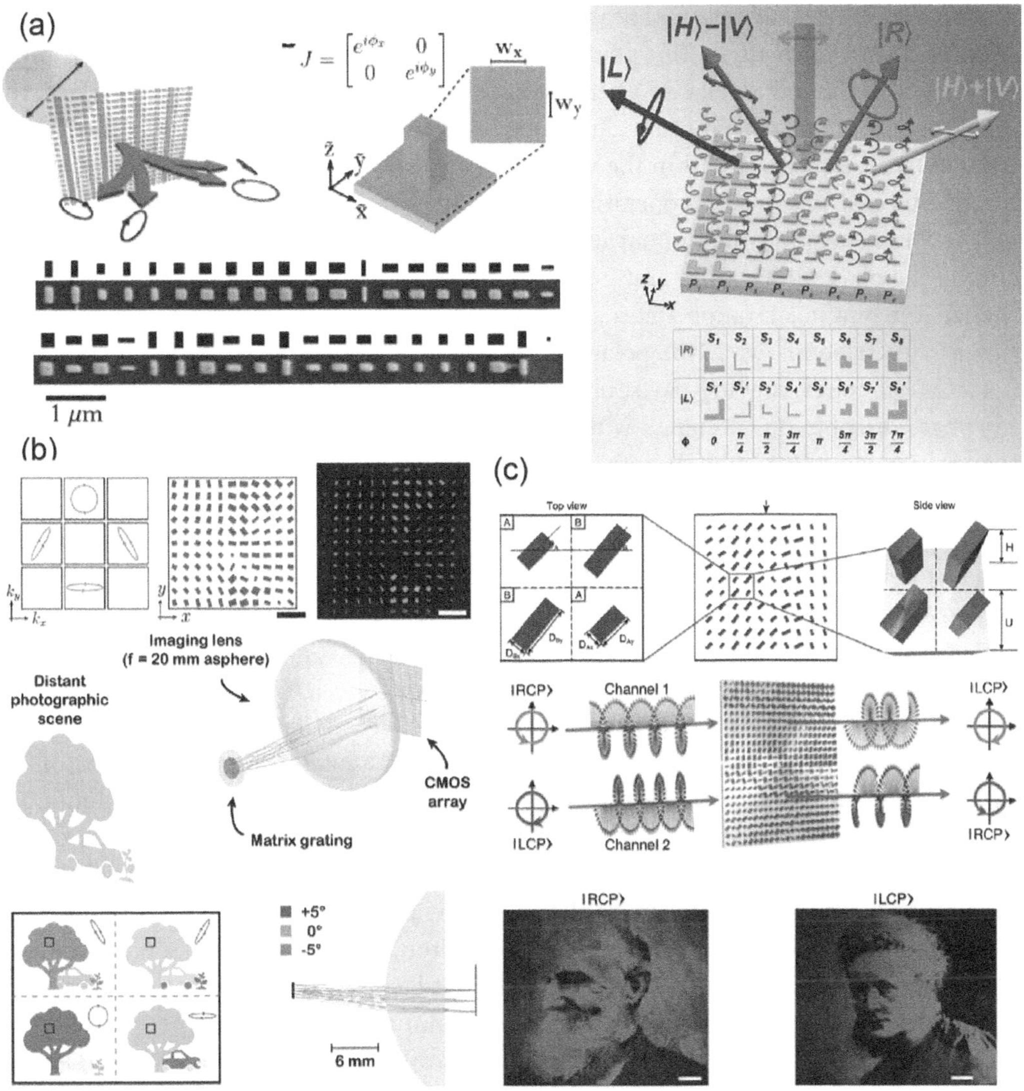

FIGURE 2.1 (a) Generation and separation of multiple polarization states in space based on multi-order diffraction birefringent meta-atoms and (b) its application in polarization imaging; (c) orthogonal polarization state modulation and application based on diatomic interference [12].

polarized light from unpolarized light and holds promise for miniaturized optical devices with higher integration than traditional polarization-converting metasurfaces.

Incorporating phase distribution design endows metasurfaces with more powerful polarization control capabilities. As reported by Rubin and Gao [18,19], as shown in Fig. 2.1(a), birefringent unit structures analogous to wave plates are utilized to independently manipulate the phase delays of x-polarized and y-polarized waves. By adjusting the size of these unit structures, independent phase delay control for two orthogonal polarization states can be achieved within the range of $0–2\pi$. Multiple structures are selected and arranged according to specific patterns to form a unit cell, which is then periodically extended into a metasurface. This enables spatial beam separation and the output of different polarization states in distinct directions. This phenomenon is fundamentally based on the multi-order diffraction of the metasurface. When incident light illuminates such a metasurface, the diffracted field at a far-field point can be expressed as the superposition of contributions from multiple structures within the unit cell. The total diffraction orders are determined by the unit cell size and the operating wavelength, while the diffraction angles depend on the arrangement of the unit structures. Since different unit structures independently control the phase delays of x- and y-polarized waves, the resulting metasurface can synthesize different polarization states at various diffraction orders. Based on these properties, as illustrated in Fig. 2.1(b), polarization imaging can be achieved without additional polarization optics, moving parts, or specialized sensors [20]. Furthermore, by combining phase distributions designed with special principles and algorithms, scalar and vector polarization holography in both real and momentum spaces can be realized [21–25]. Additionally, researchers have proposed approaches using multi-atom interference for polarization transformation [26–28]. For example, by combining two structures with independent polarization conversions into a unit cell, interference can generate a third polarization state [26] or achieve more complex functionalities such as independent amplitude and phase modulation of orthogonal polarization states [27,28], as shown in Fig. 2.1(c).

Although most metasurfaces with phase distribution designs for polarization control employ multiple unit cells with different sizes, if the phase distribution is designed properly, even if the unit cells have the same size parameter, some interesting polarization control can be achieved. For instance, as shown in Fig. 2.2(a), Zang et al. demonstrated multiple linear polarization rotation outputs based on a single rectangular bar [29]. The design principle can be succinctly described as follows: assume an incident x-polarized wave is decomposed into left-handed and right-handed circularly polarized components without a phase difference. To achieve linear polarization rotation at an angle ϕ, phase delays of ϕ and $-\phi$ are applied to the left-handed and right-handed components, respectively. By assigning multiple phase delays with the same magnitude but opposite signs to the circular components, different polarization rotation angles can be obtained. Combined with corresponding focusing phase designs, linear polarized waves with varying rotation angles can be spatially separated. The phase distribution is designed based on the PB phase principle, allowing the rotation of a single rectangular bar to meet the diverse phase delay requirements for constructing the metasurface. Similarly, Wang et al. demonstrated spatial multiple elliptical polarization state outputs based on a single Z-shaped metallic structure [30]. This structure exhibits excellent circular polarization conversion capabilities and relies on the PB phase principle to design different focusing phase distributions, creating multiple spatial transmission channels. By rotating the metallic structure, the phase delay requirements of the metasurface can be satisfied. When elliptically polarized light is incident on the metasurface, each channel outputs a different elliptical polarization state. As illustrated in Fig. 2.2(b), Deng et al., inspired by Malus's law, designed a Malus metasurface [31]. Using only one size of rectangular bar, when linearly polarized light is incident, the transmitted intensity of linearly polarized light varies with the rotation of the rectangular bars, enabling the formation of near-field intensity grayscale patterns. Simultaneously, for the

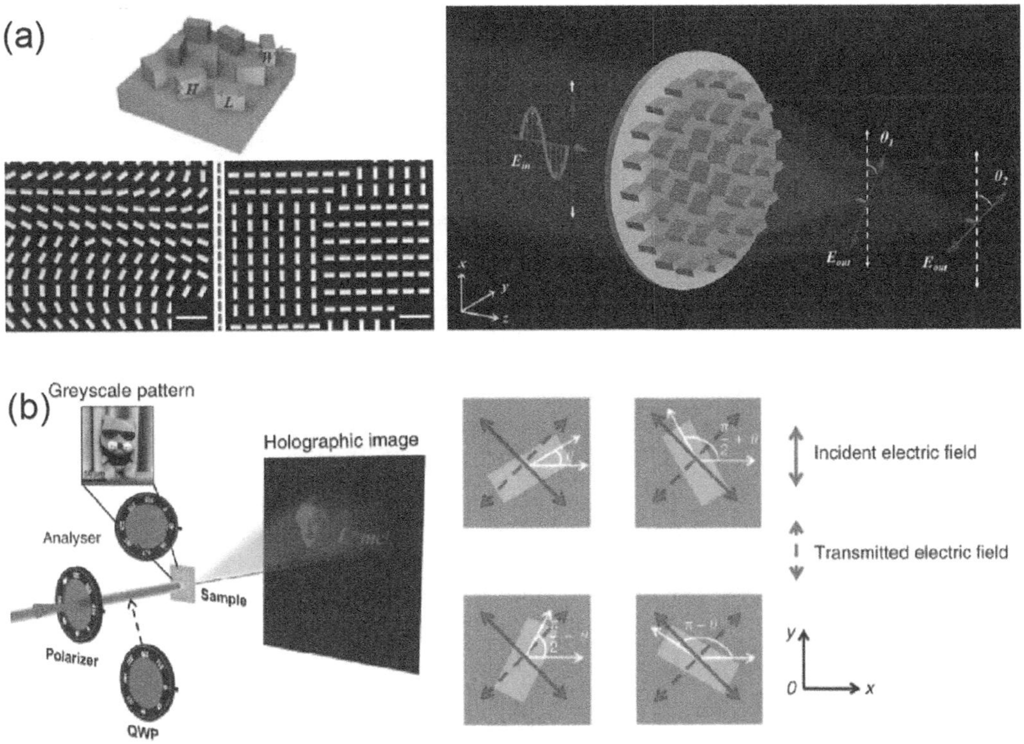

FIGURE 2.2 (a) The spatial rotation output of multiple different linear polarization states based on unit cells with the same size parameters. (b) Multi-channel and multi-polarization information metasurface based on Malus's law, where all unit cells are obtained by rotating the same rectangular bar at different angles without any size changes [12].

same output intensity, multiple rotation angles of the rectangular bars can achieve phase delays covering 0–2π, allowing the design of phase holographic metasurfaces. Through appropriate holographic phase design and diversified rectangular bar orientation selection, near-field greyscale display and far-field holographic display can be simultaneously achieved with distinct patterns. Moreover, different far-field holographic responses can be designed for circularly polarized incident light and linearly polarized incident light at various rotation angles, enabling multi-channel and multi-polarization information encoding.

Coherent superposition of left-handed and right-handed circularly polarized waves carrying opposite topological charges yields a vector beam with an inhomogeneous in-plane polarization distribution [32]. Its defining feature is the spatial variation of polarization across the plane. In 2014, Liu et al. demonstrated a straightforward method for generating vector optical fields using metasurfaces [33]. First, polarized light was obtained from a laser using a linear polarizer and a quarter-wave plate, and its ellipticity and orientation angle were adjusted by rotating these elements. This polarized light was then incident on a metasurface designed to function as a non-uniform half-wave plate, producing various vector beams. The metasurface was composed of birefringent rectangular bars of a single size, where each bar could flip the handedness of circular polarization. By rotating the bars to

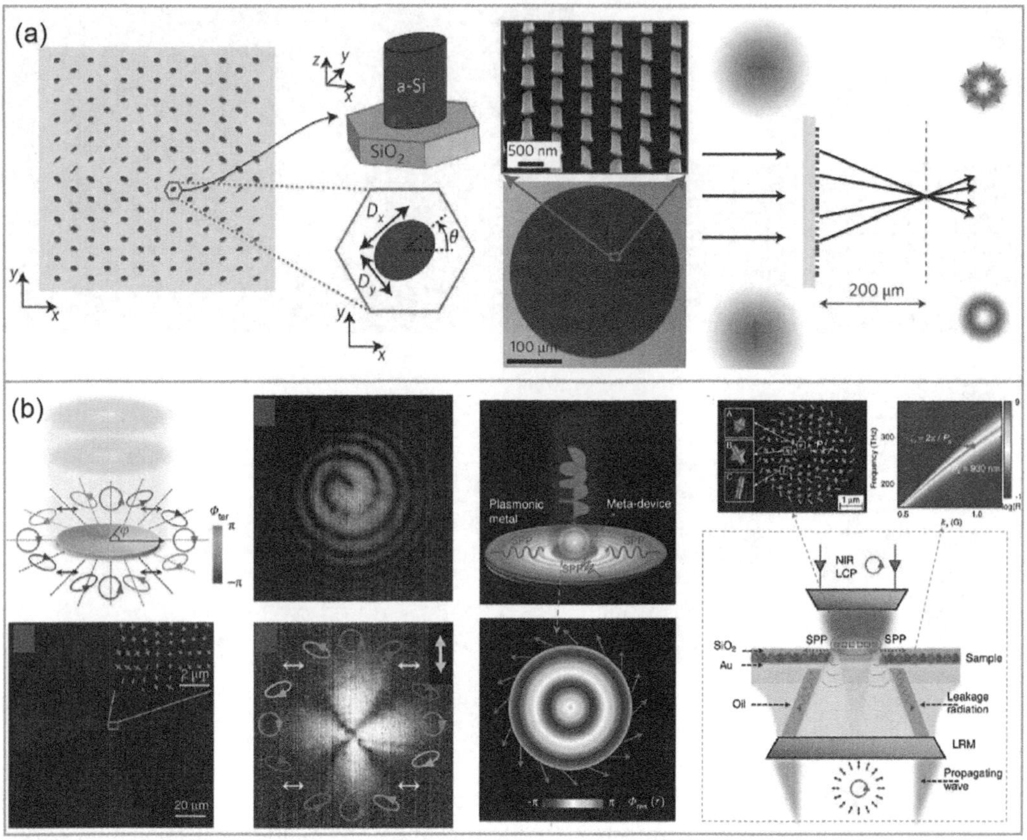

FIGURE 2.3 (a) Generation of focused vector beams based on hexagonal unit cells and birefringent elliptical bars. (b) Design of far-field and near-field vector beam generation via metasurfaces [12].

achieve phase delays covering 0–2π and arranging them according to a phase distribution containing specific topological charges, the metasurface could impart opposite topological charges to the two orthogonal circular polarization components of the incident uniformly polarized light, thereby transforming it into an inhomogeneous vector beam. Arbabi et al. designed a hexagonal unit cell [34] containing birefringent elliptical bars of varying sizes. Based on these structures, they achieved complete phase and polarization control and successfully demonstrated the ability to generate radially and azimuthally focused vector beams under orthogonal linearly polarized incidence, as shown in Fig. 2.3(a). While numerous examples of vector beam manipulation using metasurfaces exist, most operate in the far field [35,36]. To address this, in 2021, Wang et al. proposed a general strategy for generating both far-field and near-field vector beams via metasurfaces [37], as illustrated in Fig. 2.3(b). The core idea assumes that the metasurface exhibits a non-uniform and full-matrix form of the Jones matrix, enabling simultaneous control over local polarization and global wavefront. They designed a special device capable of generating cylindrical polarization vortex SP waves, and this is a unique type of near-field vortex that incorporates polarization distribution and even OAM information.

2.2 POLARIZATION MANIPULATION IN 3D SPACE

Meta-photonic devices have achieved two-dimensional in-plane inhomogeneous polarization outputs, manipulating the polarization state of light point-by-point in the transverse plane. A common approach is to locally vary the anisotropy of the structures, which independently modifies the two orthogonal polarization states of the incident wave, enabling local conversion of the incident polarization into any desired output polarization. These techniques have been widely applied in multifunctional polarization management, polarization gratings, holography, polarization imaging, and vector beam generation, significantly expanding the scope of research and applications in polarization optics. However, these metasurface devices typically focus on polarization behaviour in a single transverse plane perpendicular to the propagation direction and lack the ability to independently control polarization across multiple planes along the propagation axis, i.e. longitudinal polarization control. Such capability could be highly valuable in applications such as in vivo practices, where multiple polarized light beams may be used to operate within living tissues. A device with this capability would consolidate multiple optical polarization elements into a compact single component. In a sense, its property of altering the optical structure (intensity and polarization) along the light path can be regarded as a form of dynamic optical manipulation, with the potential to enable novel technologies such as optically tunable devices and varifocal lenses, challenging the status of static metasurfaces.

Yet achieving such a metasurface device requires addressing a key issue: the polarization of light is closely related to the photonic spin, a relationship that is generally constant. For instance, the two orthogonal circular polarization components that constitute a polarization state do not vary independently during propagation, so the resultant polarization state remains unchanged. How, then, can such a conserved quantity be altered along the spatial propagation path? Formally, the metasurface designed by Zang et al., shown in Fig. 2.2(a), also touches on the threshold of longitudinal polarization control [29]. Its physical mechanism similarly considers breaking the constant relationship between orthogonal circular polarization components during propagation, i.e. altering the phase difference between left-handed and right-handed circular polarization components at different longitudinal positions. However, this principle still lacks the ability to achieve continuous evolution of the polarization state along the longitudinal axis, as it cannot continuously modify the propagation phase difference between orthogonal circular polarization components. Before relying on metasurfaces to address this issue, strategies involved using spatial light modulators [38,39]. The general approach utilized multiple computer-generated holograms to spatially decompose the incident wave into two orthogonal polarization components, then recombined the beam polarization at the output while introducing longitudinal modulation, all while preserving its global energy and SAM. These techniques assumed the incident polarization was known in advance; once the incident polarization changed, the preset functionality of the device would be lost. In 2021, Ahmed H. Dorrah and colleagues at Harvard University achieved longitudinal polarization control based on metasurfaces [40]. The fundamental principle relies on spatial polarization beat frequency and depends on spatially interleaved meta-atom arrays, allowing an incident plane wave to undergo different polarization transformations at each propagation plane after a single interaction with the nanostructured metasurface, as shown in Fig. 2.4(a). The device demonstrated the ability to rotate a linearly polarized incident wave along the propagation axis, and the preset functionality remained effective without prior knowledge of the incident polarization. In other words, regardless of the incident polarization, the metasurface device could produce distinct polarization rotation responses along the beam axis. This work pioneered the multidimensional spatial modulation of electromagnetic polarization states based on metasurfaces in both principle and practice, potentially opening new directions for polarization-switchable devices, light-matter interactions, and three-dimensional laser microfabrication.

Using spin-decoupling techniques can also alter the relationship between orthogonal spin components during light propagation. The key to achieving photonic spin decoupling

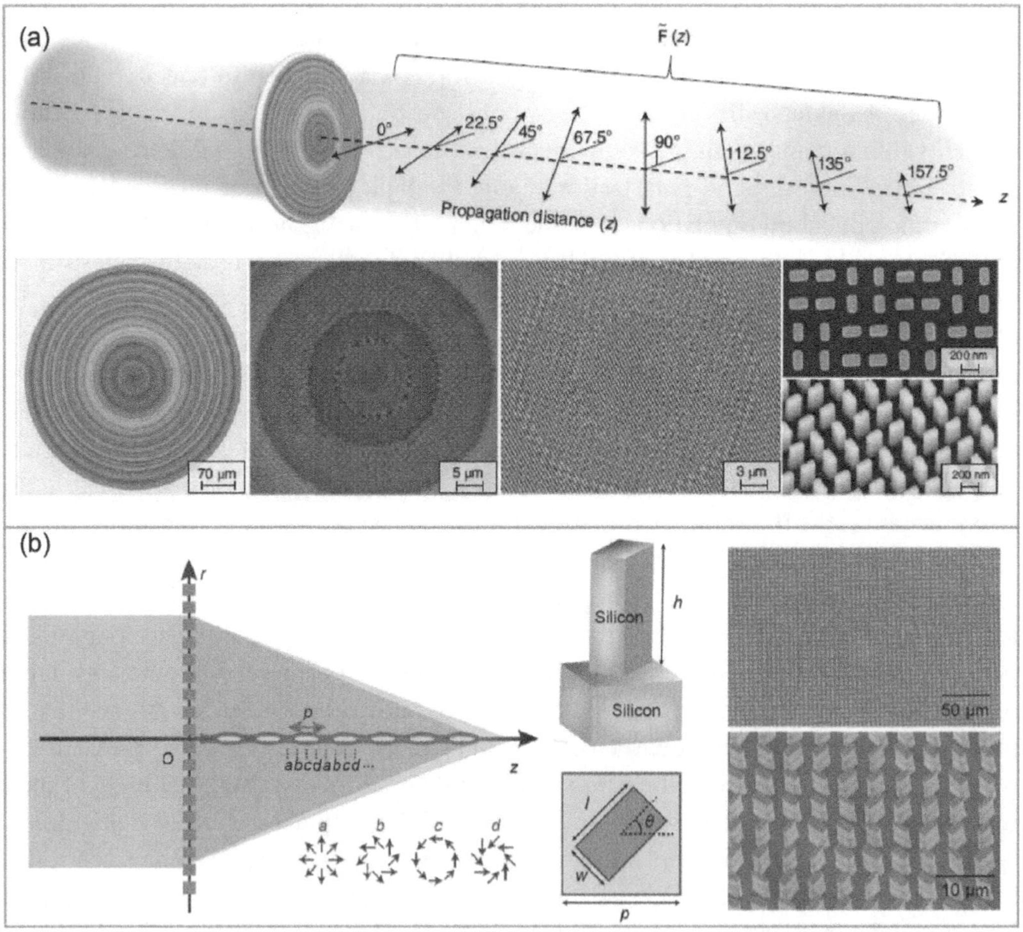

FIGURE 2.4 (a) Schematic of a longitudinal polarization control metasurface array and its function of polarization rotation along the propagation axis. The metasurface unit cells are composed of multiple spatially interleaved structures. (b) Schematic of 3D vector optical field manipulation based on spin-decoupling technology and its corresponding metasurface device. This device achieves periodic transformation from radially polarized vortex beams to azimuthally polarized vortex beams along the optical propagation path [12].

in metasurfaces lies in the phase delay design of the meta-atoms. A designed meta-atom can flip the handedness of left-handed and right-handed circularly polarized waves and produce different phase delays. Most importantly, by adjusting the structural parameters of the meta-atom, the phase delay for the LCP wave can be kept constant while achieving 0–2π phase delay coverage for the RCP wave, or vice versa. This establishes a vast database of meta-atoms. Subsequently, by independently designing the phase profiles for the two circularly polarized waves and selecting corresponding meta-atoms from the database to arrange into a metasurface array, independent manipulation of the orthogonal spin components within a polarized light beam can be realized. Based on this principle, in 2022, the team led by Zhang at the Institute of Optics and Electronics, Chinese Academy of Sciences, proposed three-dimensional vector optical field manipulation [41], as shown in Fig. 2.4(b). This achievement is based on all-silicon meta-atoms, which focus the two orthogonal spin components onto the propagation axis in the form of Bessel beams according to different phase profiles and apply opposite OAM to the two orthogonal spin components. Most critically, because the phase profiles controlling the two orthogonal spin components differ, they follow different paths to convergence, causing the phase difference between them to vary along the propagation distance. Ultimately, this work not only manipulates longitudinal polarization variation but also controls the inhomogeneous polarization distribution within the transverse plane, thereby achieving true three-dimensional optical field manipulation.

2.3 POLARIZATION MANIPULATION RELATED TO NOVEL PHYSICS

In recent years, meta-photonics has been linked to novel physical concepts such as BICs and non-Hermitian physics, greatly enriching the content of polarization manipulation.

BICs and Quasi-BICs. Meta-photonic devices exhibit a special type of electromagnetic eigenstate known as BICs. These states have frequencies within the radiation continuum but are completely localized. BICs' electromagnetic modes at the Γ-point are non-radiative and behave as polarization vortex singularities in momentum space, characterized by a non-zero polarization topological charge q. Around the Γ-point, the radiation polarization of Bloch resonant modes is linear and varies in a vortex-like manner around the Γ-point. This linear polarization vortex feature can be decomposed into the superposition of a LCP wave and a RCP wave in momentum space, each carrying opposite phase topological charges with magnitudes of $|2q|$. Therefore, BICs metadevices can perform wavefront vortex transformations on input circularly polarized waves in momentum space, a property entirely distinct from polarization manipulation in real space, as shown in Fig. 2.5(a). The generation of vortices based on BICs is different from traditional perceptions of vortex manipulation. In most cases, vortex generation primarily occurs in real space using phase assembled metasurface devices, which typically exhibit visible vortex-like geometric arrangements. However, BICs metasurfaces feature completely periodic unit cell arrangements in real space without any visible vortex geometry, making it counterintuitive to generate vortex beams; nevertheless, they indeed possess the ability to produce vortex beams in momentum space. This unique capability may give BICs metasurfaces an advantage over traditional phase metasurfaces in applications such as secure electromagnetic communication.

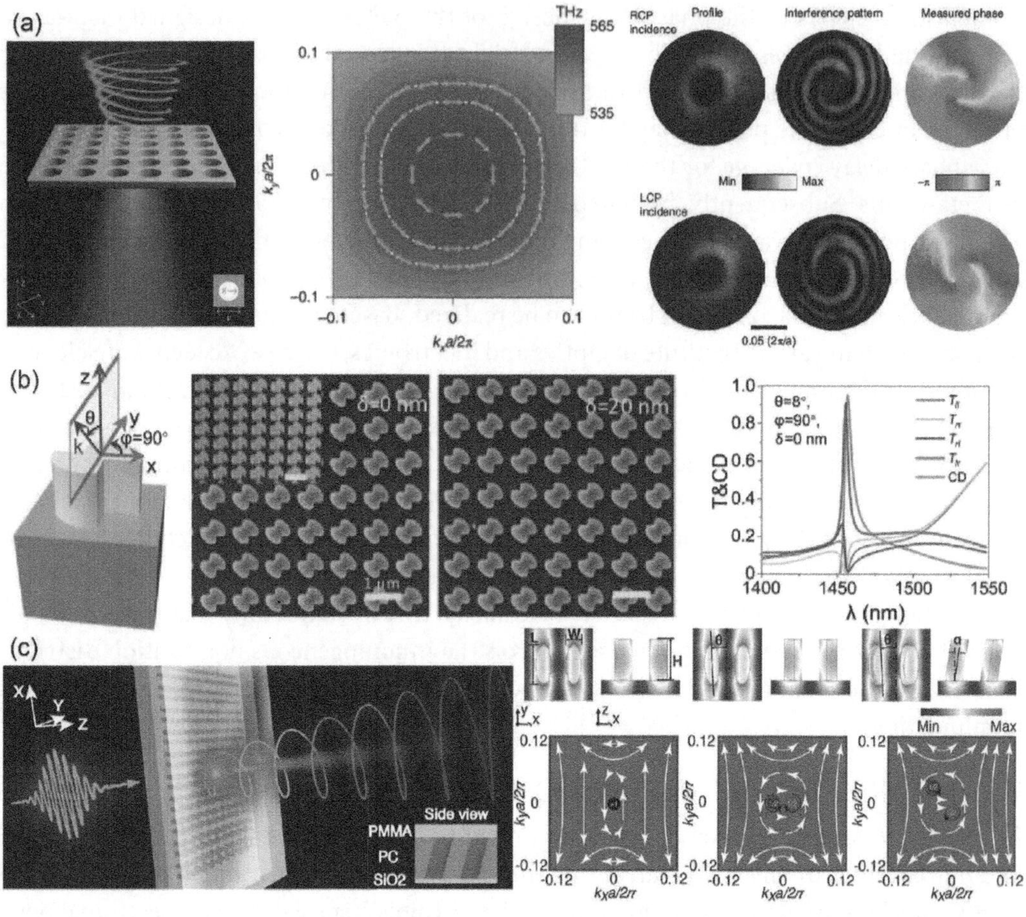

FIGURE 2.5 (a) Generation of circularly polarized wave vortex states in momentum space based on BICs metasurfaces. (b) A flat chiral quasi-BIC metasurface and its circularly polarized transmission spectrum. (c) High-quality-factor circularly polarized micro-laser emission based on chiral quasi-BICs [12].

Although Γ-BICs are polarization singularities in momentum space, when they leak and become quasi-BICs, the singularity is disrupted, and the radiation polarization state at the Γ-point in momentum space becomes well-defined [42]. This polarization state exhibits an extremely narrow frequency linewidth, a feature unattainable by traditional polarization control devices. By appropriately adjusting structural geometric parameters, the Γ-point radiation polarization of quasi-BICs is not limited to linear polarization and can also exhibit significant ellipticity, indicating decoupling between LCP and RCP states and the emergence of chirality. Adam Overvig et al., based on spatially closely stacked structures, constructed in-plane C_2 symmetry-broken quasi-BICs and theoretically demonstrated that the Γ-point radiation polarization of such quasi-BICs can be arbitrarily elliptical, proposing the concept of chiral quasi-BICs [43]. Maxim V. Gorkunov et al., using paired elliptical bars, constructed out-of-plane C_2 symmetry-broken chiral quasi-BICs with in-plane C_4 symmetry. This design theoretically enables perfect narrowband circular dichroism in scattered waves under normal incidence. Analytical expressions for circular dichroism were derived using temporal coupled-mode theory, proving that maximum circular dichroism is achieved at quasi-BIC frequencies [44]. In 2022, to enhance the feasibility of fabricating chiral quasi-BICs, flat chiral quasi-BICs were proposed [45]. These were designed in flat structures symmetric under mirror reflection about the *z*-plane, with unit cells based on in-plane C_2 symmetry-broken dielectric pillars. Experiments successfully verified the narrowband asymmetric transmission capability of chiral quasi-BICs, as shown in Fig. 2.5(b). Quasi-BICs with frequencies below the first-order diffraction limit possess only a single zeroth-order radiation channel, confining a significant portion of electromagnetic energy within the metasurface, analogous to a resonant cavity. This forms the basis for micro-lasing applications. Leveraging chiral quasi-BICs and their high-intensity resonant properties, circularly polarized micro-lasing devices can be realized. In 2022, Zhang et al. from Harbin Institute of Technology collaborated with Yuri et al. from the Australian National University [46]. By obliquely cutting a pair of titanium dioxide-doped elliptical dielectric pillars supporting BICs along the *z*-axis, they obtained out-of-plane C_2 symmetry-broken quasi-BIC modes. This quasi-BIC metasurface achieved circular polarization winding around the Γ-point in momentum space and exhibited nearly perfect high-quality-factor chirality. Under pulsed laser pumping, it emitted high-quality-factor circularly polarized light, as shown in Fig. 2.5(c).

Non-Hermitian Physics. Losses in electromagnetic metasurfaces (e.g. radiation loss and ohmic loss) are inevitable. In the design and application of electromagnetic metasurfaces, losses are often considered detrimental as they reduce efficiency, and thus efforts are made to minimize them, for example, by using dielectric materials. Interestingly, with the development of non-Hermitian physics in recent years, researchers have realized that losses can be transformed into a new degree of freedom in metasurface design, offering new directions for expanding metasurface functionalities and achieving unconventional physical effects. Broadly speaking, all open systems that exchange energy with their environment are non-Hermitian systems, and their physical quantities can be described by non-Hermitian operators. Non-Hermitian systems typically exhibit complex eigenvalues and non-orthogonal eigenstates, along with special degeneracy points. At these points, two or more eigenvalues and their corresponding eigenstates simultaneously degenerate, known as exceptional points (EPs).

Non-Hermitian metadevices utilize gain-loss mechanisms to induce and control the generation of EPs. Circulating parameters around an EP can be used for polarization manipulation. In 2014, Lawrence et al. [47] used a single-layer chiral metasurface to study the optical properties of parity-time (PT) symmetry breaking. The designed structure consisted of two mutually perpendicular split circular metal rings with identical resonance frequencies but different loss coefficients. The perpendicular orientation of the metal rings facilitated independent probing of the radiation fields from each resonant antenna, and the orthogonality of the two radiation fields ensured a real coupling coefficient. The elliptical eigenstates exhibited different principal axis orientations in the corresponding PT-symmetric and PT-symmetry-broken regions. However, when the coupling length was adjusted to transition the eigenstates through the phase transition point, the elliptical principal axis abruptly rotated by $\pi/4$, and the eigenmodes degenerated into a single circular polarization state. In 2016, Kang et al. [48] designed a non-Hermitian chiral metasurface using a pair of lossy linearly coupled antennas as unit structures, investigating the physical properties of EPs. By varying the incident light frequency and the coupling loss rate of the metasurface, they observed the emergence of EPs with chiral characteristics, linked to circularly polarized transmission eigenstates.

2.4 CHAPTER SUMMARY

This chapter provides an overview of the development of meta-photonics, with a focus on the spatial manipulation of light polarization. Spatial polarization control has evolved from simple to complex functionalities, from scalar to vector light fields, and from two-dimensional planes to three-dimensional spaces. It has further developed connections with novel physical concepts such as BICs and non-Hermitian systems. Next, Chapters 3 to 9 of this book will delve into the principles of spatial polarization manipulation and device design in detail. Given that polarization is a physical concept primarily associated with electromagnetic waves in free space, this chapter focuses on reviewing the manipulation of spatial polarization using meta-photonic devices, but it does not encompass the entire field of meta-photonic polarization control. Beyond the polarization manipulation of spatial light, meta-photonic polarization control has also extended from free space to on-chip platforms. Furthermore, by integrating artificial intelligence technologies, it is advancing into the era of inverse design. These developments will be discussed in Chapters 10 and 11.

REFERENCES

1. Ebbesen, T.W., Lezec, H.J., Ghaemi, H.F., et al. (1998) 'Extraordinary optical transmission through sub-wavelength hole arrays', Nature, 391, 667–669.
2. Schroter, U. and Heitmann, D. (1998) 'Surface-plasmon-enhanced transmission through metallic gratings', Physical Review B, 58(23), 15419–15421.
3. Beruete, M., Navarro-Cía, M., Falcone, F., et al. (2010) 'Single negative birefringence in stacked spoof plasmon metasurfaces by prism experiment', Optics Letters, 35(5), 643–645.
4. Tang, S. and Ding, F. (2020) 'High-efficiency focused optical vortex generation with geometric gap-surface plasmon metalenses', Applied Physics Letters, 117, 011103.
5. Wen, D., Yue, F. and Li, G. (2015) 'Helicity multiplexed broadband metasurface holograms', Nature Communications, 6, 8241.
6. Hu, Y., Li, L., Wang, Y., et al. (2020) 'Trichromatic and tripolarization-channel holography with noninterleaved dielectric metasurface', Nano Letters, 20(2), 994–1002.
7. Bao, Y., Jiang, Q., Kang, Y., et al. (2017) 'Enhanced optical performance of multifocal metalens with conic shapes', Light: Science & Applications, 6, e17071.
8. Wang, Q., Zhang, X., Xu, Y., et al. (2015) 'A broadband metasurface-based terahertz flat-lens array', Advanced Optical Materials, 3(6), 779–785.
9. Ding, F., Chang, B., Wei, Q., et al. (2020) 'Versatile polarization generation and manipulation using dielectric metasurfaces', Laser & Photonics Reviews, 14(11), 2000116.
10. Mykhailo, T., Nikitin, A.Y. and Luis, M.M. (2013) 'Faraday rotation due to excitation of magnetoplasmons in graphene microribbons', ACS Nano, 7(11), 9780–9787.
11. Poumirol, J.M., Liu, P., Slipchenko, T.M., et al. (2017) 'Electrically controlled terahertz magneto-optical phenomena in continuous and patterned graphene', Nature Communications, 8, 14626.

12. Dai, Z., Ye, J., Liang, Y., et al. (2026) 'Metasurfaces manipulating polarization evolution from the perspectives of 2D/3D real space and momentum space', Advanced Optical Materials, 14(6), e02952.
13. Hu, G., Hong, X., Wang, K., et al. (2019) 'Coherent steering of nonlinear chiral valley photons with a synthetic Au-WS2 metasurface', Nature Photonics, 13, 467–472.
14. Han, B., Li, S., Li, Z., et al. (2021) 'Asymmetric transmission for dual-circularly and linearly polarized waves based on a chiral metasurface', Optics Express, 29(13), 19643–19654.
15. Wu, C., Arju, N., Kelp, G., et al. (2014) 'Spectrally selective chiral silicon metasurfaces based on infrared Fano resonances', Nature Communications, 5, 3892.
16. Wang, Q., Plum, E., Yang, Q., et al. (2018) 'Reflective chiral meta-holography: multiplexing holograms for circularly polarized waves', Light: Science & Applications, 7, 25.
17. Wang, S., Deng, Z., Wang, Y., et al. (2021) 'Arbitrary polarization conversion dichroism metasurfaces for all-in-one full Poincaré sphere polarizers', Light: Science & Applications, 10, 24.
18. Rubin, N.A., Zaidi, A., Juhl, M., et al. (2018) 'Polarization state generation and measurement with a single metasurface', Optics Express, 26(17), 21455–21478.
19. Gao, Y., Xiong, X., Wang, Z., et al. (2020) 'Simultaneous generation of arbitrary assembly of polarization states with geometrical-scaling-induced phase modulation', Physical Review X, 10(3), 031035.
20. Rubin, N.A., Aversa, G.D., Chevalier, P., et al. (2019) 'Matrix Fourier optics enables a compact full-Stokes polarization camera', Science, 365, eaax1839.
21. Zhao, R., Sain, B., Wei, Q., et al. (2018) 'Multichannel vectorial holographic display and encryption', Light: Science & Applications, 7, 95.
22. Wei, Q., Sain, B., Wang, Y., et al. (2019) 'Simultaneous spectral and spatial modulation for color printing and holography using all-dielectric metasurfaces', Nano Letters, 19(12), 8964–8971.
23. Song, Q., Baroni, A., Sawant, R., et al. (2020) 'Ptychography retrieval of fully polarized holograms from geometric-phase metasurfaces', Nature Communications, 11, 2651.
24. Wen, D., Cadusch, J.J., Meng, J., et al. (2021) 'Vectorial holograms with spatially continuous polarization distributions', Nano Letters, 21(4), 1735–1741.
25. Zhao, R., Xiao, X., Geng, G., et al. (2021) 'Polarization and holography recording in real- and k-space based on dielectric metasurface', Advanced Functional Materials, 31(27), 2100406.
26. Li, J., Zheng, C., Li, J.T., et al. (2021) 'Terahertz wavefront shaping with multi-channel polarization conversion based on all-dielectric metasurface', Photonics Research, 9(10), 1939–1947.
27. Deng, Z., Deng, J., Zhuang, X., et al. (2018) 'Diatomic metasurface for vectorial holography', Nano Letters, 18(5), 2885–2892.
28. Liu, M., Zhu, W., Huo, P., et al. (2021) 'Multifunctional metasurfaces enabled by simultaneous and independent control of phase and amplitude for orthogonal polarization states', Light: Science & Applications, 10, 107.
29. Zang, X., Ding, H., Intaravanne, Y., et al. (2019) 'A multi-foci metalens with polarization-rotated focal points', Laser & Photonics Reviews, 13(12), 1900182.
30. Wang, R., Han, J., Liu, J., et al. (2020) 'Multi-foci metalens for terahertz polarization detection', Optics Letters, 45(13), 3506–3509.
31. Deng, L., Deng, J., Guan, Z., et al. (2020) 'Malus-metasurface-assisted polarization multiplexing', Light: Science & Applications, 9, 101.
32. Naidoo, D., Roux, F.S., Dudley, A., et al. (2016) 'Controlled generation of higher-order Poincaré sphere beams from a laser', Nature Photonics, 10, 327–332.
33. Liu, Y., Ling, X., Yi, X., et al. (2014) 'Realization of polarization evolution on higher-order Poincaré sphere with metasurface', Applied Physics Letters, 104, 191110.
34. Arbabi, A., Horie, Y., Bagheri, M., et al. (2015) 'Dielectric metasurfaces for complete control of phase and polarization with subwavelength spatial resolution and high transmission', Nature Nanotechnology, 10, 937–943.

35. Pfeiffer, C. and Grbic, A. (2014) 'Controlling vector Bessel beams with metasurfaces', Physical Review Applied, 2(4), 044012.
36. Wen, D., Pan, K., Meng, J., et al. (2022) 'Broadband multichannel cylindrical vector beam generation by a single metasurface', Laser & Photonics Reviews, 16(10), 2200206.
37. Wang, D., Liu, F., Liu, T., et al. (2021) 'Efficient generation of complex vectorial optical fields with metasurfaces', Light: Science & Applications, 10, 67.
38. Schulze, C., Roux, F.S., Dudley, A., et al. (2015) 'Accelerated rotation with orbital angular momentum modes', Physical Review A, 91(4), 043821.
39. Otte, E., Rosales-Guzmán, C., Ndagano, B., et al. (2018) 'Entanglement beating in free space through spin-orbit coupling', Light: Science & Applications, 7, 18009.
40. Dorrah, A.H., Rubin, N.A., Zaidi, A., et al. (2021) 'Metasurface optics for on-demand polarization transformations along the optical path', Nature Photonics, 15, 287–296.
41. Zhang, F., Pu, M., Guo, Y., et al. (2022) 'Synthetic vector optical fields with spatial and temporal tenability', Science China-Physics, Mechanics & Astronomy, 65(5), 254211.
42. Wang, J., Zhao, M., Liu, W., et al. (2021) 'Shifting beams at normal incidence via controlling momentum-space geometric phases', Nature Communications, 12, 6046.
43. Overvig, A., Yu, N. and Alù, A. (2021) 'Chiral quasi-bound states in the continuum', Physical Review Letters, 126(7), 073001.
44. Gorkunov, M.V., Antonov, A.A. and Kivshar, Y.S. (2020) 'Metasurfaces with maximum chirality empowered by bound states in the continuum', Physical Review Letters, 125(9), 093903.
45. Shi, T., Deng, Z.L., Geng, G., et al. (2022) 'Planar chiral metasurfaces with maximal and tunable chiroptical response driven by bound states in the continuum', Nature Communications, 13, 4111.
46. Zhang, X., Liu, Y., Han, J., et al. (2022) 'Chiral emission from resonant metasurfaces', Science, 377, 1215–1218.
47. Lawrence, M., Xu, N., Zhang, X., et al. (2014) 'Manifestation of PT-symmetry breaking in polarization space with terahertz metasurfaces', Physical Review Letters, 113(9), 093901.
48. Kang, M., Chen, J. and Chong, Y.D. (2016) 'Chiral exceptional points in metasurfaces', Physical Review A, 94(3), 033834.

CHAPTER 3

Fundamental Polarization Conversion

Abstract

Polarization conversion is a fundamental capability requirement for many multifunctional polarized metasurfaces. The underlying mechanisms of polarization conversion can be divided into two categories: non-resonant and resonant. The typical non-resonant effect is a birefringence-like effect (mimicking the birefringence of natural materials), which is widely used as the theoretical basis for polarization conversion. Electromagnetic resonance is the interaction between light and electromagnetic eigenstates of devices, and it is also one of the main theoretical foundations for polarization conversion. This chapter will explain the fundamental principles of birefringence-like and electromagnetic resonance effects and demonstrate the design of polarization conversion devices based on them.

3.1 BIREFRINGENCE-INDUCED POLARIZATION CONVERSION

Birefringence-like effect is the most widely relied upon physical effect when performing polarization conversion on metasurfaces. The primitive birefringence effect was discovered during the study of optical phenomena in crystals. The phenomenon where a beam of light is incident on an anisotropic crystal and decomposes into two beams of light that refract in different directions. When light propagates in a non-uniform medium, its propagation speed and refractive index value change with different vibration directions, and there is more than one refractive index value. When light waves enter a non-uniform matter, except in special directions, they undergo birefringence and decompose into two types of polarized light with perpendicular vibration directions, different propagation speeds, and unequal refractive indices. This phenomenon is called birefringence. In metasurfaces, the birefringence phenomenon of anisotropic crystals can be imitated to achieve fundamental polarization conversion between

DOI: 10.1201/9781003784821-3

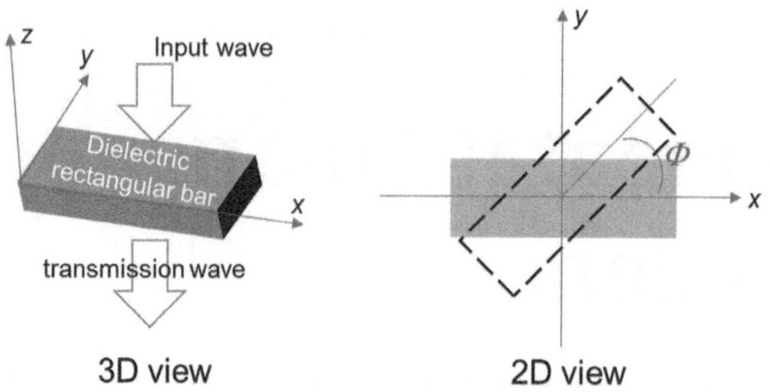

FIGURE 3.1 The structure schematic for a dielectric rectangular bar as a unit cell used in a metasurface.

various linearly polarized (LP) states and between linearly and circularly polarized (CP) states [1].

3.1.1 Theoretical Deduction of Birefringent Polarization Conversion

Herein, borrow a dielectric rectangular bar structure (Fig. 3.1) – the most commonly used unit cell in metasurfaces – to deduce the theory of birefringent polarization conversion. First, let us see the polarization conversion property of a rectangular bar. After an electric field is incident on the dielectric rectangular bar, the transmission electric field is as follows:

$$\begin{pmatrix} E_x^t \\ E_y^t \end{pmatrix} = \begin{pmatrix} t_{xx} & t_{xy} \\ t_{yx} & t_{yy} \end{pmatrix} \begin{pmatrix} E_x^i \\ E_y^i \end{pmatrix} = \boldsymbol{T} \begin{pmatrix} E_x^i \\ E_y^i \end{pmatrix}, \tag{3.1}$$

where E_x^i, E_y^i are incident electric fields on x-axis and y-axis, respectively, and E_x^t, E_y^t are transmission electric fields on x-axis and y-axis, respectively. $\boldsymbol{T}$ is the transfer Jones matrix and consists of transmission coefficients t_{ab} (a, $b \in \{x, y\}$), where t_{ab} means the coefficient of transmission of a-polarized wave under an incident b-polarized wave.

When an x-polarized or a y-polarized wave passes through a dielectric rectangle bar parallel to the x-axis or the y-axis, there are no cross-polarization components in transmission, and hence the transfer Jones matrix $\boldsymbol{T}$ can be expressed as follows:

$$\boldsymbol{T} = \begin{pmatrix} t_{xx} & 0 \\ 0 & t_{yy} \end{pmatrix} = \begin{pmatrix} |t_{xx}| e^{i\varphi_{xx}} & 0 \\ 0 & |t_{yy}| e^{i\varphi_{yy}} \end{pmatrix}. \tag{3.2}$$

When the dielectric rectangle bar is rotated with an angle of Φ, the matrix $\boldsymbol{T}$ will be changed under a coordinate transformation base matrix $\Lambda_1 = \begin{pmatrix} \cos\Phi & -\sin\Phi \\ \sin\Phi & \cos\Phi \end{pmatrix}$ (the derivation of the coordinate transformation matrix will be seen in Section 3.1.2), and the general form of the new matrix $\boldsymbol{T}_0$ is as follows:

$$\boldsymbol{T}_0 = \Lambda_1 \boldsymbol{T} \Lambda_1^{-1} = \begin{pmatrix} \cos\Phi & -\sin\Phi \\ \sin\Phi & \cos\Phi \end{pmatrix} \begin{pmatrix} |t_{xx}|e^{i\varphi_{xx}} & 0 \\ 0 & |t_{yy}|e^{i\varphi_{yy}} \end{pmatrix} \begin{pmatrix} \cos\Phi & \sin\Phi \\ -\sin\Phi & \cos\Phi \end{pmatrix}$$
$$= \begin{pmatrix} \cos^2\Phi \cdot |t_{xx}|e^{i\varphi_{xx}} + \sin^2\Phi \cdot |t_{yy}|e^{i\varphi_{yy}} & \cos\Phi\sin\Phi \cdot \left(|t_{xx}|e^{i\varphi_{xx}} - |t_{yy}|e^{i\varphi_{yy}}\right) \\ \cos\Phi\sin\Phi \cdot \left(|t_{xx}|e^{i\varphi_{xx}} - |t_{yy}|e^{i\varphi_{yy}}\right) & \sin^2\Phi \cdot |t_{xx}|e^{i\varphi_{xx}} + \cos^2\Phi \cdot |t_{yy}|e^{i\varphi_{yy}} \end{pmatrix}. \tag{3.3}$$

3.1.1.1 Conversion from LP Waves to Other Polarized Waves

When an x-polarized wave $[1\ 0]^{\mathrm{T}}$ is incident on the dielectric rectangle bar, the transmission electric field can be expressed as follows:

$$\boldsymbol{E}_0^t = \boldsymbol{T}_0 \begin{pmatrix} 1 \\ 0 \end{pmatrix}$$
$$= \begin{pmatrix} \cos^2\Phi \cdot |t_{xx}|e^{i\varphi_{xx}} + \sin^2\Phi \cdot |t_{yy}|e^{i\varphi_{yy}} & \cos\Phi\sin\Phi \cdot \left(|t_{xx}|e^{i\varphi_{xx}} - |t_{yy}|e^{i\varphi_{yy}}\right) \\ \cos\Phi\sin\Phi \cdot \left(|t_{xx}|e^{i\varphi_{xx}} - |t_{yy}|e^{i\varphi_{yy}}\right) & \sin^2\Phi \cdot |t_{xx}|e^{i\varphi_{xx}} + \cos^2\Phi \cdot |t_{yy}|e^{i\varphi_{yy}} \end{pmatrix} \begin{pmatrix} 1 \\ 0 \end{pmatrix}. \tag{3.4}$$

Let $\Phi = \pi/4$, i.e. $\cos^2\Phi = \sin^2\Phi = 1/2$, and Let $|t_{xx}| = |t_{yy}| = t$, where t is the transmission coefficient and determined by the physical properties of the dielectric material. If reflection and absorption of dielectric material are neglected, the t equals 1. Making $\varphi_{yy} - \varphi_{xx} = 2\theta$ (for the convenience of discussion, the range of θ is limited to 0–π so that 2θ is between 0 and 2π), Eq. (3.4) is simplified as:

$$\boldsymbol{E}_0^t = \frac{t}{2}e^{i\varphi_{xx}} \begin{pmatrix} 1+e^{i2\theta} & 1-e^{i2\theta} \\ 1-e^{i2\theta} & 1+e^{i2\theta} \end{pmatrix} \begin{pmatrix} 1 \\ 0 \end{pmatrix}. \tag{3.5}$$

Then, making $\varphi_{xx} = \varphi - \theta$, there is:

$$\boldsymbol{E}_0^t = \frac{t}{2}e^{i\varphi} \begin{pmatrix} 2\cos\theta & -2i\sin\theta \\ -2i\sin\theta & 2\cos\theta \end{pmatrix} \begin{pmatrix} 1 \\ 0 \end{pmatrix} = \frac{t}{2}e^{i\varphi} \begin{pmatrix} 2\cos\theta \\ -2i\sin\theta \end{pmatrix}. \tag{3.6}$$

It is easy to observe that when $\theta = \pi/2$, the above equation represents a y-polarized wave. Otherwise, the above equation can be transformed into:

$$\boldsymbol{E}_0^t = \frac{t}{2}e^{i\varphi} \begin{pmatrix} 2\cos\theta \\ -2i\sin\theta \end{pmatrix} = te^{i\varphi}\cos\theta \begin{pmatrix} 1 \\ -i\tan\theta \end{pmatrix}. \tag{3.7}$$

It can be seen that when $\theta = 0$ or π, the above equation represents an x-polarized wave. When $\theta = \pi/4$ or $3\pi/4$, the above equation represents a left-handed circularly polarized (LCP) light or right-handed circularly polarized (RCP) wave. Otherwise, the above equation represents an elliptically polarized (EP) wave, where if the x-direction amplitude is normalized to 1, the absolute value of the y-direction amplitude is $|\tan\theta|$, and the phase difference between them is $\pm\pi/2$. The above deduction indicates that an x-polarized wave can theoretically be converted into a y-polarized wave, an EP, or a CP wave after passing through a rectangular bar rotated at an angle of $\pi/4$.

Next, deduce whether it is possible to convert an x-polarized wave into any LP wave. Making $|t_{xx}| = |t_{yy}| = t$, $\varphi_{yy} - \varphi_{xx} = 2\theta$, and Φ to be pending, Eq. (3.4) can be transformed into:

$$\begin{aligned} E_0^t = T_0\begin{pmatrix} 1 \\ 0 \end{pmatrix} &= te^{i\varphi_{xx}}\begin{pmatrix} \cos^2\Phi + \sin^2\Phi \cdot e^{i2\theta} & \cos\Phi\sin\Phi\cdot\left(1-e^{i2\theta}\right) \\ \cos\Phi\sin\Phi\cdot\left(1-e^{i2\theta}\right) & \cos^2\Phi + \sin^2\Phi \cdot e^{i2\theta} \end{pmatrix}\begin{pmatrix} 1 \\ 0 \end{pmatrix} \\ &= te^{i\varphi_{xx}}\begin{pmatrix} \cos^2\Phi + \sin^2\Phi \cdot e^{i2\theta} \\ \cos\Phi\sin\Phi\cdot\left(1-e^{i2\theta}\right) \end{pmatrix} \end{aligned} . \quad (3.8)$$

If the above equation needs to be expressed as an LP wave, the imaginary term needs to be eliminated and can be taken as $2\theta = 0, \pi, 2\pi$. But it is not difficult to find that if $2\theta = 0$ or 2π, the above equation will be fully expressed as an x-polarized state, that is, without polarization conversion. Therefore, in order to obtain other linear polarizations, the above equation can only take $2\theta = \pi$:

$$E_0^t = te^{i\varphi_{xx}}\begin{pmatrix} \cos^2\Phi - \sin^2\Phi \\ 2\cos\Phi\sin\Phi \end{pmatrix} = \begin{pmatrix} \cos 2\Phi \\ \sin 2\Phi \end{pmatrix}. \quad (3.9)$$

The above equation represents an LP wave of an electric vector with an angle of 2Φ to the x-axis. Since the value of Φ can be arbitrary, the incident an x-polarized wave can theoretically be converted into an arbitrary LP wave.

In summary, based on a birefringent rectangular bar, an x-polarized wave can be converted into an EP, a CP, or any other LP state. The conversion conditions are related to the co-polarized transmission phase difference between the x- and y-polarized waves ($\varphi_{yy} - \varphi_{xx}$), the absolute value of the co-polarized transmission coefficient ($|t_{xx}|$, $|t_{yy}|$), and the rotation angle Φ of the rectangular bar, as summarized in Table 3.1.

3.1.1.2 Conversion from EP Waves to Other Polarized Waves

The deduction for converting an incident EP wave into any other polarization state is as follows. The electric vector of the incident EP wave can be represented as $[1 \pm y_0 i]^{\mathrm{T}}$, where the coefficient $y_0 > 0$ and $y_0 \neq 1$ of the y-polarized components and the symbol "±" correspond to different spin directions. Make $\Phi = \pi/4$, i.e. $\cos^2\Phi = \sin^2\Phi = 1/2$, and take

TABLE 3.1 The Conversion Conditions from LP Waves to Other Polarized Waves

Input Wave	Transmission Wave	Conversion Conditions: Co-polarized Transmission Phase Difference "$\theta = (\varphi_{yy} - \varphi_{xx})/2$"	Co-polarized Transmission Coefficient "t_{xx}, t_{yy}"	Rotation Angle of Rectangle Bar "Φ"
x-polarization	y-polarization	$\theta = \pi/2$	$\lvert t_{xx}\rvert = \lvert t_{yy}\rvert$	$\Phi = \pi/4$
	Other LP	$\theta = \pi/2$	$\lvert t_{xx}\rvert = \lvert t_{yy}\rvert$	$\Phi \neq 0, \pi/4, \pi/2$
	LCP/RCP	$\theta = \pi/4, 3\pi/4$	$\lvert t_{xx}\rvert = \lvert t_{yy}\rvert$	$\Phi = \pi/4$
	EP	$\theta \neq 0, \pi/4, \pi/2, 3\pi/4, \pi$	$\lvert t_{xx}\rvert = \lvert t_{yy}\rvert$	$\Phi = \pi/4$

$|t_{xx}| = |t_{yy}| = t$, $\varphi_{yy} - \varphi_{xx} = 2\theta$, $\varphi_{xx} = \varphi - \theta$. Furthermore, when $\theta \neq \pi/2$, i.e. $\cos\theta \neq 0$, the transfer Jones matrix can be extracted from Eq. (3.3) as:

$$T' = \frac{t}{2}e^{i\varphi}\begin{pmatrix} 2\cos\theta & -2i\sin\theta \\ -2i\sin\theta & 2\cos\theta \end{pmatrix} = te^{i\varphi}\cos\theta\begin{pmatrix} 1 & -i\tan\theta \\ -i\tan\theta & 1 \end{pmatrix}. \tag{3.10}$$

If a left-handed elliptically polarized (LEP) wave is incident, the transmitted field is:

$$T'\begin{pmatrix} 1 \\ -y_0 i \end{pmatrix} = te^{i\varphi}\cos\theta\begin{pmatrix} 1 & -i\tan\theta \\ -i\tan\theta & 1 \end{pmatrix}\begin{pmatrix} 1 \\ -y_0 i \end{pmatrix} = te^{i\varphi}\cos\theta\begin{pmatrix} 1 - y_0\tan\theta \\ -i\left(\tan\theta + y_0\right) \end{pmatrix}. \tag{3.11}$$

After simple calculation, it can be found that when $\tan\theta = (1 - y_0)/(1 + y_0)$, that is, $\theta = \arctan[(1 - y_0)/(1 + y_0)]$, the x- and y-polarized components in Eq. (3.11) have equal magnitudes but different signs, which is represented as an LCP wave. When $\tan\theta = (1 + y_0)/(y_0 - 1)$, that is, $\theta = \arctan[(1 + y_0)/(y_0 - 1)]$, the x- and y-polarized components have equal magnitudes and signs, which is represented as an RCP wave. The analysis of the transmitted wave when a right-handed elliptically polarized (REP) wave is incident is similar to the above (the derivation process will not be repeated here). Specifically, when $y_0 = -\tan\theta$ or $1/\tan\theta$, i.e. $\theta = \arctan(-y_0)$ or $\arctan(1/y_0)$, Eq. (3.11) will only retain one x- or y-polarized component, thus representing an x- or y-polarized wave.

Now, let $\Phi = \pi/2$, i.e. $\cos\Phi = 0$, $\sin\Phi = 1$, and take $|t_{xx}| = |t_{yy}| = t$, $\varphi_{yy} - \varphi_{xx} = 2\theta = \pi$. At this point, it can be observed that the transmission matrix Eq. (3.3) is transformed into:

$$T'' = te^{i\varphi_{xx}}\begin{pmatrix} -1 & 0 \\ 0 & 1 \end{pmatrix}. \tag{3.12}$$

Based on Eq. (3.12), it is easy to calculate that regardless of whether the incident wave is an LEP or REP, the transmitted wave will be an EP wave with opposite spin direction. In summary, an EP wave can be converted into a CP, a cross-EP, or a LP wave, with the conditions summarized in Table 3.2.

3.1.1.3 Conversion from CP Waves to Other Polarized Waves

The deduction for converting an incident CP wave into any other polarization state is as follows. The electric vector of the incident CP wave can be expressed as $[1 \pm i]^{T}$. Still let

TABLE 3.2 The Conversion Conditions from LP Waves to Other Polarized Waves

Input Wave	Transmission Wave	Conversion Conditions: Co-polarized Transmission Phase Difference "$\theta = (\varphi_{yy} - \varphi_{xx})/2$"	Co-polarized Transmission Coefficient "t_{xx}, t_{yy}"	Rotation Angle of Rectangle Bar "Φ"
LEP	LCP/RCP	$\theta = \arctan[(1 - y_0)/(1 + y_0)]$, $\arctan[(1 + y_0)/(y_0 - 1)]$	$\lvert t_{xx}\rvert = \lvert t_{yy}\rvert$	$\Phi = \pi/4$
	x- or y-polarization	$\theta = \arctan(-y_0)$, $\arctan(1/y_0)$	$\lvert t_{xx}\rvert = \lvert t_{yy}\rvert$	$\Phi = \pi/4$
	REP	$\theta = \pi/2$	$\lvert t_{xx}\rvert = \lvert t_{yy}\rvert$	$\Phi = \pi/2$

$\Phi = \pi/4$, i.e. $\cos^2\Phi = \sin^2\Phi = 1/2$, and take $|t_{xx}| = |t_{yy}| = t$, $\varphi_{yy} - \varphi_{xx} = 2\theta$, $\varphi_{xx} = \varphi - \theta$, $\theta \neq \pi/2$ so that the transmission matrix represented by Eq. (3.10) can be used again. When an LCP wave is incident, the transmitted wave electric field is:

$$T'\begin{pmatrix} 1 \\ -i \end{pmatrix} = te^{i\varphi}\cos\theta \begin{pmatrix} 1 & -i\tan\theta \\ -i\tan\theta & 1 \end{pmatrix}\begin{pmatrix} 1 \\ -i \end{pmatrix} = te^{i\varphi}\cos\theta \begin{pmatrix} 1-\tan\theta \\ -i(1+\tan\theta) \end{pmatrix}. \tag{3.13}$$

Since $|\tan\theta| \neq 0$ and 1, i.e. $\theta \neq 0$ and π, and $\theta \neq \pi/2$, the above equation represents a transmitted EP wave. Easy to calculate, when an RCP wave is incident, the transmitted electric field also represents an EP wave (the derivation process will not be repeated here). Specifically, if $|\tan\theta| = 1$, that is, $\theta = \pi/4$ or $3\pi/4$, Eq. (3.13) represents an LP wave in the x- or y-direction. In addition, if we want to obtain a CP wave with opposite spin direction, which only needs to use the conditions required by Eq. (3.12), that is, $\Phi = \pi/2$, $|t_{xx}| = |t_{yy}| = t$, and $\theta = \pi/2$. In summary, a CP wave can be converted into an EP, a cross-CP, or a LP wave, and the required conditions are summarized in Table 3.3.

3.1.2 Design and Analysis of Birefringent Polarization Conversion Devices

The above content deduces the theory of interchange between different electromagnetic polarization states, providing a systematic theoretical foundation for polarization conversion. In practical applications, linear and circular polarization states are the most common; therefore, the focus here will further narrow to the physical realization and application of linear-circular polarization state interchange. To verify the birefringence effect and phase modulation characteristics of metasurfaces, this section will

TABLE 3.3 The Conversion Conditions from EP Waves to Other Polarized Waves

Input Wave	Transmission Wave	Conversion Conditions: Co-polarized Transmission Phase Difference "$\theta = (\varphi_{yy} - \varphi_{xx})/2$"	Co-polarized Transmission Coefficient "t_{xx}, t_{yy}"	Rotation Angle of Rectangle Bar "Φ"
LCP	EP	$\theta \neq 0, \pi/2, \pi$	$\lvert t_{xx}\rvert = \lvert t_{yy}\rvert$	$\Phi = \pi/4$
	x- or y-polarization	$\theta = \pi/4, 3\pi/4$	$\lvert t_{xx}\rvert = \lvert t_{yy}\rvert$	$\Phi = \pi/4$
	RCP	$\theta = \pi/2$	$\lvert t_{xx}\rvert = \lvert t_{yy}\rvert$	$\Phi = \pi/2$

demonstrate a design method for a metalens capable of linear-circular polarization conversion. If certain conditions are met, including $\Phi = \pi/4$, $|t_{xx}| = |t_{yy}| = t$, and $\varphi_{yy} - \varphi_{xx} = \pi/2$, then Eq. (3.4) can be simplified to:

$$E_0^t = T_0\begin{pmatrix} 1 \\ 0 \end{pmatrix} = \frac{t}{2}e^{i\varphi_{xx}}\begin{pmatrix} 1+e^{i\frac{\pi}{2}} & 1-e^{i\frac{\pi}{2}} \\ 1-e^{i\frac{\pi}{2}} & 1+e^{i\frac{\pi}{2}} \end{pmatrix}\begin{pmatrix} 1 \\ 0 \end{pmatrix}, \tag{3.14}$$

where, for computational convenience, let t equal 1 (meaning the reflection and absorption of the dielectric material are neglected). Furthermore, φ_{xx} can be regarded as the sum of a phase φ and an additional phase $-\pi/4$, and thus Eq. (3.14) simplifies to:

$$E_0^t = T_0\begin{pmatrix} 1 \\ 0 \end{pmatrix}$$

$$= \frac{t}{2}e^{i\left(\varphi-\frac{\pi}{4}\right)}\begin{pmatrix} 1+e^{i\frac{\pi}{2}} & 1-e^{i\frac{\pi}{2}} \\ 1-e^{i\frac{\pi}{2}} & 1+e^{i\frac{\pi}{2}} \end{pmatrix}\begin{pmatrix} 1 \\ 0 \end{pmatrix} = \frac{t\sqrt{2}}{2}e^{i\varphi}\begin{pmatrix} 1 & -i \\ -i & 1 \end{pmatrix}\begin{pmatrix} 1 \\ 0 \end{pmatrix} = \frac{t\sqrt{2}}{2}e^{i\varphi}\begin{pmatrix} 1 \\ -i \end{pmatrix}, \tag{3.15}$$

The vector $[1\ -i]^{\mathrm{T}}$ in Eq. (3.15) is a typical form of the LCP electric field, meaning that when conditions such as $\Phi = \pi/4$, $|t_{xx}| = |t_{yy}| = t$, and $\varphi_{yy} - \varphi_{xx} = \pi/2$ are satisfied, an LCP wave output is obtained. According to Eq. (3.15), the simple form of the transfer Jones matrix T_0 can be written as:

$$T_0 = \frac{t\sqrt{2}}{2}e^{i\varphi}\begin{pmatrix} 1 & -i \\ -i & 1 \end{pmatrix} = \beta\begin{pmatrix} 1 & -i \\ -i & 1 \end{pmatrix}, \beta = \frac{t\sqrt{2}}{2}e^{i\varphi}. \tag{3.16}$$

Equations (3.15) and (3.16) describe the birefringence effect using the Jones matrix. An x-polarized wave passing through this unit cell structure will split into a co-polarized component (T_x) and a cross-polarized component (T_y). These two mutually perpendicular transmission components (T_x and T_y) have a phase difference of $\pi/2$ and the same amplitude, ultimately combining into an LCP wave.

The theory of birefringence-induced polarization conversion can be verified by the following numerical simulations and experiments. The metasurface model was established in the commercial software CST Microwave Studio, and all numerical simulation results were obtained by the time-domain solver. A simple silicon rectangular bar was used as the basic structure to construct the metalens. The structural parameters of the unit cell (height, length, and width of the silicon rectangular bar) were selected from a large number of parameter scanning results. The metalens sample was fabricated using a high-resistivity silicon (Si) wafer (relative permittivity $\varepsilon = 11.9$, resistivity $\rho > 5000\ \Omega{\cdot}\mathrm{cm}$) with a thickness of 500 μm, prepared using a plasma etching method (ICP ion etcher using SF_6 as the etching gas). The silicon substrate thickness was $h_1 = 300$ μm, the silicon bar height was

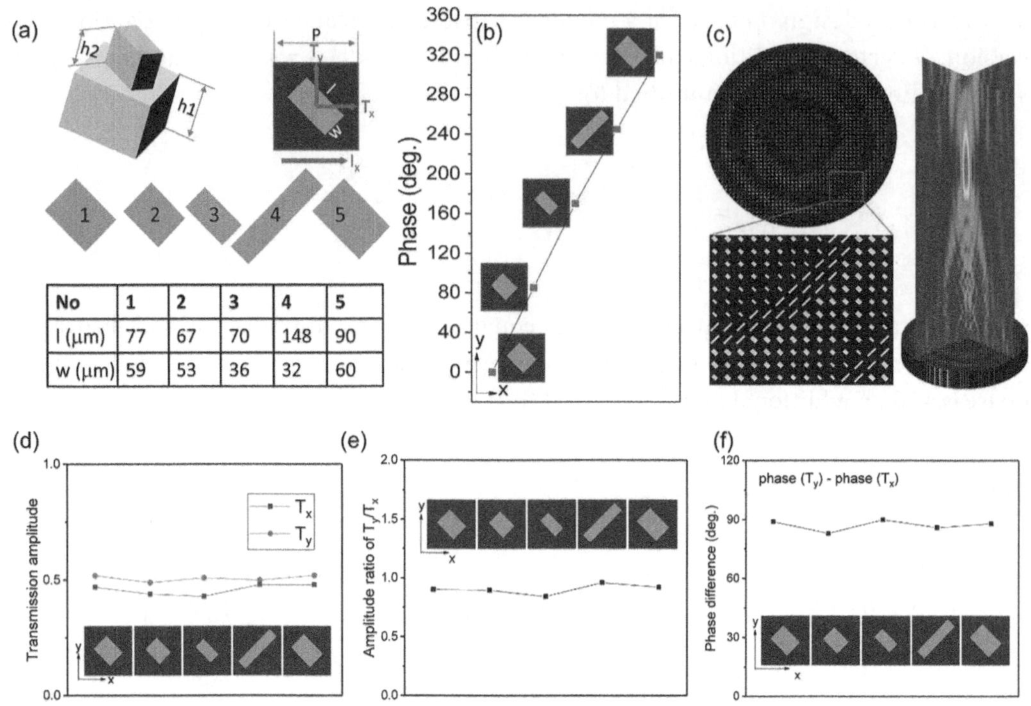

FIGURE 3.2 (a) Schematic diagrams of the five basic structures; parameters in the inset table are length "*l*" and width "*w*"; I_x – incident *x*-polarized wave, T_x – co-polarized transmission along the *x*-axis, and T_y – cross-polarized transmission along the *y*-axis. (b) Phase delay based on T_x for the five basic structures. (c) Schematic diagram of the metalens device. (d) Amplitudes of T_x and T_y under incident *x*-polarized wave. (e) Amplitude ratio of T_y/T_x. (f) Phase difference between T_x and T_y.

$h_2 = 200$ μm, and the lattice constant of the square unit cell structure was $P = 160$ μm. By changing the geometric parameters of the silicon bar length and width, a phase variation of 0–2π was obtained. Five basic structures were identified, whose shapes and parameters are shown in Fig. 3.2(a). The phase (based on T_x) of these five basic structures varies approximately from 0 to 2π, as shown in Fig. 3.2(b). Furthermore, using these five basic structures to construct an array satisfying the phase distribution $\varphi = \frac{2\pi}{\lambda}\left(f - \sqrt{x^2 + y^2 + f^2}\right)$ yields a metalens, as shown in Fig. 3.2(c), where the wavelength λ corresponds to 0.96 THz, and (x, y) are the coordinates of any point within the metalens (with the metalens centre as the coordinate origin (0, 0)), and $f = 8$ mm is the focal length of the metalens. Simultaneously, to achieve complete and stable LCP conversion, it is required that the two components (T_x and T_y) of each silicon bar have almost identical amplitudes and a stable $\pi/2$ phase difference, as shown in Fig. 3.2(d–f).

When the metasurface is rotated by angles θ = 0°, 45°, 90°, 135°, respectively, each unit cell structure will rotate accordingly, as shown in Fig. 3.3(a). Consequently, the transfer Jones matrix $T_0 = \beta\begin{pmatrix} 1 & -i \\ -i & 1 \end{pmatrix}$ will change via the basis transformation matrix

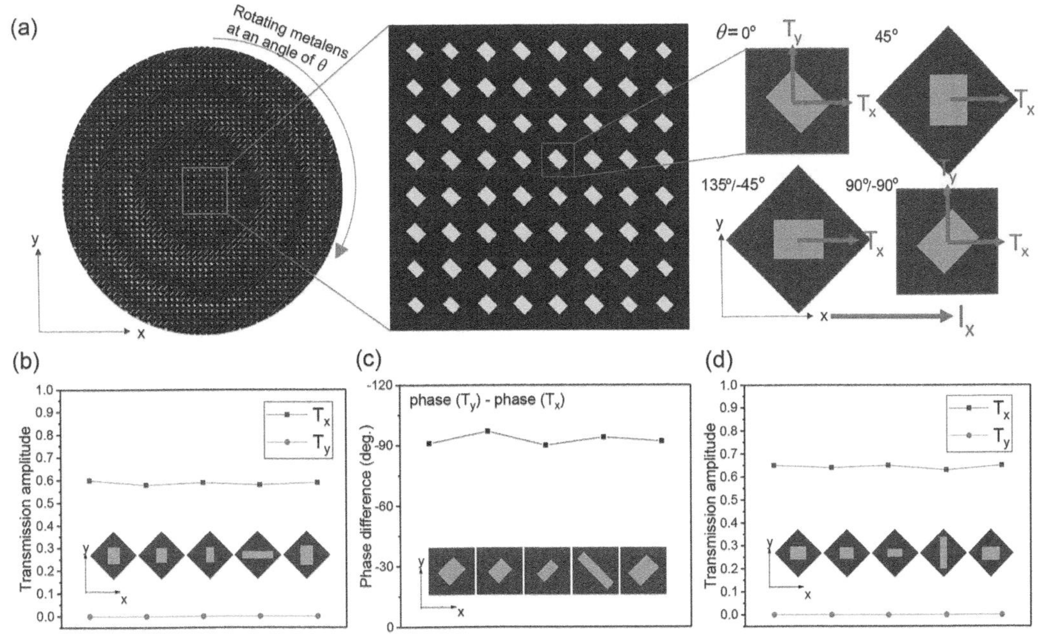

FIGURE 3.3 (a) Schematic change of the unit structure when the entire metalens is rotated by angle θ. (b) Transmission amplitudes T_x and T_y under incident x-polarized wave when the entire metalens is rotated by 45°. (c) Phase difference between T_x and T_y under incident x-polarized wave when the entire metalens is rotated by ±90°. (d) Phase difference between T_x and T_y under incident x-polarized wave when the entire metalens is rotated by 135°/−45°.

$\Lambda_2 = \begin{pmatrix} \cos\theta & -\sin\theta \\ \sin\theta & \cos\theta \end{pmatrix}$ (derivation of the coordinate transformation matrix will be seen in the following text) as follows.

For $\theta = 45°$, there is:

$$\boldsymbol{T}_1 = \Lambda_2 \beta \begin{pmatrix} 1 & -i \\ -i & 1 \end{pmatrix} \Lambda_2^{-1} = \frac{\sqrt{2}}{2}\beta \begin{pmatrix} 1 & -1 \\ 1 & 1 \end{pmatrix} \begin{pmatrix} 1 & -i \\ -i & 1 \end{pmatrix} \begin{pmatrix} 1 & 1 \\ -1 & 1 \end{pmatrix} = \sqrt{2}\beta \begin{pmatrix} 1+i & 0 \\ 0 & 1-i \end{pmatrix}. \quad (3.17)$$

For $\theta = 90°/-90°$, there is:

$$\boldsymbol{T}_2 = \Lambda_2 \beta \begin{pmatrix} 1 & -i \\ -i & 1 \end{pmatrix} \Lambda_2^{-1} = \beta \begin{pmatrix} 0 & -1 \\ 1 & 0 \end{pmatrix} \begin{pmatrix} 1 & -i \\ -i & 1 \end{pmatrix} \begin{pmatrix} 0 & 1 \\ -1 & 0 \end{pmatrix} = \beta \begin{pmatrix} 1 & i \\ i & 1 \end{pmatrix}. \quad (3.18)$$

For $\theta = 135°/-45°$, there is:

$$\boldsymbol{T}_3 = \Lambda_2 \beta \begin{pmatrix} 1 & -i \\ -i & 1 \end{pmatrix} \Lambda_2^{-1} = \frac{\sqrt{2}}{2}\beta \begin{pmatrix} 1 & 1 \\ -1 & 1 \end{pmatrix} \begin{pmatrix} 1 & -i \\ -i & 1 \end{pmatrix} \begin{pmatrix} 1 & -1 \\ 1 & 1 \end{pmatrix} = \sqrt{2}\beta \begin{pmatrix} 1-i & 0 \\ 0 & 1+i \end{pmatrix}. \quad (3.19)$$

Correspondingly, under x-polarized incidence, the electric field of the output wave is expressed as follows.

For $\theta = 45°$, there is:

$$\mathbf{E}_1^t = \mathbf{T}_1\begin{pmatrix} 1 \\ 0 \end{pmatrix} = \sqrt{2}\beta(1+i)\begin{pmatrix} 1 \\ 0 \end{pmatrix}. \tag{3.20}$$

For $\theta = 90°/-90°$, there is:

$$\mathbf{E}_2^t = \mathbf{T}_2\begin{pmatrix} 1 \\ 0 \end{pmatrix} = \beta\begin{pmatrix} 1 \\ i \end{pmatrix}. \tag{3.21}$$

For $\theta = 135°/-45°$, there is:

$$\mathbf{E}_3^t = \mathbf{T}_3\begin{pmatrix} 1 \\ 0 \end{pmatrix} = \sqrt{2}\beta(1-i)\begin{pmatrix} 1 \\ 0 \end{pmatrix}. \tag{3.22}$$

Equations (3.20) and (3.22) imply that when the metalens is rotated by ±45°, the transmitted wave remains in an x-polarized state without polarization conversion. Meanwhile, the five basic structures show almost identical transmission amplitudes, as shown in Figs. 3.3(b, d). Under these conditions, the metalens will focus the wave as an x-polarized field without polarization conversion. When the metalens is rotated by ±90°, Eq. (3.21) shows that transmitted RCP conversion will occur. Clearly, the transmission amplitudes of the two components (T_x and T_y) are the same as before rotating the metalens, but the two components (T_x and T_y) will have a phase difference of $\Delta\varphi = -\pi/2$, as shown in Fig. 3.3(c), meaning the metalens converts the incident x-polarized wave into an RCP wave.

The above analysis indicates that we only need to rotate this metasurface by ±90° to achieve focusing with arbitrary circular polarization conversion for the incident x-polarized wave, respectively. In fact, it is easy to understand that any linear polarized wave in the x-o-y plane passing through the metasurface can also be focused and undergo arbitrary circular polarization conversion. For example, if the incident linear polarized wave is at an angle ϕ to the x-axis, we can rotate the metasurface by an additional angle ϕ on the basis of the rotation angle θ. This would then have the same significance as the aforementioned cases. Therefore, focusing and conversion from any linear polarization to any circular polarization can be achieved. Here, we can give their general forms.

Any incident linear polarized wave in the x-o-y plane can be obtained by rotating the x-polarized wave by an angle ϕ and described based on the coordinate transformation principle. As shown in Fig. 3.4, when a vector $\mathbf{P}(x_1, y_1)$ is rotated by an angle ϕ, the new vector $\mathbf{P}'(x_2, y_2)$ can be expressed by Eq. (3.23) and Eq. (3.24):

$$\begin{cases} x_2 = x_1\cos\phi - y_1\sin\phi \\ y_2 = x_1\sin\phi + y_1\cos\phi \end{cases}. \tag{3.23}$$

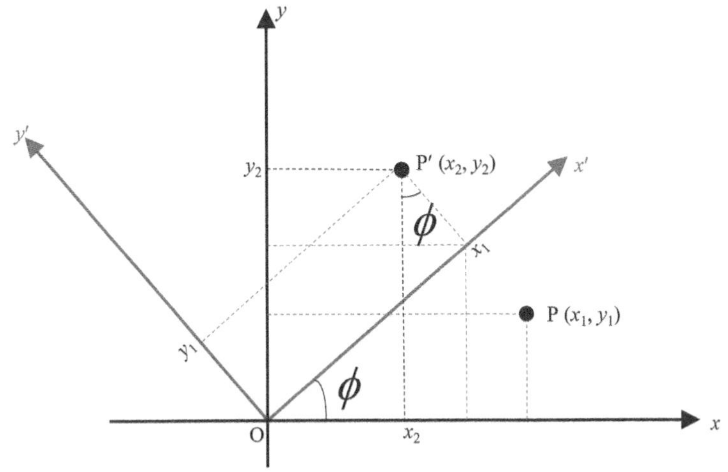

FIGURE 3.4 Change of a vector in the coordinate transformation.

$$\begin{pmatrix} x_2 \\ y_2 \end{pmatrix} = \begin{pmatrix} \cos\phi & -\sin\phi \\ \sin\phi & \cos\phi \end{pmatrix} \begin{pmatrix} x_1 \\ y_1 \end{pmatrix}. \tag{3.24}$$

According to Eq. (3.24), the coordinate transformation matrix $\boldsymbol{H}$ is as follows:

$$\boldsymbol{H} = \begin{pmatrix} \cos\phi & -\sin\phi \\ \sin\phi & \cos\phi \end{pmatrix}. \tag{3.25}$$

Therefore, the arbitrary incident LP $\boldsymbol{E}_A^i$ can be written as follows:

$$\boldsymbol{E}_A^i = \boldsymbol{H} \begin{pmatrix} 1 \\ 0 \end{pmatrix} = \begin{pmatrix} \cos\phi & -\sin\phi \\ \sin\phi & \cos\phi \end{pmatrix} \begin{pmatrix} 1 \\ 0 \end{pmatrix} = \begin{pmatrix} \cos\phi \\ \sin\phi \end{pmatrix}. \tag{3.26}$$

Subsequently, as shown in Fig. 3.5, we rotate the metasurface by an additional angle ϕ on the basis of the rotation angle θ. This will be equivalent to the case of an x-polarized

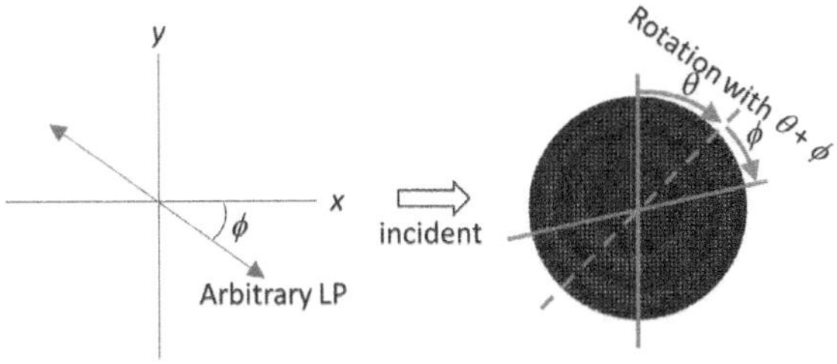

FIGURE 3.5 Schematic diagram of an arbitrary incident linear polarized wave and the metasurface rotated by an additional angle ϕ on the basis of the rotation angle θ.

wave passing through the metasurface without the additional ϕ rotation. At this point, the transfer Jones matrix $\boldsymbol{T}_n'$ based on the $\boldsymbol{H}$ transformation is:

$$\boldsymbol{T}_n' = \boldsymbol{H}\boldsymbol{T}_n\boldsymbol{H}^{-1} = \begin{pmatrix} \cos\phi & -\sin\phi \\ \sin\phi & \cos\phi \end{pmatrix} \boldsymbol{T}_n \begin{pmatrix} \cos\phi & \sin\phi \\ -\sin\phi & \cos\phi \end{pmatrix}, \quad n = 0,1,2,3. \tag{3.27}$$

For example, for the original metalens $n = 0$, the $\boldsymbol{T}_0'$ is:

$$\begin{aligned} \boldsymbol{T}_0' &= \begin{pmatrix} \cos\phi & -\sin\phi \\ \sin\phi & \cos\phi \end{pmatrix} \beta \begin{pmatrix} 1 & -i \\ -i & 1 \end{pmatrix} \begin{pmatrix} \cos\phi & \sin\phi \\ -\sin\phi & \cos\phi \end{pmatrix} \\ &= \beta \begin{pmatrix} \cos\phi + i\sin\phi & -i\cos\phi - \sin\phi \\ \sin\phi - i\cos\phi & -i\sin\phi + \cos\phi \end{pmatrix} \begin{pmatrix} \cos\phi & \sin\phi \\ -\sin\phi & \cos\phi \end{pmatrix} \\ &= \beta \begin{pmatrix} 1 + i\sin 2\phi & -i\cos 2\phi \\ -i\cos 2\phi & 1 - i\sin 2\phi \end{pmatrix} \end{aligned}. \tag{3.28}$$

And the transmitted electric field $\boldsymbol{E}_0^{t'}$ can be further expressed as:

$$\begin{aligned} \boldsymbol{E}_0^{t'} &= \boldsymbol{T}_0'\boldsymbol{E}_A^i = \boldsymbol{T}_0' \begin{pmatrix} \cos\phi \\ \sin\phi \end{pmatrix} = \beta \begin{pmatrix} 1 + i\sin 2\phi & -i\cos 2\phi \\ -i\cos 2\phi & 1 - i\sin 2\phi \end{pmatrix} \begin{pmatrix} \cos\phi \\ \sin\phi \end{pmatrix} \\ &= \beta \begin{pmatrix} \cos\phi + i\sin 2\phi\cos\phi - i\cos 2\phi\sin\phi \\ -i\cos 2\phi\cos\phi + \sin\phi - i\sin 2\phi\sin\phi \end{pmatrix} \\ &= \beta \begin{pmatrix} \cos\phi + i\sin\phi \\ -i\cos\phi + \sin\phi \end{pmatrix} = \beta \begin{pmatrix} \cos\phi & -\sin\phi \\ \sin\phi & \cos\phi \end{pmatrix} \begin{pmatrix} 1 \\ -i \end{pmatrix} \end{aligned}. \tag{3.29}$$

Further, substituting Eq. (3.15) and Eq. (3.25) into Eq. (3.29), we obtain:

$$\boldsymbol{E}_0^{t'} = \boldsymbol{H}\boldsymbol{E}_0^t. \tag{3.30}$$

From Eq. (3.30), it can be seen that the final transmitted electric field $\boldsymbol{E}_0^{t'}$ is equivalent to rotating the original transmitted electric field $\boldsymbol{E}_0^t$ by an additional angle ϕ. Therefore, the general form of the transmitted electric field $\boldsymbol{E}_n^{t'}$ can be easily derived as follows:

$$\boldsymbol{E}_n^{t'} = \boldsymbol{T}_n'\boldsymbol{E}_A^i = \begin{pmatrix} \cos\phi & -\sin\phi \\ \sin\phi & \cos\phi \end{pmatrix} \boldsymbol{E}_n^t = \boldsymbol{H}\boldsymbol{E}_n^t, \quad n = 0,1,2,3. \tag{3.31}$$

The scanning electron microscope (SEM) image of the original metalens is shown in Fig. 3.6(a). Figure 3.6(b) shows the simulation and experimental results of the incident x-polarized wave converted into a transmitted circular polarized wave and focused on the focal plane. Overall, these experimental results agree well with the simulation results.

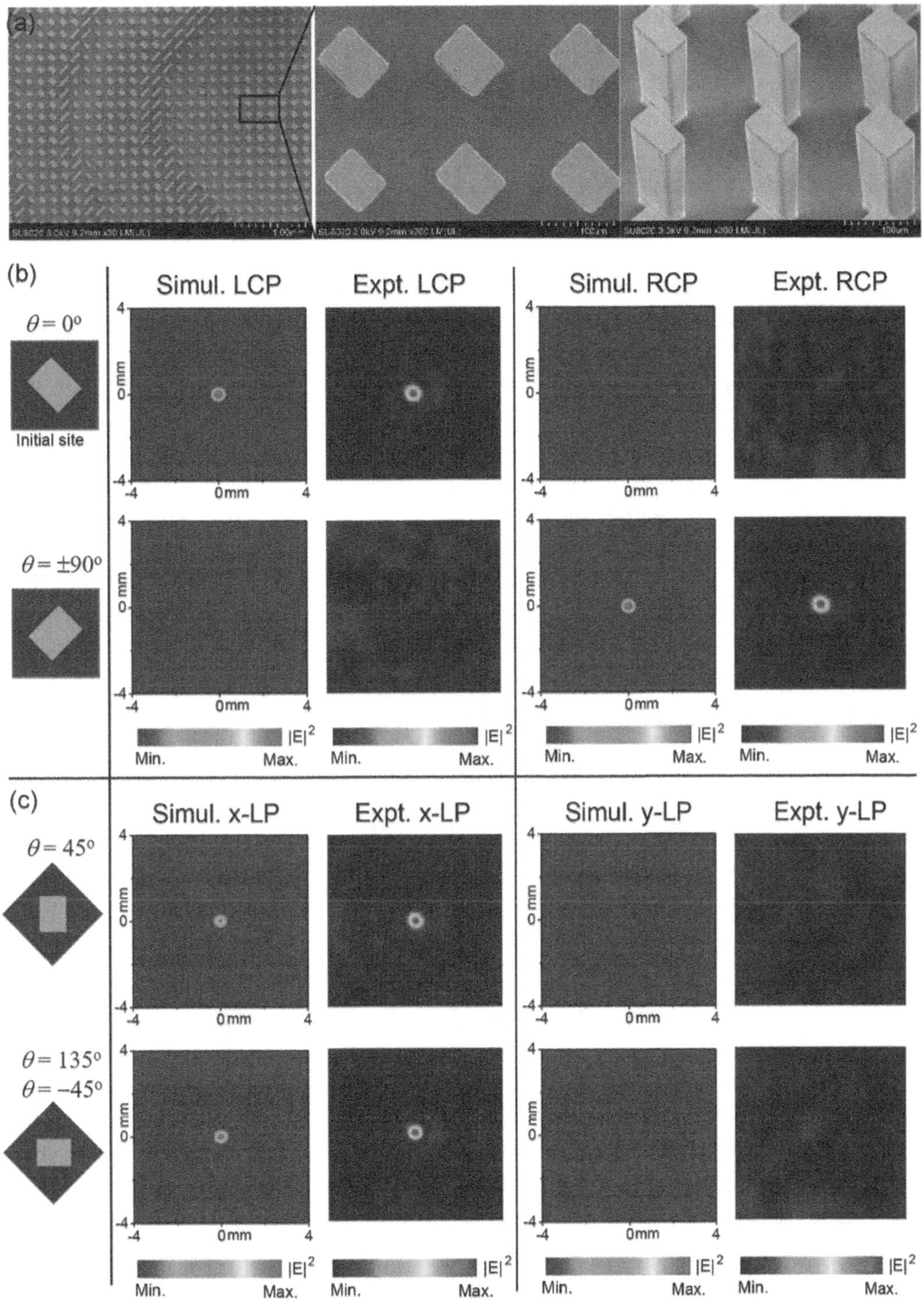

FIGURE 3.6 (a) SEM image of the metalens sample. (b) Simulation and experimental results of the incident x-polarized wave converted into a transmitted circular polarized wave and focused. (c) Simulation and experimental results of the incident x-polarized wave transmitted without polarization conversion.

The original metalens without any rotation ($\theta = 0°$) functions to convert the x-polarized wave into an LCP wave. When the metalens is rotated by $\theta = \pm 90°$, conversion from the x-polarized wave to an RCP wave can be achieved. Furthermore, if we wish to maintain x-polarized transmission without polarization conversion, the metalens can be rotated by $\theta = \pm 45°$, as shown in Fig. 3.6(c).

Reversible polarization conversion is another characteristic of this metalens, meaning it can not only convert an incident linear polarized wave into a circular polarized wave but also convert an incident circular polarized wave into a linear polarized wave. For the original metalens without any rotation ($\theta = 0°$), the incident RCP and LCP are transmitted by the transfer Jones matrix $\boldsymbol{T}_0$, and the transmitted fields $\boldsymbol{E}_n^t$ ($n = 4, 5$) are as follows:

$$\boldsymbol{E}_4^t = \boldsymbol{T}_0 \begin{pmatrix} 1 \\ i \end{pmatrix} = \beta \begin{pmatrix} 1 & -i \\ -i & 1 \end{pmatrix} \begin{pmatrix} 1 \\ i \end{pmatrix} = 2\beta \begin{pmatrix} 1 \\ 0 \end{pmatrix}. \tag{3.32}$$

$$\boldsymbol{E}_5^t = \boldsymbol{T}_0 \begin{pmatrix} 1 \\ -i \end{pmatrix} = \beta \begin{pmatrix} 1 & -i \\ -i & 1 \end{pmatrix} \begin{pmatrix} 1 \\ -i \end{pmatrix} = -2\beta \begin{pmatrix} 0 \\ 1 \end{pmatrix}. \tag{3.33}$$

The above two formulas indicate that RCP passing through the original metalens will be converted into an x-polarized wave, while LCP passing through the metalens will be converted into a y-polarized wave.

Furthermore, we can rotate the metalens to change the direction of the transmitted linear polarized wave. Based on the above derivation, it is easy to understand that when the metalens is rotated by an arbitrary angle θ, the transfer Jones matrix is:

$$\boldsymbol{T}' = \beta \begin{pmatrix} 1 + i\sin 2\theta & -i\cos 2\theta \\ -i\cos 2\theta & 1 - i\sin 2\theta \end{pmatrix}. \tag{3.34}$$

When an RCP wave passes through the metalens, the transmission electric field $\boldsymbol{E}_4^{t'}$ is:

$$\boldsymbol{E}_4^{t'} = \boldsymbol{T}' \begin{pmatrix} 1 \\ i \end{pmatrix} = \beta \begin{pmatrix} 1 + i\sin 2\theta & -i\cos 2\theta \\ -i\cos 2\theta & 1 - i\sin 2\theta \end{pmatrix} \begin{pmatrix} 1 \\ i \end{pmatrix} = \beta \begin{pmatrix} 1 + e^{i2\theta} \\ i\left(1 - e^{i2\theta}\right) \end{pmatrix} = 2\beta e^{i\theta} \begin{pmatrix} \cos\theta \\ \sin\theta \end{pmatrix}. \tag{3.35}$$

Equation (3.35) shows that when the metalens is rotated by an angle θ, the RCP wave passing through the metalens will be converted into a linear polarized wave at an angle θ to the x-axis, as shown in Fig. 3.7(a). Similarly, when an LCP passes through the metalens, the transmitted electric field $\boldsymbol{E}_5^{t'}$ is:

$$\begin{aligned} \boldsymbol{E}_5^{t'} = \boldsymbol{T}' \begin{pmatrix} 1 \\ -i \end{pmatrix} &= \beta \begin{pmatrix} 1 + i\sin 2\theta & -i\cos 2\theta \\ -i\cos 2\theta & 1 - i\sin 2\theta \end{pmatrix} \begin{pmatrix} 1 \\ -i \end{pmatrix} = \beta \begin{pmatrix} 1 - e^{-i2\theta} \\ -i\left(1 + e^{-i2\theta}\right) \end{pmatrix} \\ &= -2\beta i e^{-i\theta} \begin{pmatrix} -\sin\theta \\ \cos\theta \end{pmatrix} = -2\beta e^{i\left(\frac{\pi}{2} - \theta\right)} \begin{pmatrix} -\sin\theta \\ \cos\theta \end{pmatrix} \end{aligned} . \tag{3.36}$$

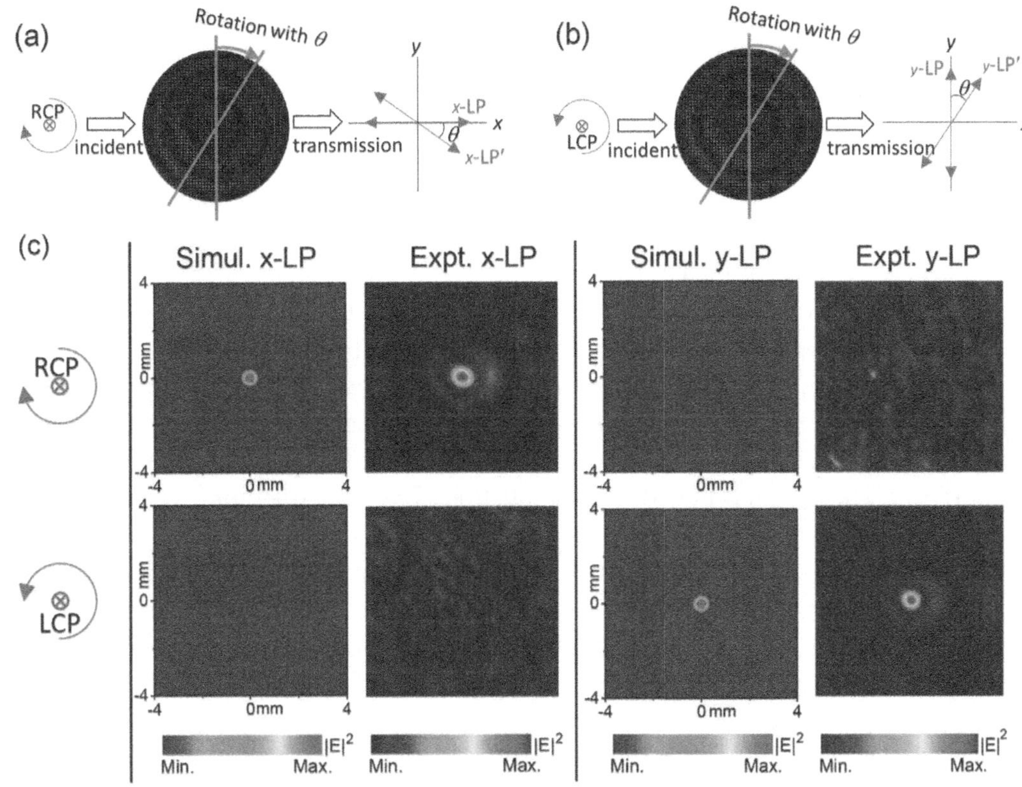

FIGURE 3.7 (a, b) RCP and LCP passing through the rotated metalens will be converted into linear polarized waves at angles θ to the x-axis and y-axis, respectively. The x-LP and y-LP correspond to the case of the unrotated metalens, and the x-LP′ and y-LP′ correspond to the case of the rotated metalens. (c) Simulation and experimental results of the incident circular polarized wave focused and converted into a transmitted linear polarized wave.

Equation (3.36) shows that when the metalens is rotated by an angle θ, the LCP passing through the metalens will be converted into a linear polarized wave at an angle θ to the y-axis, as shown in Fig. 3.7(b).

To verify the above theoretical analysis, Fig. 3.7(c) shows the simulation and experimental results of the incident circular polarized wave converted into a transmitted linear polarized wave and focused. Since there is a correspondence between the transmitted linear polarized waves before and after rotating the metalens, only the results for the original metalens ($\theta = 0°$) are shown here. From Fig. 3.7(c), it can be seen that the original metalens functions to convert RCP into an x-polarized wave and LCP into a y-polarized wave, which well verifies the above theoretical analysis.

3.2 RESONANCE-DRIVEN POLARIZATION CONVERSION

3.2.1 Resonance Photonic Band and Electromagnetic Multipoles

This section will introduce another commonly used polarization conversion method, namely electromagnetic resonance. There are electromagnetic eigenstates in optical structures, and the electromagnetic structures of these eigenstates may be simple or quite complex, but they all originate from several basic multipoles, such as electric dipoles, magnetic dipoles, toroidal dipoles, and so on. They can be excited by light waves with specific frequencies and polarization states and coupled with the incident wave, thereby changing the polarization state of the output wave. When light and matter undergo electromagnetic resonance, it can cause a phase transition, which can also facilitate the polarization conversion process. The phase change caused by light-matter resonance is called "resonance phase", which is different from the birefringence effect. In the birefringence effect, different polarized waves are affected by the refractive index of the medium in different directions, resulting in different phase delays during propagation – this phase delay is a type of "propagation phase".

The electromagnetic behaviour in all optical devices can be described by Maxwell's equations. For periodic photonic systems, this has been elaborated in detail in the book "*Photonic Crystals: Molding the Flow of Light*" by John D. Joannopoulos et al. [2] and will not be repeated here. Simply put, Maxwell's equations for periodic photonic systems possess a series of Bloch eigen solutions. Each eigen solution corresponds to an angular frequency ω and a wavevector k, along with an associated electromagnetic mode. Representing the angular frequency and wavevector in ω-k coordinates yields the photonic band structure. In photonic crystals, the dispersion of light can be understood from a perspective similar to the free electron approximation. In a uniform, infinitely extended medium, light exhibits the dispersion relation $\omega = |k|c/n$, where c is the speed of light in vacuum and n is the refractive index of the medium. When a periodic potential is introduced, breaking the continuous translational symmetry of the system and replacing it with discrete translational symmetry, the Bloch boundary condition makes k and $k + G$ (where G is the reciprocal lattice vector) equivalent. In other words, the original dispersion relation becomes folded at the boundaries of the Brillouin zone, and band gaps open at these boundaries, forming the photonic band structure. In photonic crystal slabs, the folded dispersion originates from the guided modes within the dielectric slab. A uniform dielectric slab, due to the boundary conditions at its upper and lower surfaces, supports propagating modes confined between the surfaces via total internal reflection – these are known as waveguide modes, as shown in Fig. 3.8(a). When the slab is periodically patterned, these modes become folded due to the Bloch boundary condition, forming the band structure, as illustrated in Fig. 3.8(b).

The eigen electromagnetic modes play a decisive role in determining the far-field polarization states of photonic devices and can be further analysed through multipole decomposition techniques. Multipole decomposition represents an arbitrary field distribution as a superposition of fields generated by a set of multipoles, mathematically expressed as a series expansion, with the contributions of different multipoles reflected in the mathematical results of the multipole expansion. Multipoles are primarily categorized into three series [3]: electric multipoles, magnetic multipoles, and toroidal multipoles. Each series can be further divided into dipole, quadrupole, and octupole orders, as illustrated intuitively in Fig. 3.8(c). Electric dipoles and magnetic dipoles are the two most fundamental types of multipoles proposed earliest in electromagnetism. An electric dipole consists of a pair of positive and negative charges, with the charges oscillating along the dipole moment direction. A magnetic dipole is formed by an oscillating circular current loop, with its magnetic

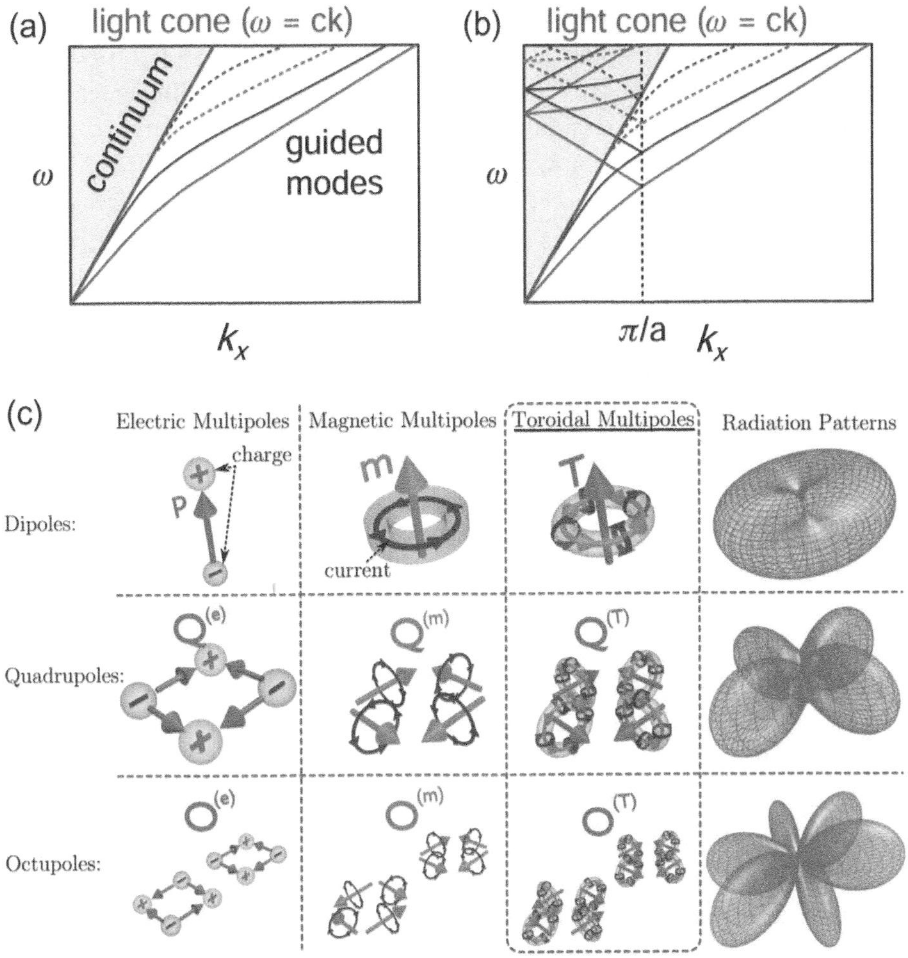

FIGURE 3.8 (a) The photonic band structure for a uniform dielectric slab. (b) The photonic band structure for a photonic crystal slab. (c) Fundamental electromagnetic multipoles.

moment located at the centre of the loop and perpendicular to the loop's plane. A toroidal dipole is generated by an oscillating poloidal current, which flows along the meridian of a cylindrical torus. Intuitively, it can be visualized as an infinite series of magnetic dipoles connected head-to-tail, with the toroidal dipole moment located at the centre of the torus and perpendicular to its plane. These three types of multipoles form the foundation for the higher-order members in their respective series. For example, a quadrupole is created by two anti-aligned dipoles, and an octupole by anti-aligned quadrupoles. All multipoles exhibit no electromagnetic radiation field along their dipole moment direction; instead, the electromagnetic field radiates in directions other than the direction of the dipole moment.

3.2.2 Design and Analysis of Resonant Polarization Conversion Devices

Both metal materials and dielectric materials can produce electromagnetic resonance with light waves. Generally speaking, resonance-driven polarization conversion based on metal material systems has significant losses, which is not conducive to efficient polarization conversion. Dielectric material systems have low losses and are commonly used for efficient polarization conversion. Next, we will demonstrate the benefits and implementation methods of resonance-driven polarization conversion through a case study – a meta-mirror preserving the handedness of arbitrarily circular polarization waves.

Mirror reflection is a fundamental phenomenon in electromagnetic wave propagation. Mirror reflection perfectly preserves the polarization states of the incident LP waves, and the phase shifts upon reflection are identical for all LP waves [4]. However, when a CP wave is incident, the reflected wave viewed along the propagation direction undergoes a cross-polarization conversion (handedness reversal) – a limitation inherent to conventional mirror reflection [5, 6]. In optical system designs involving mirrors and CP waves, conventional mirror reflection may complicate the design process due to the need to account for the abrupt variable of handedness reversal. If a new type of mirror reflection could maintain the handedness of CP waves, the adverse effects induced by conventional mirror reflection might be avoided, potentially simplifying the design complexity of optical systems.

Metasurfaces have demonstrated remarkable capabilities in manipulating the polarization states of electromagnetic waves [7–17], holding the potential to achieve co-polarized reflection for arbitrarily polarized incident waves. Herein, we provide an investigation into the theoretical principles and design methodology of the CP wave reflector. Starting from the reflective behaviour of the device for LP waves, we theoretically derived the conditions for efficient co-polarized reflection of CP waves with arbitrary handedness. It is discovered that the device must exhibit co-polarized x- and y-polarized reflections with equal amplitude and opposite phase, while suppressing all cross-polarized LP components. Through numerical simulations, a meta-mirror composed of a high-resistance silicon-based metasurface and a metal plate is designed. The electromagnetic resonance of the metasurface can only be excited by a y-polarized wave and remains unresponsive to an x-polarized wave, generating only co-polarized x- and y-polarized reflection with opposite phases. The metal plate serves to eliminate transmission losses, ensuring efficient reflection of co-polarized x- and y-polarized waves while maintaining equal amplitudes. Ultimately, the device successfully achieves efficient co-polarized reflection for arbitrarily handed CP waves.

We begin by analysing the reflection behaviour of the metasurface for LP waves and then derive the theoretical formulation for the reflection of co-polarized CP waves. The relationship between the incident and reflected LP waves is generally expressed as:

$$\boldsymbol{E}^r_{x,y} = \boldsymbol{R}_{x,y}\boldsymbol{E}^i_{x,y} \Rightarrow \begin{pmatrix} E^r_x \\ E^r_y \end{pmatrix} = \begin{pmatrix} r_{xx} & r_{xy} \\ r_{yx} & r_{yy} \end{pmatrix}\begin{pmatrix} E^i_x \\ E^i_y \end{pmatrix} = \begin{pmatrix} A_{xx}e^{i\Psi_{xx}} & A_{xy}e^{i\Psi_{xy}} \\ A_{yx}e^{i\Psi_{yx}} & A_{yy}e^{i\Psi_{yy}} \end{pmatrix}\begin{pmatrix} E^i_x \\ E^i_y \end{pmatrix}, \quad (3.37)$$

where $\boldsymbol{E}^i_{x,y} = [E_x^i\ E_y^i]^{\mathrm{T}}$ and $\boldsymbol{E}^r_{x,y} = [E_x^r\ E_y^r]^{\mathrm{T}}$ denote the incident and reflected electric field vectors in the x- and y-polarizations, respectively, and r_{kj} ($k, j \in \{x, y\}$) denotes the reflection coefficient of k component under j component incidence. All reflection coefficients r_{kj} in matrix $\boldsymbol{R}_{x,y}$ are complex numbers containing amplitude A_{kj} and phase ψ_{kj}. Using a basis matrix Λ can transfer a CP electric field vector to the LP electric field vector $\boldsymbol{E}_{x,y}$, and the transfer relationship is written as $\boldsymbol{E}_{x,y} = \Lambda \boldsymbol{E}_{+,-}$, where $\boldsymbol{E}_{+,-}$ denotes the CP electric field vector, and Λ is:

$$\Lambda = \frac{1}{\sqrt{2}} \begin{pmatrix} 1 & 1 \\ -i & i \end{pmatrix}, \tag{3.38}$$

where subscripts "+" and "−" represent the clockwise and counterclockwise rotating CP electric field vector, respectively, when viewed along the "−z" direction. Equation (3.37) can be changed as a CP-based reflection matrix via the transfer relationship $\boldsymbol{E}_{x,y} = \boldsymbol{\Lambda E}_{+,-}$, that is:

$$\begin{aligned} &\boldsymbol{E}^r_{x,y} = \boldsymbol{R}_{x,y} \boldsymbol{E}^i_{x,y} \Rightarrow \Lambda \boldsymbol{E}^r_{+,-} = \boldsymbol{R}_{x,y} \Lambda \boldsymbol{E}^i_{+,-} \\ &\Rightarrow \Lambda^{-1} \Lambda \boldsymbol{E}^r_{+,-} = \Lambda^{-1} \boldsymbol{R}_{x,y} \Lambda \boldsymbol{E}^r_{+,-} \\ &\Rightarrow \boldsymbol{E}^r_{+,-} = \boldsymbol{R}_{+,-} \boldsymbol{E}^r_{+,-} \end{aligned} \quad . \tag{3.39}$$

It can be found from Eq. (3.39) that the relationship between coefficient matrix $\boldsymbol{R}_{+,-}$ of CP waves and coefficient matrix $\boldsymbol{R}_{x,y}$ of LP waves is:

$$\boldsymbol{R}_{+,-} = \Lambda^{-1} \boldsymbol{R}_{x,y} \Lambda = \begin{pmatrix} r_{++} & r_{+-} \\ r_{-+} & r_{--} \end{pmatrix} = \frac{1}{2} \begin{pmatrix} r_{xx} + r_{yy} + i\left(r_{xy} - r_{yx}\right) & r_{xx} - r_{yy} - i\left(r_{xy} + r_{yx}\right) \\ r_{xx} - r_{yy} + i\left(r_{xy} + r_{yx}\right) & r_{xx} + r_{yy} - i\left(r_{xy} - r_{yx}\right) \end{pmatrix}. \tag{3.40}$$

Generally, waves with clockwise and counterclockwise rotating electric field vectors when viewed along the direction of wave propagation are defined as right-handed and left-handed CP waves (RCP, LCP), respectively. To conform to this definition, $\boldsymbol{R}_{+,-}$ need to be adjusted accordingly as (in this work, the incident direction of waves is "$-z$", and the reflected direction is "z"):

$$\boldsymbol{R}_{+,-} = \begin{pmatrix} r_{++} & r_{+-} \\ r_{-+} & r_{--} \end{pmatrix} = \begin{pmatrix} r_{LR} & r_{LL} \\ r_{RR} & r_{RL} \end{pmatrix}, \tag{3.41}$$

where r_{mn} ($m, n \in \{R, L\}$) denotes the reflection coefficient of the m component under the n component incidence, and subscripts "R" and "L" represent RCP and LCP components, respectively. By rearranging the coefficients in Eq. (3.41), the reflection Jones matrix of CP waves can be obtained as $\boldsymbol{R}_{R,L}$:

$$\boldsymbol{R}_{R,L} = \begin{pmatrix} r_{RR} & r_{RL} \\ r_{LR} & r_{LL} \end{pmatrix} = \frac{1}{2} \begin{pmatrix} r_{xx} - r_{yy} + i\left(r_{xy} + r_{yx}\right) & r_{xx} + r_{yy} - i\left(r_{xy} - r_{yx}\right) \\ r_{xx} + r_{yy} + i\left(r_{xy} - r_{yx}\right) & r_{xx} - r_{yy} - i\left(r_{xy} + r_{yx}\right) \end{pmatrix}. \tag{3.42}$$

When electromagnetic waves are at normal incidence, conventional mirrors exhibit identical reflection characteristics (both in amplitude and phase) for any LP wave and do not generate cross-polarized components. Consequently, in Eq. (3.42), there are $r_{xx} = r_{yy}$

and $r_{xy} = r_{yx} = 0$. Under this condition, the reflection matrix $\boldsymbol{R}_{R,L}$ retains only the cross-polarized CP reflection terms as non-zero, specifically $r_{LR} = r_{RL} = (r_{xx} + r_{yy})/2$. This explains why conventional mirrors flip the handedness of CP waves upon reflection compared to the incident waves (when viewed along the wave propagation direction), as shown in Fig. 3.9(a). However, observing Eq. (3.42) reveals that when the x- and y-polarized waves have no cross-polarized reflection components ($r_{xy} = r_{yx} = 0$), and the co-polarized reflections exhibit the same reflection amplitude but opposite reflection phases ($r_{xx} = -r_{yy}$), the metasurface can then achieve CP reflected waves that maintain the same handedness as the incident waves (when viewed along the wave propagation direction). When $r_{xx} = -r_{yy}$, by setting $A_{xx} = A_{yy} = A$ and $\psi_{xx} = -\psi_{yy} = \psi$, the following equation can be obtained:

$$\boldsymbol{R}_{R,L} = \begin{pmatrix} r_{RR} & r_{RL} \\ r_{LR} & r_{LL} \end{pmatrix} = \frac{1}{2}\begin{pmatrix} r_{xx} - r_{yy} & 0 \\ 0 & r_{xx} - r_{yy} \end{pmatrix} = \frac{1}{2}\begin{pmatrix} A\left(e^{i\Psi} - e^{-i\Psi}\right) & 0 \\ 0 & A\left(e^{i\Psi} - e^{-i\Psi}\right) \end{pmatrix}. \quad (3.43)$$

Equations (3.42) and (3.43) demonstrate that simple mathematical constraints can enable the reflected wave to maintain the same circular polarization state as the incident wave – a capability completely opposite to that of a conventional reflective mirror. This work will subsequently demonstrate how to physically implement these mathematical constraints ($r_{xy} = r_{yx} = 0$, $A_{xx} = A_{yy} = A$, $\psi_{xx} = -\psi_{yy} = \psi$).

To avoid absorption loss of electromagnetic waves, this study employs high-resistivity silicon (Si) wafer (relative permittivity $\varepsilon = 11.9$, resistivity $\rho > 5000\ \Omega \cdot \text{cm}$) for metasurface design. The rectangular unit cells with dual symmetry about both x- and y-axes can readily achieve zero response to cross-polarized components of incident LP waves, thus easily satisfying the reflection constraint $r_{xy} = r_{yx} = 0$. The key challenge of this work lies in achieving both $A_{xx} = A_{yy}$ and $\psi_{xx} = -\psi_{yy}$. The phase relationship $\psi_{xx} = -\psi_{yy}$ implies a $\pm\pi$ phase difference

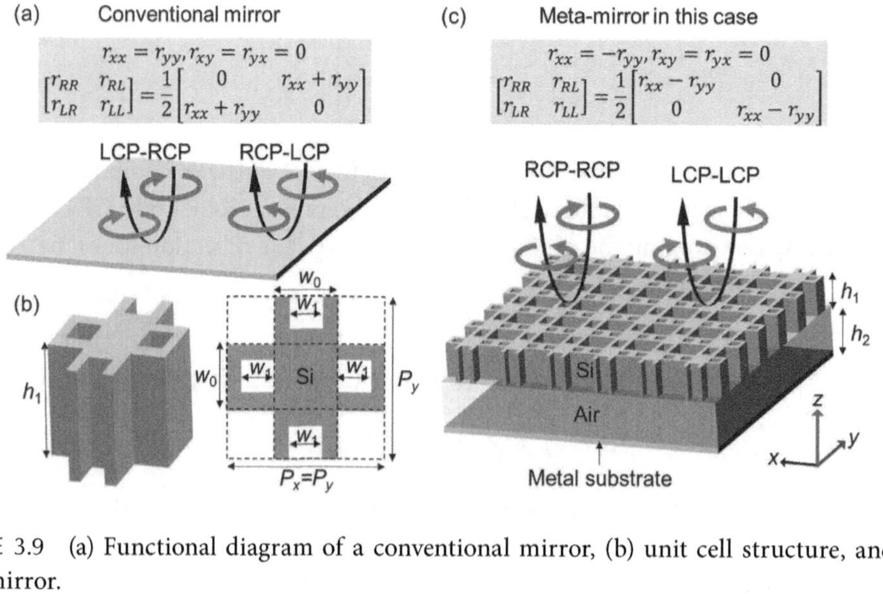

FIGURE 3.9 (a) Functional diagram of a conventional mirror, (b) unit cell structure, and (c) a meta-mirror.

between the co-polarized reflection coefficients r_{xx} and r_{yy}. While such phase characteristics are unattainable with conventional materials, metasurfaces can realize this condition. As shown in Fig. 3.9(b), the designed metasurface operates at a central frequency of 0.702 THz. The unit cell consists of two orthogonally crossed rectangular strips with hollow patterns. Key structural parameters include: unit cell thickness $h_1 = 200$ μm; period constants $P_x = P_x =$ 200 μm. Identical dimensions for both crossed strips: length matching the period constant (200 μm); width $w_0 = 80$ μm. The rectangular strips contain four square hollow regions with side length $w_1 = 40$ μm. The two crossed strips feature differently positioned hollow regions to control the metasurface's distinct responses to x- and y-polarized waves.

For the stand-alone metasurface, the incident x-polarized wave does not excite resonant interactions with the metasurface, while the y-polarized wave effectively induces electromagnetic resonance. Consequently, the reflected x-polarized wave has no resonance-induced phase shifts, whereas the reflected y-polarized wave undergoes a significant phase discontinuity due to resonance, achieving a $\pm\pi$ phase difference between the two orthogonal polarizations, i.e. the required phase constraint $\psi_{xx} = -\psi_{yy}$ is satisfied. Nevertheless, a stand-alone metasurface inherently permits partial electromagnetic transmission, making it virtually impossible to simultaneously achieve high-efficiency reflection for both co-polarized LP components while satisfying $A_{xx} = A_{yy}$. To address this, as shown in Fig. 3.9(c), a highly conductive metal plate is positioned adjacent to the transmission side of the metasurface to form a reflective metasurface device, with a separation distance of $h_2 = 280$ μm between the metal substrate and the metasurface's bottom surface. The reflective metasurface device completely suppresses electromagnetic wave transmission. Given the negligible absorption loss in high-resistivity silicon and the total elimination of transmission loss, energy conservation dictates that all incident wave energy is converted into reflected waves. As a result, the required amplitude constraint is further satisfied and gets high amplitude value $A_{xx} = A_{yy} = 1$.

The numerical simulations in this work were performed using the frequency domain solver in CST Microwave Studio. The metasurface's response to normally incident LP waves is investigated, with results presented in Fig. 3.10. Figure 3.10(a–c) first demonstrates the reflection coefficients and phase variations of the metasurface without a metal substrate across the 0.67–0.74 THz range. Since the symmetry axes of the metasurface unit cells perfectly align with the x-y coordinate axes, this configuration eliminates cross-polarized components; i.e. $|r_{xy}|$ and $|r_{yx}|$ remain zero throughout the entire observation band (0.67–0.74 THz). Notably, the y-polarized co-polarized reflection $|r_{yy}|$ exhibits a distinct resonant lineshape between 0.7 and 0.705 THz, while the x-polarized co-polarized reflection $|r_{xx}|$ remains smooth (Fig. 3.10(a)). This confirms that the electromagnetic resonance in the metasurface responds strongly to y-polarized waves while remaining inactive to x-polarized waves. Correspondingly, the y-polarized reflection phase ψ_{yy} undergoes abrupt changes within 0.7–0.705 THz, whereas the x-polarized reflection phase ψ_{xx} varies smoothly (Fig. 3.10(b)). The phase difference $\psi_{yy} - \psi_{xx}$, a critical parameter, is magnified for detailed observation in Fig. 3.10(c). Near 0.702 THz (with slight rightward offset), $\psi_{yy} - \psi_{xx}$ reaches a maximum of π (marked by the black dashed circle in Fig. 3.10(c)), indicating that the phase constraint $\psi_{xx} = -\psi_{yy}$ is satisfied near 0.702 THz. Nevertheless, returning to Fig. 3.10(a), there exists a significant disparity between the reflection coefficients

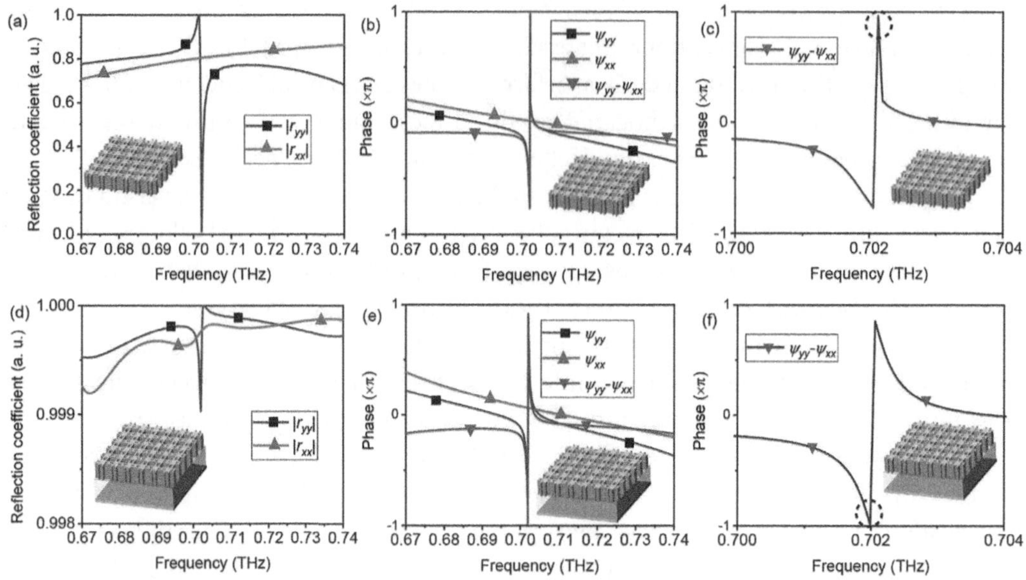

FIGURE 3.10 Reflection coefficient and phase of co-polarized LP waves: (a–c) for stand-alone metasurface and (d–f) for entire device composed of metasurface and metal plate.

of co-polarized y- and x-polarized waves, indicating that another crucial amplitude constraint $A_{xx} = A_{yy}$ remains unsatisfied.

The next issue to consider is how to further satisfy the constraint $A_{xx} = A_{yy}$. From Fig. 3.10(a), the reflection coefficients generally are between 0.7 and 0.8 due to transmission loss. Furthermore, near the resonance, the y-polarized reflection spectrum exhibits a sharp dip, with the minimum value approaching zero, indicating that the electromagnetic wave energy is almost entirely transmitted. Obviously, eliminating the transmitted wave would completely prevent the transmission loss of the metasurface, which would not only improve the reflection coefficient but also ensure that the constraint $A_{xx} = A_{yy}$ is satisfied. Therefore, this work introduces a metal substrate at the bottom of the metasurface to reflect the electromagnetic energy transmitted through the metasurface. It is important to note that to maintain the resonance position unchanged, the dielectric environment near the metasurface must not be disturbed. Thus, a sufficient distance (280 μm) is required between the metal substrate and the bottom surface of the metasurface. In this configuration, the entire system will only have reflected waves, and the reflection coefficient can approach 1 under the principle of energy conservation. Figure 3.10(d–f) presents the reflection response of the device after adding the metal substrate. Figure 3.10(d) demonstrates that the reflection coefficients $|r_{xx}|$ and $|r_{yy}|$ of the device are both greater than 0.999, very close to 1, resulting in A_{xx} approximately equal to A_{yy}, which means the amplitude constraint $A_{xx} = A_{yy}$ is satisfied. The bottom metal substrate leads to slight modifications in the reflection phase spectrum (Fig. 3.10(e, f)), where the co-polarization phase difference $\psi_{yy} - \psi_{xx}$ reaches $-\pi$ at 0.702 THz (marked by the black dashed circle in Fig. 3.10(f)), indicating that the phase constraint $\psi_{xx} = -\psi_{yy}$ is satisfied at 0.702 THz.

The above results indicate that the designed device can meet the key constraints ($r_{xy} = r_{yx} = 0$, $A_{xx} = A_{yy}$, $\psi_{xx} = -\psi_{yy}$) under 0.702 THz LP waves incidence. Next, we will convert the LP waves into CP waves using theoretical formulas and compare this theoretical result with the simulation results obtained under direct CP wave excitation. The conversion from LP waves to CP waves is achieved using the aforementioned Eq. (3.42). Since $r_{xy} = r_{yx} = 0$ across the entire observation band (0.67–0.74 THz), the theoretically derived cross-polarized CP reflectivity ($|r_{LR}|^2$ and $|r_{RL}|^2$) is given by $|r_{xx} + r_{yy}|^2/4$, while the co-polarized CP reflectivity ($|r_{RR}|^2$ and $|r_{LL}|^2$) is given by $|r_{xx} - r_{yy}|^2/4$. Figure 3.11(a) presents the reflectivity results of the CP waves converted through theoretical formulas. In the non-resonant region, where $|r_{xx} + r_{yy}|^2/4$ value is high and $|r_{xx} - r_{yy}|^2/4$ value is low, the device still follows conventional mirror reflection; i.e. a CP wave incident on the device will be primarily reflected as a cross-polarized CP wave. However, at the resonant frequency of 0.702 THz, the constraints $A_{xx} = A_{yy}$ and $\psi_{xx} = -\psi_{yy}$ work. Consequently, $|r_{xx} + r_{yy}|^2/4$ value diminishes to 0 while $|r_{xx} - r_{yy}|^2/4$ value increases to 1, achieving complete co-polarized CP reflection at 0.702 THz.

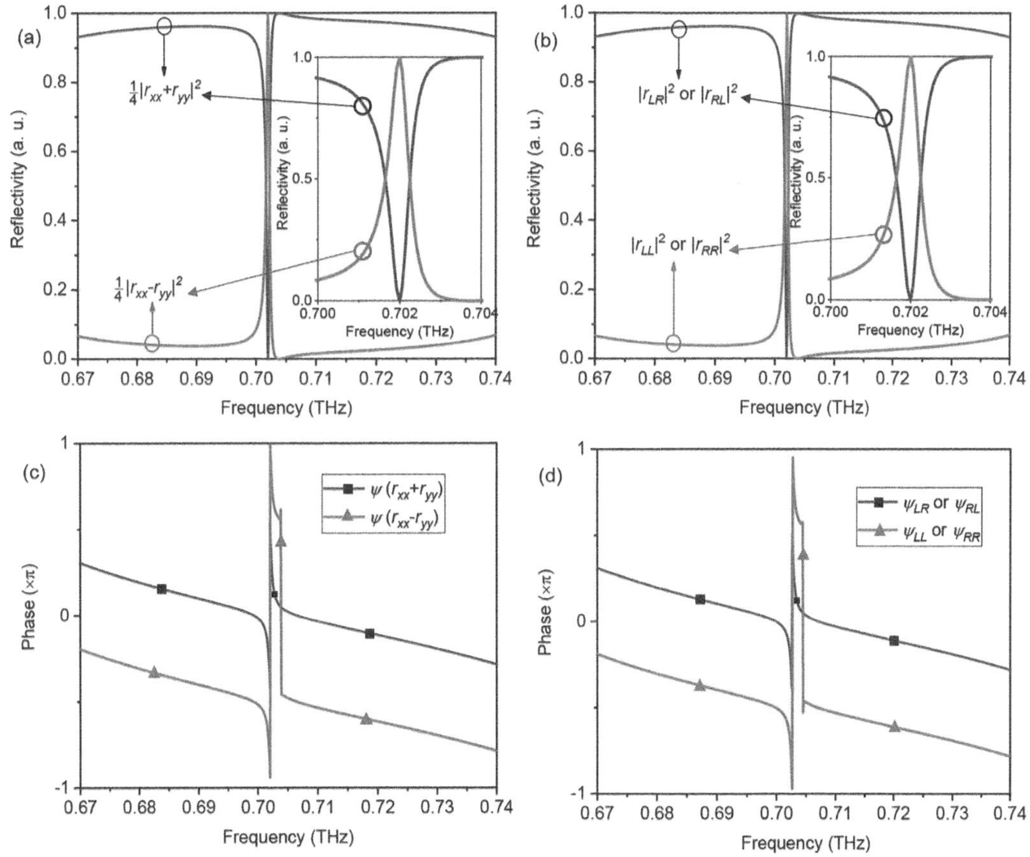

FIGURE 3.11 (a, c) The CP waves reflectivity and phase derived from LP waves reflection data through theoretical conversion formulas. (b, d) The CP waves reflectivity and phase obtained by directly using CP waves excitation.

As previously mentioned, the CP waves reflection results presented in Fig. 3.11(a) were derived from LP waves reflection data through theoretical conversion formulas, rather than obtained from direct CP waves excitation. To validate the theoretical formulation, we performed additional simulations in CST Microwave Studio using direct CP wave excitation, with results shown in Fig. 3.11(b). It can be seen that the simulated reflectivity results obtained by directly using CP waves excitation (Fig. 3.11(b)) are completely consistent with those derived from converting LP waves to CP waves using theoretical formulas (Fig. 3.11(a)), i.e. $|r_{LR}|^2 = |r_{RL}|^2 = |r_{xx} + r_{yy}|^2/4$ and $|r_{LL}|^2 = |r_{RR}|^2 = |r_{xx} - r_{yy}|^2/4$. Additionally, the reflection phases are extracted from the theoretical reflection coefficients and the simulated reflection coefficients, respectively, as shown in Fig. 3.11(c, d). Similarly, the reflection phases under direct CP waves excitation (Fig. 3.11(d)) agree with that obtained through the theoretical conversion formula (Fig. 3.11(c)), i.e. $\psi_{LR} = \psi_{RL} = \psi(r_{xx} + r_{yy})$ and $\psi_{LL} = \psi_{RR} = \psi(r_{xx} - r_{yy})$. Obviously, the simulation results obtained through direct CP wave excitation further validate the correctness of the theoretical design.

Analysis of resonant electromagnetic modes facilitates a deeper understanding of the interaction between CP waves and the device. Figure. 3.12(a–e) presents the monitored electromagnetic field distributions under *y*-polarized wave excitation at 0.702 THz. The observed electromagnetic field distributions reveal the presence of a transverse electric (TE) mode, characterized by the magnetic fields H_z along the *z*-axis and electric fields E_x and E_y in the *x*-*y* plane. The magnetic fields H_z (Fig. 3.12(a, b)) exhibit strong localization in two square hollows arranged along the *x*-axis. Notably, magnetic fields in two hollows are opposite phases, which collectively establish a closed magnetic loop in the *z*-*x* plane, confirmed by the magnetic field line distribution presented in Fig. 3.12(c). Furthermore, a pair of toroidal electric fields in the *x*-*y* plane form and exhibit mirror symmetry about the *y*-axis (Fig. 3.12(d, e)). These features indicate that the electromagnetic resonance mode excited by the *y*-polarized wave at 0.702 THz corresponds to a toroidal dipole (TD) mode, with its dipole moment M_{TD} oriented along the *y*-axis, as schematically illustrated in

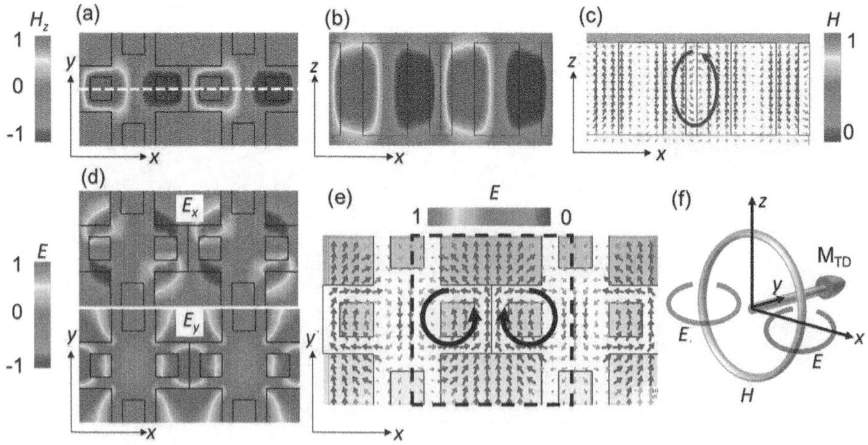

FIGURE 3.12 (a, b) The magnetic field H_z and (c) magnetic field line distributions under *y*-polarized wave excitation at 0.702 THz, corresponding to (d) electric fields E_x, E_y and (e) electric field line distributions, and (f) a schematic multipole structure of electromagnetic resonance.

Fig. 3.12(f). The TD resonant mode enables a high-quality factor [18], significantly enhancing the interaction between electromagnetic waves and the device. This directly results in an extremely narrow resonance linewidth in the reflection spectra (see Figs. (3.10) and (3.11)).

Considering that structural parameter errors may disrupt resonance and lead to the failure of the device's functions if the device is actually fabricated, the effects of various key structural parameters on resonance are further analysed. The simulation results in Fig. 3.13 show that limited deviations in unit cell size parameters primarily affect the resonance frequency without destroying the device's functions, and the reflection intensity of co-polarized CP waves remains almost unaffected. When the size w_1 of the square hollows arranged along the x-axis is varied while maintaining other parameters, the resonance shifts towards higher frequencies as w_1 increases (Fig. 3.13(a)). When the distance d of the square hollows from the central y-axis is adjusted while maintaining other parameters, the resonance shifts towards higher frequencies as d increases (Fig. 3.13(b)). When the thickness h_1 of the Si wafer is modified while maintaining other parameters, the resonance shifts towards lower frequencies as h_1 increases (Fig. 3.13(c)). Since the wafer thickness used in this work refers to mature products already available in reality, the thickness h_1 was predetermined before simulation design, eliminating the need to consider potential thickness variations during device fabrication. Subsequently, only careful adjustments of parameters such as the size w_1 and distance d, along with high-precision fabrication techniques, are required to ensure that the resonance frequency drifts within a controllable range.

Furthermore, in actual fabrication, it is difficult to control precisely the distance h_2 between the metal substrate and the bottom surface of the metasurface, with potential errors reaching several micrometres. Therefore, it is necessary to investigate the effect of h_2 on resonance. The co-polarized CP waves reflections under different wafer thicknesses h_1 are examined as h_2 varies. The spectral dependences on h_2 for three typical wafer thicknesses (h_1 = 195, 200, and 215 μm) are presented in Fig. 3.14(a, d), (b, e), and (c, f), respectively. It can be observed that both the reflection intensity and frequency of the co-polarized CP waves remain nearly unaffected by h_2. Within the wafer thickness range of 195–215 μm, the allowable variation range for h_2 is relatively large (260–290 μm) regardless of which specific wafer thickness is employed, indicating that the device's resonance exhibits its strong robustness against h_2. Thus, in this work, significant deviations in the distance between the metal substrate and the bottom surface of the metasurface are permissible.

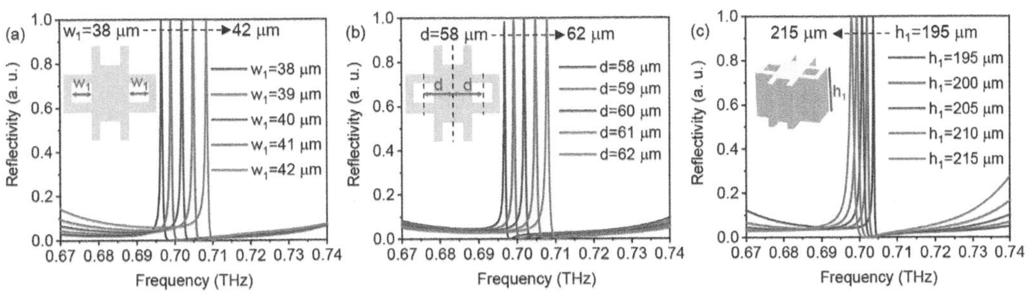

FIGURE 3.13 The effects of several key unit cell size parameters on co-polarized CP waves reflectivity ($|r_{RR}|^2$ or $|r_{LL}|^2$) (a) for size w_1 of the square hollows, (b) for distance d, and (c) wafer thickness h_1.

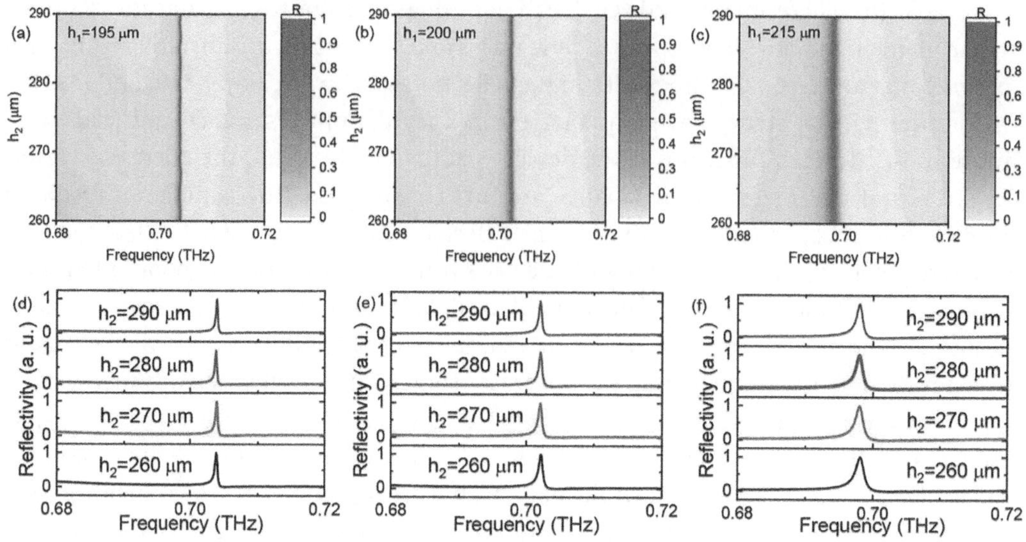

FIGURE 3.14 (a–c) The co-polarized CP waves reflectivity ($|r_{RR}|^2$ or $|r_{LL}|^2$) under continuous h_2 parameter scanning for three wafer thickness $h_1 = 195$, 200, and 215 μm, respectively; corresponding to several discrete reflectivity spectra related to h_2 shown in (d–f).

Although the above device in Fig. 3.9(c) is designed to protect the handedness feature of CP reflected waves, it can also perform linear polarization conversion. Assuming an incident LP wave with an electric vector oscillation direction at a $\pi/4$ angle to the x-axis, it can be decomposed into x- and y-polarized components. Due to the phase difference π between the reflected x-polarized component and the y-polarized component, a LP reflected wave with an electric field vector at a $3\pi/4$ angle to the x-axis will be synthesized, which means that the original LP incident wave undergoes a polarization rotation with the angle of $\pi/2$ after reflection. Equivalently, we can rotate the structure at a $\pi/4$ angle, where an incident x- or y-polarized wave will be converted into a y- or x-polarized wave, which is easy to understand intuitively. For theoretical analysis, let's return to Eq. (3.37). By substituting $r_{xy} = r_{yx} = 0$, $r_{xx} = -r_{yy} = r$, into Eq. (3.37), we can obtain the reflection Jones matrix

$$\boldsymbol{R}_{x,y} = \begin{pmatrix} r & 0 \\ 0 & -r \end{pmatrix}. \tag{3.44}$$

When the device is rotated with a $\pi/4$ angle, the reflection Jones matrix will be transformed into:

$$\begin{aligned} \boldsymbol{R}'_{x,y} &= \begin{pmatrix} \cos(\pi/4) & -\sin(\pi/4) \\ \sin(\pi/4) & \cos(\pi/4) \end{pmatrix} \begin{pmatrix} r & 0 \\ 0 & -r \end{pmatrix} \begin{pmatrix} \cos(\pi/4) & \sin(\pi/4) \\ -\sin(\pi/4) & \cos(\pi/4) \end{pmatrix} \\ &= \begin{pmatrix} \sqrt{2}/2 & -\sqrt{2}/2 \\ \sqrt{2}/2 & \sqrt{2}/2 \end{pmatrix} \begin{pmatrix} r & 0 \\ 0 & -r \end{pmatrix} \begin{pmatrix} \sqrt{2}/2 & \sqrt{2}/2 \\ -\sqrt{2}/2 & \sqrt{2}/2 \end{pmatrix} = \begin{pmatrix} 0 & r \\ r & 0 \end{pmatrix} \end{aligned}. \tag{3.45}$$

With an x- or a y-polarized wave incidence, the reflected wave will be:
For incident x-polarized wave,

$$\boldsymbol{E}^{r'}_{x,y} = \boldsymbol{R}'_{x,y}\boldsymbol{E}^{i}_{x} = \begin{pmatrix} 0 & r \\ r & 0 \end{pmatrix}\begin{pmatrix} 1 \\ 0 \end{pmatrix} = \begin{pmatrix} 0 \\ r \end{pmatrix}. \tag{3.46}$$

For incident y-polarized wave,

$$\boldsymbol{E}^{r'}_{x,y} = \boldsymbol{R}'_{x,y}\boldsymbol{E}^{i}_{y} = \begin{pmatrix} 0 & r \\ r & 0 \end{pmatrix}\begin{pmatrix} 0 \\ 1 \end{pmatrix} = \begin{pmatrix} r \\ 0 \end{pmatrix}. \tag{3.47}$$

From Eqs. (3.46) and (3.47), it can be seen that when an x- or a y-polarized wave is incident, the reflected wave is converted into a y- or an x-polarized wave.

In summary, through the above case study about a meta-mirror preserving the handedness of arbitrarily circular polarization waves, we have seen that resonance-driven polarization conversion can achieve functions that are difficult to achieve with conventional optical devices. The use of electromagnetic resonance can also achieve various polarization conversions. Although we cannot list them one by one here, the key role of resonance-driven polarization conversion has been demonstrated; that is, the phase transition induced by resonance promotes the realization of polarization conversion.

3.3 CHAPTER SUMMARY

This chapter introduces two types of fundamental physics that induce polarization conversion, namely the birefringence-like effect and the electromagnetic resonance effect. Specifically, for the conversion between linear polarization, circular polarization, and elliptical polarization, the Jones matrix is used as a tool to derive a detailed mathematical description based on birefringent rectangular bars. Subsequently, the design of multifunctional birefringent polarization devices is demonstrated for the commonly used linear-circular polarization conversion. Further, the polarization conversion caused by electromagnetic resonance is introduced. Still using the Jones matrix as a tool, the electromagnetic resonance between light and the device is demonstrated through polarization device simulation, and the role of this effect in polarization conversion is demonstrated.

REFERENCES

1. Li, J., Wang, G., Yue, Z., et al. (2022) 'Dynamic phase assembled terahertz metalens for reversible conversion between linear polarization and arbitrary circular polarization', Opto-Electronic Advances, 5, 210062.
2. Joannopoulos, J.D., Johnson, S.G., Winn, J.N., et al. (2011) Photonic crystals: Molding the flow of light, 2nd edn. Princeton: Princeton University Press.
3. Yao, J., Li, J., Zhang, Y., et al. (2023) 'Bound states in continuum in periodic optical systems', Chinese Optics, 16, 1–23.
4. Fedotov, V.A., Rogacheva, A.V. and Zheludeva, N.I. (2006) 'Mirror that does not change the phase of reflected waves', Applied Physics Letters, 88(9), 091119.
5. Plum, E. and Zheludev, N.I. (2015) 'Chiral mirrors', Applied Physics Letters, 106(22), 221901.

6. Kang, L., Rodrigues, S.P., Taghinejad, M., et al. (2017) 'Preserving spin states upon reflection: linear and nonlinear responses of a chiral meta-mirror', Nano Letters, 17(11), 7102–7109.
7. Wu, P.C., Tsai, W.-Y., Chen, W.T., et al. (2017) 'Versatile polarization generation with an aluminum plasmonic metasurface', Nano Letters, 17(1), 445–452.
8. Zhang, F., Pu, M., Li, X., et al. (2017) 'All-dielectric metasurfaces for simultaneous giant circular asymmetric transmission and wavefront shaping based on asymmetric photonic spin-orbit interactions', Advanced Functional Materials, 27(47), 1704295.
9. Chang, C.-C., Zhao, Z., Li, D., et al. (2019) 'Broadband linear-to-circular polarization conversion enabled by birefringent off-resonance reflective metasurfaces', Physical Review Letters, 123, 237401.
10. Ding, F., Chang, B., Wei, Q., et al. (2020) 'Versatile polarization generation and manipulation using dielectric metasurfaces', Laser & Photonics Reviews, 14(11), 2000116.
11. Wang, S., Deng, Z.-L., Wang, Y., et al. (2021) 'Arbitrary polarization conversion dichroism metasurfaces for all-in-one full Poincaré sphere polarizers', Light: Science & Applications, 10, 24.
12. Chen, Y., Deng, H., Sha, X., et al. (2023) 'Observation of intrinsic chiral bound states in the continuum', Nature, 613, 474–478.
13. Sinev, I., Richter, F.U., Toftul, I., et al. (2025) 'Chirality encoding in resonant metasurfaces governed by lattice symmetries', Nature Communications, 16, 6091.
14. Li, W., Coppens, Z.J., Besteiro, L.V., et al. (2015) 'Circularly polarized light detection with hot electrons in chiral plasmonic metamaterials', Nature Communications, 6, 8379.
15. Wang, Q., Plum, E., Yang, Q., et al. (2018) 'Reflective chiral meta-holography: multiplexing holograms for circularly polarized waves', Light: Science & Applications, 7, 25.
16. Li, J., Li, J., Zheng, C., et al. (2021) 'Dynamic control of reflective chiral terahertz metasurface with a new application developing in full grayscale near field imaging', Carbon, 172, 189–199.
17. Li, J., Yue, Z., Li, J., et al. (2023) 'Wavefront-controllable all-silicon terahertz meta-polarizer', Science China Materials, 66(1), 300–308.
18. Zhang, Y., Wang, L., He, H., et al. (2024) 'High-Q magnetic toroidal dipole resonance in all-dielectric metasurfaces', APL Photonics, 9, 076106.

CHAPTER 4

Polarization Generation from Unpolarized Waves

Abstract

Generally, the light emitted by the light source may be unpolarized light and partially polarized light, and the primary task of the optical system is to obtain specific polarized light from these light sources, which requires the use of polarization-generating devices. The conventional polarization conversion metasurface depends on a specific incident polarization state, meaning the output polarization state is limited by the input polarization state. When the input polarization state changes, the output polarization state of the conventional polarization conversion metasurface will also change accordingly. In contrast, polarization-generating metasurfaces are required to have the capability to convert any arbitrary polarization state into a specific polarization state, thereby exhibiting the characteristics of a polarizer that directly extracts stable polarized waves from unpolarized waves. This chapter will first demonstrate the general principle of generating arbitrary polarization states without input polarization state limitations. Then, for the two most commonly used polarization states, namely linear polarization and circular polarization, the principle of polarization-generating metasurfaces is derived and the design method is demonstrated. It should be noted in advance that the device functions mentioned in this chapter will be implemented using the principle of two meta-atoms beam interference, but their physical roots and foundations can still be linked to the birefringence-like effect introduced in Chapter 3. Based on special resonance effects, such as bound states in continuum, polarization generation can also be achieved, which will be demonstrated in detail in Chapter 8.

DOI: 10.1201/9781003784821-4

4.1 THE GENERAL PRINCIPLE OF AN ARBITRARY POLARIZATION GENERATION

To obtain polarized waves, the most widely used is the optical linear polarizer, which is the most basic optical element for filtering polarized waves from unpolarized or partially polarized waves. It is generally made of special optical materials, such as metal wire grids, polarizing materials, nano films, or birefringent prisms, and in some cases uses the principle of Brewster angle reflection. It is easy to obtain linearly polarized light using traditional polarizers, with only one linearly polarized filter required. However, for other polarization states, such as circular polarization and elliptical polarization generation, the flexibility of current optical devices is insufficient, and multiple optical elements need to be cascaded to achieve the goal, which is extremely unfriendly to the miniaturization of polarization optical systems. The acquisition of circularly polarized waves can rely on anisotropic materials represented by birefringent crystals. By combining a linear polarization filter with birefringent crystal elements, circularly polarized waves can be obtained from unpolarized waves. By performing simple geometric operations on anisotropic materials (such as rotating them to change the phase difference of the output waves along the long and short axes), it is also possible to obtain elliptically polarized waves with specific orientation angles. However, to obtain elliptically polarized waves with arbitrary orientation angles, more complex combinations of optical elements are required. Simplifying the volume of optical systems is an important issue in optical design. The emergence of metasurfaces provides convenience for directly generating arbitrary polarized waves from unpolarized or partially polarized waves. The direct generation of linearly polarized and circularly polarized waves is easy for a metasurface, and even elliptically polarized waves with arbitrary orientation angles can be achieved using a thin metasurface device, providing development opportunities for miniaturization and high integration of polarization optical systems.

Most of the naturally formed light is unpolarized or partially polarized. Therefore, traditional polarized optical devices follow an intuitive development path, which is to first obtain polarized light from unpolarized light and then utilize polarized light (including further conversion of the obtained polarized light). However, the development path for polarized metasurfaces is exactly the opposite of that of traditional polarized optical devices. In the development history of polarized metasurfaces, limited by the physical principles used, studying polarization manipulation under specific input polarization state conditions is more mature and extensive than studying obtaining directly polarized light from unpolarized light.

In early reports, researchers used Brewster metasurfaces to generate linearly polarized light from unpolarized light [1–4]. Y. Tamayama et al. [1] in 2006 used a subwavelength split ring structure to construct an artificial medium with a relative permeability not equal to unity and achieved manipulation of transverse-electric (TE) waves in the microwave range. Ryosuke Watanabe et al. [2] in 2008 constructed a metasurface in which aluminium oxide and aluminium (Al_2O_3-Al) thin films were periodically stacked along the z-axis and also achieved manipulation of TE waves. In 2019, Chun Wang et al. [4] proposed dispersion engineering at each incident angle to strictly match the longitudinal wave impedance on both sides of the surface, resulting in all-angle TE waves manipulation.

The Brewster angle is an important fundamental concept in optics. When natural light is incident on a medium at the Brewster angle, the *p* component light (linearly polarized light with an electric field parallel to the incident plane) does not reflect but is completely transmitted into the medium. The reflected light only contains the *s* component (with the electric field perpendicular to the incident plane), thus obtaining linearly polarized light, and the reflected light and refracted light are perpendicular to each other. This effect was discovered by Scottish physicist David Brewster (1781–1868) and is therefore named the Brewster effect.

The physical principle of the Brewster phenomenon can be explained by the electric dipole model. Electric dipoles with linear electric vector vibration cannot radiate in the direction of vibration. Due to the electric dipole of the *p*-wave (transverse-magnetic (TM) wave) being located within the incident plane, when the Brewster condition is met, the electric vector vibration direction of the *p*-wave is consistent with the wave vector of the reflected wave. Therefore, no wave is output in the reflected direction because it does not conform to the spatial propagation rule of a wave vector perpendicular to the electric field. The electric dipole of *s*-wave (TE wave) is perpendicular to the incident plane, so there is no extraordinary reflection angle.

In mathematics, the reflection coefficient of optical materials is described by Fresnel's law as "$r_s = (n_i\cos i - n_t\cos t)/(n_i\cos i + n_t\cos t) = -\sin(i - t)/\sin(i + t)$" and "$r_p = (n_t\cos i - n_i\cos t)/(n_t\cos i + n_i\cos t) = \tan(i - t)/\tan(i + t)$", where n_i and n_t are the refractive indices of the incident and refractive media, respectively, and i and t are the incident angle and refractive angle, respectively. As can be seen, for the *p* component, when $i + t = \pi/2$, the reflection coefficient r_p is equal to zero, indicating complete transmission, which is the performance in traditional optical materials. However, the Brewster metasurface can manipulate both the *p* and *s* components, attributed to its ability to construct relative permeability and relative permittivity that natural materials do not possess [5]. In addition, generalized Brewster effect and anomalous Brewster effect have been reported in metasurface devices [6–8].

From the perspective of polarization generation, the limitation of the Brewster metasurface is quite obvious; that is, the output object is limited to linearly polarized waves. The study of directly obtaining an arbitrary polarized light from unpolarized light emerged after 2015 [9]; at that time, the theoretical system of polarization conversion under specific polarization inputs was generally mature. That's also why this book first discusses polarization conversion under specific input polarization states in the previous chapter and then starts to talk about polarization generation from unpolarized waves in this chapter. Before discussing the general principle of polarization generation metasurfaces, it should be noted that the principles proposed in this section will be based on the birefringence-like effect of meta-atoms. Of course, this is not the only solution for polarization generation (other options, e.g. the Brewster metasurface, mentioned above). Some special resonance effects, such as bound states in continuum (BICs), can also achieve polarization generation. However, in terms of flexibility, versatility, and the ability to generate arbitrary polarization in device design, the principles proposed in this section are still the optimal solution as compared to other candidate solutions. Polarization generation based on special resonance effects will be discussed in Chapter 8.

Metasurfaces with polarization manipulation capabilities have undergone development from polarization metasurface converters to metasurface polarizers [10–16]. The unit cell of a polarization metasurface converter usually contains only one meta-atom, so the device can only serve the conversion between two specific polarization states. In practical use, additional polarizers need to be used to give a determined input polarization [17–23]. The meta-polarizers described in this section refer to devices that can generate arbitrary stable polarization states without input polarization state limitations. The currently used meta-polarizer schemes generally include at least two different meta-atoms in a unit cell, which can directly generate polarized waves from unpolarized waves by using the beam interference constructive or destructive effects of multiple meta-atoms, without the need for additional polarizers. Due to the fact that the unpolarized waves can be seen as an irregular collection of various polarized waves, the role of a meta-polarizer is essentially to induce the conversion of an arbitrary polarization to a specific polarization. Zhang et al. proposed circularly polarized asymmetric transmission based on asymmetric photonic spin-orbit interactions in 2017, which was the earliest demonstration of the idea of multiple meta-atom beam interference [13]. Subsequently, Fan et al. established the selective transmission mechanism for orthogonal polarization states based on multiple meta-atom beam interference in 2020, achieving the generation of linear, elliptical, and circularly polarized waves [14]. Wang et al. proposed a more universal theory and application in 2021, allowing arbitrary linear, elliptical, and circularly polarized waves to be directly generated from unpolarized light [9].

In this section, we present the general principle of generating arbitrary stable polarization states in metasurfaces based on the Jones matrix operation [9]. As is well known, any polarized wave can be decomposed into the superposition of two orthogonal polarizations, which can be x and y polarizations, RCP and LCP, or two orthogonal elliptical polarizations. In Chapter 1, we mentioned that an arbitrary elliptical polarization wave can be fully described by its orientation angle Φ and ellipticity angle χ. In the global linear polarization base defined in the Cartesian coordinates (x-o-y coordinate system), the Jones vector of elliptical polarization $\boldsymbol{\alpha}$ can be explicitly written as:

$$\alpha = \mathbf{R}(\Phi)\begin{pmatrix} \cos(\chi) \\ -i\sin(\chi) \end{pmatrix} = \frac{\sqrt{2}}{2}\mathbf{R}\left(\Phi - \frac{\pi}{4}\right)\begin{pmatrix} e^{i\chi} \\ e^{-i\chi} \end{pmatrix}, \tag{4.1}$$

where the rotation matrix $\mathbf{R}(\Phi) = \begin{pmatrix} \cos(\Phi) & -\sin(\Phi) \\ \sin(\Phi) & \cos(\Phi) \end{pmatrix}$ has been explained in the previous chapter. The Jones vector of its orthogonal polarization $\boldsymbol{\beta}$ is as follows:

$$\boldsymbol{\beta} = \mathbf{R}\left(\Phi - \frac{\pi}{2}\right)\begin{pmatrix} \cos(\chi) \\ i\sin(\chi) \end{pmatrix} = \frac{\sqrt{2}}{2}\mathbf{R}\left(\Phi - \frac{\pi}{4}\right)\begin{pmatrix} e^{i\chi} \\ -e^{-i\chi} \end{pmatrix}. \tag{4.2}$$

The geometric relationship between the orthogonal elliptical polarization states $\boldsymbol{\alpha}$ and $\boldsymbol{\beta}$ represented by Eq. (4.1) and Eq. (4.2) is shown in Fig. 4.1. Orthogonal elliptical polarization

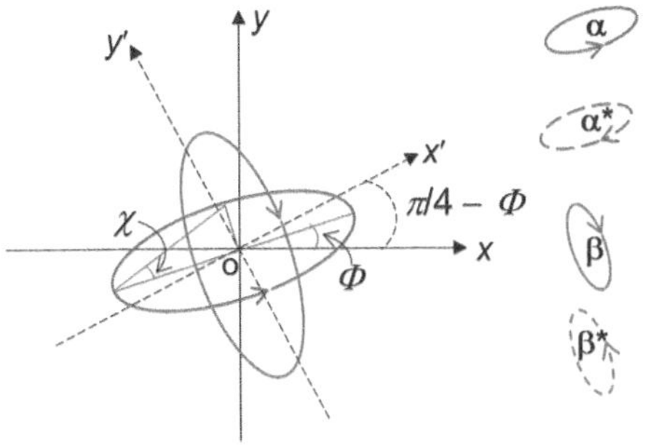

FIGURE 4.1 The geometric schematic diagram of the orthogonal elliptical polarization bases.

states not only have orientation angles perpendicular to each other but also have opposite spin directions. It is easy to understand that when the ellipticity angle χ is zero, $\boldsymbol{\alpha}$ and $\boldsymbol{\beta}$ will represent a pair of orthogonal linear polarization bases, while when χ is $\pi/4$, $\boldsymbol{\alpha}$ and $\boldsymbol{\beta}$ will represent a pair of circular polarization bases (i.e. LCP and RCP). Therefore, mathematically, elliptical polarization bases are a more universal form than linear and circular polarization bases, which is why we first discuss the elliptical polarization bases here.

The following derivation is based on birefringent dielectric rectangular bars. In the *x-o-y* coordinate, the transmission field **t** and the incident field **i** are connected by the transfer Jones matrix **J**, written as:

$$\begin{pmatrix} t_x \\ t_y \end{pmatrix} = \mathbf{J} \begin{pmatrix} i_x \\ i_y \end{pmatrix}, \tag{4.3}$$

where, $\mathbf{t} = [t_x\ t_y]^{\mathrm{T}}$ and $\mathbf{i} = [i_x\ i_y]^{\mathrm{T}}$ denote complex amplitude vectors of the transmitted field and the incident field, respectively. Equation (4.3) is described in a pair of orthogonal linear polarization bases. For studying the arbitrary polarization generation, we need to perform a base transformation, i.e. converting Eq. (4.3) to a new form in a pair of orthogonal elliptical polarization bases. Herein, two Jones vectors $\hat{\mathbf{i}}$ and $\hat{\mathbf{t}}$ in the polarization pair bases of ($\boldsymbol{\alpha}$, $\boldsymbol{\beta}$) and ($\boldsymbol{\alpha}^*$, $\boldsymbol{\beta}^*$) are, respectively, defined, and they are related to Jones vectors **i** and **t** as:

$$\begin{cases} \mathbf{i} = \Lambda_i \hat{\mathbf{i}} = \begin{pmatrix} \alpha & \beta \end{pmatrix} \hat{\mathbf{i}} \\ \mathbf{t} = \Lambda_t \hat{\mathbf{t}} = \begin{pmatrix} \alpha^* & \beta^* \end{pmatrix} \hat{\mathbf{t}} \end{cases}, \tag{4.4}$$

where the Λ_k ($k = i, t$) is the base transformation matrix. It should be noted that the two pairs of elliptical polarization bases used in the transmission field and incident field are mathematically conjugate, and their physical meaning lies in the handedness flipping of ($\boldsymbol{\alpha}^*$, $\boldsymbol{\beta}^*$) relative to ($\boldsymbol{\alpha}$, $\boldsymbol{\beta}$). The cleverness of these definitions will be seen in subsequent discussions, which is the key to achieving stable polarization generation. Substituting Eq. (4.4) into Eq. (4.3), we have:

$$\mathbf{t} = \mathbf{J}\mathbf{i} \rightarrow \Lambda_t \hat{\mathbf{t}} = \mathbf{J}\Lambda_i \hat{\mathbf{i}} \rightarrow \hat{\mathbf{t}} = \Lambda_t^{-1}\mathbf{J}\Lambda_i \hat{\mathbf{i}}. \tag{4.5}$$

Further, we can get another Jones matrix $\mathbf{J}^{\#}$:

$$\mathbf{J}^{\#} = \Lambda_t^{-1}\mathbf{J}\Lambda_i. \tag{4.6}$$

Therefore, in two pairs of orthogonal elliptical polarization bases of ($\boldsymbol{\alpha}$, $\boldsymbol{\beta}$) and ($\boldsymbol{\alpha}^*$, $\boldsymbol{\beta}^*$), the complex amplitudes of the incident and transmitted field can be connected as follows:

$$\begin{pmatrix} t_{\alpha^*} \\ t_{\beta^*} \end{pmatrix} = \mathbf{J}^{\#} \begin{pmatrix} i_\alpha \\ i_\beta \end{pmatrix}, \tag{4.7}$$

where the $[t_{\alpha^*}\ t_{\beta^*}]^{\mathrm{T}}$ and $[i_\alpha\ i_\beta]^{\mathrm{T}}$ denote the transmitted and incident field in elliptical polarization bases of ($\boldsymbol{\alpha}^*$, $\boldsymbol{\beta}^*$) and ($\boldsymbol{\alpha}$, $\boldsymbol{\beta}$), respectively.

Returning to Eqs. (4.1) and (4.2), it can be found that elliptical polarization bases ($\boldsymbol{\alpha}$, $\boldsymbol{\beta}$) contain a rotation term of $\mathbf{R}(\Phi - \pi/4)$, which makes the mathematical form of ($\boldsymbol{\alpha}$, $\boldsymbol{\beta}$) complex. Hence, let us eliminate the rotation term of $\mathbf{R}(\Phi - \pi/4)$ via a coordinate transformation. Here, define a new local coordinate system x'-o-y' that coordinate is anticlockwise rotating by "$\pi/4 - \Phi$" with respect to the original x-o-y system, as shown in Fig. 4.1. In the local x'-o-y' coordinate, the rotation term of $\mathbf{R}(\Phi - \pi/4)$ will be eliminated, and Jones vectors will be as follows:

$$\begin{cases} \alpha' = R\left(\dfrac{\pi}{4} - \Phi\right)\alpha = \dfrac{\sqrt{2}}{2}\begin{pmatrix} e^{i\chi} \\ e^{-i\chi} \end{pmatrix} \\ \beta' = R\left(\dfrac{\pi}{4} - \Phi\right)\beta = \dfrac{\sqrt{2}}{2}\begin{pmatrix} e^{i\chi} \\ -e^{-i\chi} \end{pmatrix} \end{cases}. \tag{4.8}$$

Thus, the handedness-flipped elliptical polarization bases can be written as follows:

$$\begin{cases} \alpha'^* = \dfrac{\sqrt{2}}{2}\begin{pmatrix} e^{-i\chi} \\ e^{i\chi} \end{pmatrix} \\ \beta'^* = \dfrac{\sqrt{2}}{2}\begin{pmatrix} e^{-i\chi} \\ -e^{i\chi} \end{pmatrix} \end{cases}. \tag{4.9}$$

Following the same way to Eq. (4.3), the relationship of the transmitted field $\mathbf{t}' = [t_x'\ t_y']^{\mathrm{T}}$, the incident field $\mathbf{i}' = [i_x'\ i_y']^{\mathrm{T}}$, and the Jones matrix $\mathbf{J}'$ in the local x'-o-y' coordinate can be expressed as:

$$\begin{pmatrix} t_x' \\ t_y' \end{pmatrix} = \mathbf{J}' \begin{pmatrix} i_x' \\ i_y' \end{pmatrix}. \tag{4.10}$$

According to the base vector transformation under coordinate rotation, $\mathbf{t} = \mathbf{J}\mathbf{i} \rightarrow \mathbf{R}(\pi/4 - \Phi)\mathbf{t} = \mathbf{J}'\mathbf{R}(\pi/4 - \Phi)\mathbf{i} \rightarrow \mathbf{t} = \mathbf{R}(\Phi - \pi/4)\mathbf{J}'\mathbf{R}(\pi/4 - \Phi)\mathbf{i}$, we get

$$\mathbf{J} = \mathbf{R}\left(\Phi - \frac{\pi}{4}\right)\mathbf{J}'\mathbf{R}\left(\frac{\pi}{4} - \Phi\right). \tag{4.11}$$

For planar structures with up-down mirror symmetry (i.e. mirror symmetry in the propagation direction), the Jones matrix $\mathbf{J}'$ is a symmetric matrix [24]:

$$\mathbf{J}' = \begin{pmatrix} a & b \\ b & c \end{pmatrix}. \tag{4.12}$$

According to Eq. (4.6), the Jones matrix $\mathbf{J}^{\#}$ in the elliptical polarization base can be obtained by combining $\mathbf{J}'$, $(\boldsymbol{\alpha}', \boldsymbol{\beta}')$ and $(\boldsymbol{\alpha}'^{*}, \boldsymbol{\beta}'^{*})$, as follows:

$$\begin{aligned} \mathbf{J}^{\#} &= \begin{pmatrix} \mathbf{A} & \mathbf{B} \\ \mathbf{C} & \mathbf{D} \end{pmatrix} = \begin{pmatrix} \boldsymbol{\alpha}'^{*} & \boldsymbol{\beta}'^{*} \end{pmatrix}^{-1} \mathbf{J}' \begin{pmatrix} \boldsymbol{\alpha}' & \boldsymbol{\beta}' \end{pmatrix} \\ &= \begin{pmatrix} e^{-i\chi} & e^{-i\chi} \\ e^{i\chi} & -e^{i\chi} \end{pmatrix}^{-1} \mathbf{J}' \begin{pmatrix} e^{i\chi} & e^{i\chi} \\ e^{-i\chi} & -e^{-i\chi} \end{pmatrix} \\ &= \frac{\mathbf{1}}{\mathbf{2}} \begin{pmatrix} ae^{2i\chi} + 2b + ce^{-2i\chi} & ae^{2i\chi} - ce^{-2i\chi} \\ ae^{2i\chi} - ce^{-2i\chi} & ae^{2i\chi} - 2b + ce^{-2i\chi} \end{pmatrix}. \end{aligned} \tag{4.13}$$

Since it is only necessary to change the ellipticity angle χ to change the elliptical polarization base to any other polarization bases, Eq. (4.13) is a universal transfer Jones matrix form applicable to any polarization bases. Perfect polarization generation requires the output wave to have stable polarization characteristics that are not affected by the input polarization state. This requires $\mathbf{J}^{\#} = \begin{pmatrix} 1 & 0 \\ 0 & 0 \end{pmatrix}$, namely:

$$\begin{cases} A = \frac{1}{2}\left(ae^{2i\chi} + 2b + ce^{-2i\chi}\right) = 1 \\ B = C = \frac{1}{2}\left(ae^{2i\chi} - ce^{-2i\chi}\right) = 0 \\ D = \frac{1}{2}\left(ae^{2i\chi} - 2b + ce^{-2i\chi}\right) = 0 \end{cases}, \tag{4.14}$$

and easy to get:

$$a = \frac{1}{2}e^{-2i\chi}, b = \frac{1}{2}, c = \frac{1}{2}e^{2i\chi}. \tag{4.15}$$

At this point, it can be seen that, the cleverness of using two pairs of conjugate elliptical polarization bases in previous Eq. (4.4) to define the incident and transmitted fields separately: if only a pair of elliptical polarization bases are used to define the incident and transmitted fields, the solution that makes Eq. (4.14) hold will be lost, which readers can prove. Substituting Eq. (4.15) into Eq. (4.12), the form of the Jones matrix $\mathbf{J}'$ in the x'-o-y' coordinate system can be given as:

$$\mathbf{J}' = \frac{1}{2}\begin{pmatrix} e^{-2i\chi} & 1 \\ 1 & e^{2i\chi} \end{pmatrix}. \tag{4.16}$$

Further, substituting Eq. (4.16) into Eq. (4.11), the form of the Jones matrix $\mathbf{J}$ in the original x-o-y coordinate system can be obtained:

$$\mathbf{J} = \frac{1}{2}\mathbf{R}\left(\Phi - \frac{\pi}{4}\right)\begin{pmatrix} e^{-2i\chi} & 1 \\ 1 & e^{2i\chi} \end{pmatrix}\mathbf{R}\left(\frac{\pi}{4} - \Phi\right). \tag{4.17}$$

Jones matrix $\mathbf{J}$ described in Eq. (4.17) cannot be realized by a birefringent meta-atom, but Eq. (4.17) can be split into two Jones matrices, which can be realized by two corresponding birefringent meta-atoms. The Jones matrix $\mathbf{J}$ is split as follows:

$$\begin{aligned} \mathbf{J} &= \frac{1}{2}\mathbf{R}\left(\Phi - \frac{\pi}{4}\right)\begin{pmatrix} e^{-2i\chi} & 0 \\ 0 & e^{2i\chi} \end{pmatrix}\mathbf{R}\left(\frac{\pi}{4} - \Phi\right) + \frac{1}{2}\mathbf{R}(\Phi)\mathbf{R}\left(-\frac{\pi}{4}\right)\begin{pmatrix} 0 & 1 \\ 1 & 0 \end{pmatrix}\mathbf{R}\left(\frac{\pi}{4}\right)\mathbf{R}(-\Phi) \\ &= \frac{1}{2}\mathbf{R}\left(\Phi - \frac{\pi}{4}\right)\begin{pmatrix} e^{-2i\chi} & 0 \\ 0 & e^{2i\chi} \end{pmatrix}\mathbf{R}\left(\frac{\pi}{4} - \Phi\right) + \frac{1}{2}\mathbf{R}(\Phi)\begin{pmatrix} 1 & 0 \\ 0 & -1 \end{pmatrix}\mathbf{R}(-\Phi) \end{aligned}, \tag{4.18}$$

The above equation contains two terms, which can be respectively implemented by two birefringent meta-atoms. The first term is to apply a phase delay of -2χ and -2χ on the fast and slow axes of the meta-atom, respectively, to the output wave (when the fast axis (or slow axis) is parallel (or perpendicular) to a certain coordinate axis) and then rotate counterclockwise by an angle of "$\Phi - \pi/4$". The second term is to apply a phase delay of 0 and π to the output wave on the fast and slow axes of the meta-atom, respectively (when the fast (or slow) axis is parallel (or perpendicular) to a certain coordinate axis), and then rotate it counterclockwise by an angle of Φ. Regardless of the polarization state of the input light, it can be decomposed into a pair of orthogonal polarization states determined by the elliptical orientation angle Φ and ellipticity χ (including orthogonal linear polarization and circular polarization). Furthermore, under the influence of the Jones matrix $\mathbf{J}^{\#} = \begin{pmatrix} 1 & 0 \\ 0 & 0 \end{pmatrix}$, only one polarization component is output. Once a pair of orthogonal polarization states is given, we can select two birefringent meta-atoms that meet the phase delay requirements of the two objectives mentioned above, and then combine the two meta-atoms as a meta-molecule (a unit cell) for further optimization. Finally, we obtained the metasurface design that can generate specific polarized waves at the output without being affected by changes in the input polarization state.

4.2 DESIGN AND ANALYSIS OF META-POLARIZERS

Next, we will demonstrate the principle and design method of polarization meta-devices for circular polarization and linear polarization generation, which is based on the principle of two meta-atomic beam interference mentioned above. In addition, phase gradient design will be further integrated to synchronously implement wavefront manipulation to enhance the functionality of polarizers [25]. The synchronous manipulation of polarization and wavefront is a further requirement for some optical applications, such as imaging and spatially oriented communication. Obviously, in optical systems that require wavefront processing, directly using wavefront-controllable meta-polarizers has higher integration than using polarizers and a wavefront shaper alone.

4.2.1 The Principle Theory of Meta-Polarizers

The unpolarized waves can be seen as an irregular collection of countless polarized waves, so the problem of generating wavefront-controllable fixed polarization states from unpolarized states is essentially a problem of converting arbitrary polarization states to a fixed polarization state and implementing wavefront manipulation. However, for a unit cell containing only one birefringent meta-atom, the polarization state of the transmitted wave will be completely dependent on the polarization state of the incident wave, which means that the transition from arbitrary polarization states to a fixed polarization state cannot be achieved. Next, we will explore the conditions required to obtain the transition from arbitrary polarization states to a fixed polarization state starting from a common transfer equation.

In the Cartesian basis, an arbitrary polarization state of the electric vector $\boldsymbol{E}_{cart\text{-}base}$ is represented as the superposition of polarization components along the x- and y-axis. The transmission relationship for an electromagnetic wave with a certain polarization state incident on a unit cell is expressed as follows:

$$\boldsymbol{E}^t = \boldsymbol{J}\boldsymbol{E}^i \rightarrow \begin{bmatrix} E_x^t \\ E_y^t \end{bmatrix} = \begin{bmatrix} t_{xx} & t_{xy} \\ t_{yx} & t_{yy} \end{bmatrix} \begin{bmatrix} E_x^i \\ E_y^i \end{bmatrix}, \tag{4.19}$$

where $\boldsymbol{E}^t = \begin{bmatrix} E_x^t & E_y^t \end{bmatrix}$ and $\boldsymbol{E}^i = \begin{bmatrix} E_x^i & E_y^i \end{bmatrix}$ represent the electric vectors of the transmitted and incident waves in the Cartesian basis, respectively. And the $t_{\alpha\beta}$ ($\alpha, \beta \in \{x, y\}$) denotes the transmission coefficient of the α-component electric field under the incidence of the β-component electric field. Equation (4.19) is a general description of the transmission characteristics of a birefringent unit cell. For a system possessing mirror symmetry with respect to the $z = 0$ plane, the cross-polarized transmission coefficients are equal, i.e. $t_{xy} = t_{yx}$, and thus the transfer matrix can be expressed as:

$$\boldsymbol{J} = \begin{bmatrix} t_{xx} & t_{xy} \\ t_{yx} & t_{yy} \end{bmatrix} = \begin{bmatrix} a & b \\ b & c \end{bmatrix}. \tag{4.20}$$

For a unit cell containing only one birefringent meta-atom, the polarization state of the transmitted wave will completely depend on the polarization state of the incident wave, meaning the conversion of arbitrary polarization states to a fixed polarization state cannot be achieved. It is easy to find that to obtain a fixed linear polarization output, the following conditions are required:

$$\begin{cases} a=1 \\ b=c=0 \end{cases}, \tag{4.21a}$$

$$\begin{cases} a=b=0 \\ c=1 \end{cases}. \tag{4.21b}$$

Equation (4.21a) will only allow the x-component of the incident wave to be output while completely blocking the y-component. Equation (4.21b) will only allow the y-component of the incident wave to be output while completely blocking the x-component. Considering only the case of x-polarization output here, we obtain:

$$\boldsymbol{J}=\begin{bmatrix} 1 & 0 \\ 0 & 0 \end{bmatrix}=\frac{1}{2}\begin{bmatrix} 1 & 0 \\ 0 & 1 \end{bmatrix}+\frac{1}{2}\begin{bmatrix} 1 & 0 \\ 0 & -1 \end{bmatrix}. \tag{4.22}$$

The first term in the above equation is physically realized by a birefringent meta-atom with both fast and slow axis phase shifts equal to zero. The second term requires that the co-polarized transmitted waves along the x- and y-axis of the meta-atom are imparted with phase shifts of 0 and $\pm\pi$, respectively.

Next, we seek the conditions required for converting arbitrary polarization states to a fixed circular polarization state. To achieve this, the polarization state of the electromagnetic wave first needs to be represented in the circular polarization basis. In the circular basis, an arbitrary electric vector $\boldsymbol{E}_{cir\text{-}base}$ can be represented as the superposition of RCP and LCP components. The transformation matrix from the circular basis to the Cartesian basis is:

$$\boldsymbol{\Lambda}=\frac{\sqrt{2}}{2}\begin{bmatrix} 1 & 1 \\ i & -i \end{bmatrix}. \tag{4.23}$$

It allows any polarization state in the Cartesian basis to be represented by the circular basis, i.e. $\boldsymbol{E}_{cart\text{-}base}=\boldsymbol{\Lambda}\boldsymbol{E}_{cir\text{-}base}$. Therefore, Eq. (4.19) can be expressed as:

$$\boldsymbol{E}^t=\boldsymbol{J}\boldsymbol{E}^i \rightarrow \boldsymbol{\Lambda}\hat{\boldsymbol{E}}^t=\boldsymbol{J}\boldsymbol{\Lambda}\hat{\boldsymbol{E}}^i \rightarrow \hat{\boldsymbol{E}}^t=\boldsymbol{\Lambda}^{-1}\boldsymbol{J}\boldsymbol{\Lambda}\hat{\boldsymbol{E}}^i \rightarrow \hat{\boldsymbol{E}}^t=\boldsymbol{J}^{\#}\hat{\boldsymbol{E}}^i, \tag{4.24}$$

where $\hat{\boldsymbol{E}}^t=\begin{bmatrix} E_R^t & E_L^t \end{bmatrix}^{\mathrm{T}}$ and $\hat{\boldsymbol{E}}^i=\begin{bmatrix} E_R^i & E_L^i \end{bmatrix}^{\mathrm{T}}$ represent the electric vectors of the transmitted and incident waves in the circular basis, respectively. Here, R denotes RCP and L denotes LCP. The new transfer matrix is:

$$\boldsymbol{J}^{\#}=\boldsymbol{\Lambda}^{-1}\boldsymbol{J}\boldsymbol{\Lambda}=\begin{bmatrix} t_{RR} & t_{RL} \\ t_{LR} & t_{LL} \end{bmatrix}=\frac{1}{2}\begin{bmatrix} a+c & a-c-2ib \\ a-c+2ib & a+c \end{bmatrix}, \tag{4.25}$$

where t_{jk} (j, $k \in \{R, L\}$) denotes the transmission coefficient of the j-component electric field under the incidence of the k-component electric field. To achieve a unique circular polarization output, the following is required:

$$\begin{cases} a+c=0 \\ a-c+2ib=0 \\ a-c-2ib=2 \end{cases}, \tag{4.26a}$$

$$\begin{cases} a+c=0 \\ a-c+2ib=2 \\ a-c-2ib=0 \end{cases}. \tag{4.26b}$$

The Eq. (4.26a) will allow the conversion and transmission of the LCP component of the incident wave to an RCP wave, completely blocking the RCP component of the incident wave. The Eq. (4.26b) will allow the conversion and transmission of the RCP component of the incident wave to an LCP wave, completely blocking the LCP component of the incident wave. Solving Eq. (4.26) yields: $a=-c=-ib, b=\frac{1}{2}i$ (for outputting only RCP wave), or $a=-c=ib, b=-\frac{1}{2}i$ (for outputting only LCP wave). Substituting the solution of Eq. (4.26) into Eq. (4.20) gives:

$$J=\pm\frac{1}{2}i\begin{bmatrix} \mp i & 1 \\ 1 & \pm i \end{bmatrix}. \tag{4.27}$$

In the above equation, taking the upper sign yields only RCP output and taking the lower sign yields only LCP output. Furthermore, $\boldsymbol{J}$ can be written as:

$$\boldsymbol{J}=e^{\pm i\frac{\pi}{2}}\boldsymbol{J}'\rightarrow\boldsymbol{J}'=\frac{1}{2}\begin{bmatrix} e^{\mp i\frac{\pi}{2}} & 1 \\ 1 & e^{\pm i\frac{\pi}{2}} \end{bmatrix}. \tag{4.28}$$

The Eq. (4.28) indicates that $\boldsymbol{J}$ can be regarded as $\boldsymbol{J}'$ with an additional $\pi/2$ Pancharatnam-Berry (PB) phase shift factor. It is obtained by performing a rotation operation on $\boldsymbol{J}'$ by an angle of $\theta=-\pi/4$. Therefore, Eq. (4.28) can be written as:

$$\boldsymbol{J}=\boldsymbol{R}\left(-\frac{\pi}{4}\right)\boldsymbol{J}'\boldsymbol{R}^{-1}\left(-\frac{\pi}{4}\right)=\frac{1}{2}\boldsymbol{R}\left(-\frac{\pi}{4}\right)\begin{bmatrix} e^{\mp i\frac{\pi}{2}} & 1 \\ 1 & e^{\pm i\frac{\pi}{2}} \end{bmatrix}\boldsymbol{R}\left(\frac{\pi}{4}\right), \tag{4.29a}$$

$$\boldsymbol{R}(\theta)=\begin{bmatrix} \cos\theta & -\sin\theta \\ \sin\theta & \cos\theta \end{bmatrix}, \tag{4.29b}$$

where $\boldsymbol{R}(\theta)$ is a rotation matrix by angle of θ, and its inverse matrix is $\boldsymbol{R}^{-1}(\theta) = \boldsymbol{R}(\theta)$. The Eq. (4.29a) can be further written as:

$$\begin{aligned}\boldsymbol{J} &= \frac{1}{2}\boldsymbol{R}\left(-\frac{\pi}{4}\right)\begin{bmatrix} e^{\mp i\frac{\pi}{2}} & 0 \\ 0 & e^{\pm i\frac{\pi}{2}} \end{bmatrix}\boldsymbol{R}\left(\frac{\pi}{4}\right) + \frac{1}{2}\boldsymbol{R}\left(-\frac{\pi}{4}\right)\begin{bmatrix} 0 & 1 \\ 1 & 0 \end{bmatrix}\boldsymbol{R}\left(\frac{\pi}{4}\right) \\ &= \frac{1}{2}\boldsymbol{R}\left(-\frac{\pi}{4}\right)\begin{bmatrix} e^{\mp i\frac{\pi}{2}} & 0 \\ 0 & e^{\pm i\frac{\pi}{2}} \end{bmatrix}\boldsymbol{R}\left(\frac{\pi}{4}\right) + \frac{1}{2}\begin{bmatrix} 1 & 0 \\ 0 & -1 \end{bmatrix}\end{aligned} \quad . \tag{4.30}$$

The first term in the equation is physically realized by a birefringent meta-atom rotated by $-\pi/4$. Co-polarized linear transmissions along the x- and y-axis through this meta-atom (unrotated) will be imparted phase shifts of $-\pi/2$ and $\pi/2$ (corresponding to RCP output wave) or $\pi/2$ and $-\pi/2$ (corresponding to LCP output wave), respectively. The second term is physically realized by an unrotated birefringent meta-atom, where co-polarized transmitted waves along the x- and y-axis are imparted phase shifts of 0 and $\pm\pi$, respectively. Furthermore, observing the forms of Eq. (4.22) and Eq. (4.30), a more general expression can be summarized:

$$\boldsymbol{J} = \frac{1}{2}\boldsymbol{R}\left(-\frac{\pi}{4}\right)\begin{bmatrix} e^{i\varphi_{xx}} & 0 \\ 0 & e^{i\varphi_{yy}} \end{bmatrix}\boldsymbol{R}\left(\frac{\pi}{4}\right) + \frac{1}{2}\begin{bmatrix} 1 & 0 \\ 0 & -1 \end{bmatrix} = \boldsymbol{J}_A + \boldsymbol{J}_B, \tag{4.31}$$

where $|\varphi_{xx}| = |\varphi_{yy}|$. When both φ_{xx} and φ_{yy} are zero, it corresponds to a linear polarization output wave; when φ_{xx} and φ_{yy} are $-\pi/2$ and $\pi/2$, respectively, it corresponds to an RCP output wave; and when φ_{xx} and φ_{yy} are $\pi/2$ and $-\pi/2$, respectively, it corresponds to an LCP output wave. Therefore, a meta-molecule as a complete unit cell comprising two birefringent meta-atoms can achieve the conversion of arbitrary polarization states to a fixed polarization state.

When a phase delay is applied to the entire transfer matrix, it is equivalent to applying the same phase delay to each meta-atom, i.e.:

$$\boldsymbol{J}^D = e^{i\phi}\boldsymbol{J} = e^{i\phi}\boldsymbol{J}_A + e^{i\phi}\boldsymbol{J}_B. \tag{4.32}$$

Clearly, the transmitted wave is also imparted the same phase delay without changing its amplitude, i.e. $\boldsymbol{E}^{t^{*}} = e^{i\phi}\boldsymbol{E}^{t^{*}}$. This can be physically achieved by adjusting the structural size parameters of each meta-atom. When a phase delay covering 0–2π is obtained, wavefront manipulation of the output wave can be achieved. Particularly, when the output wave is circularly polarized, a rich phase delay can be obtained by simply applying a rotation factor to the transfer matrix, explained as follows: according to $\boldsymbol{E}_{cart\text{-}base} = \boldsymbol{\Lambda}\boldsymbol{E}_{cir\text{-}base}$, it is easy to convert the polarization state from the circular basis to the Cartesian basis. Since the transmitted wave has only one circular polarization state, we can assume the transmitted wave is either purely RCP or purely LCP. The transmitted wave $\hat{\boldsymbol{E}}^t$ in the Cartesian basis, converted from the circular basis transmitted wave $\boldsymbol{E}^t$, can be expressed, respectively, as:

$$\boldsymbol{E}_{RCP}^t = \boldsymbol{\Lambda}\hat{\boldsymbol{E}}^t = \frac{\sqrt{2}}{2}\begin{bmatrix} 1 & 1 \\ i & -i \end{bmatrix}\begin{bmatrix} E_R^t \\ 0 \end{bmatrix} = \frac{\sqrt{2}}{2}\begin{bmatrix} E_R^t \\ iE_R^t \end{bmatrix}, \tag{4.33a}$$

$$\boldsymbol{E}_{LCP}^t = \boldsymbol{\Lambda}\hat{\boldsymbol{E}}^t = \frac{\sqrt{2}}{2}\begin{bmatrix} 1 & 1 \\ i & -i \end{bmatrix}\begin{bmatrix} 0 \\ E_L^t \end{bmatrix} = \frac{\sqrt{2}}{2}\begin{bmatrix} E_L^t \\ -iE_L^t \end{bmatrix}. \tag{4.33b}$$

The Eq. (4.33a) represents the case where the transmitted wave is purely RCP, meaning the LCP component is 0, i.e. $E_L^t = 0$ in the circular basis $\hat{\boldsymbol{E}}^t$. The Eq. (4.33b) represents the case where the transmitted wave is purely LCP, meaning the RCP component is 0, i.e. $E_R^t = 0$ in the circular basis $\hat{\boldsymbol{E}}^t$. Next, a rotation operation is performed on $\boldsymbol{J}$ to explore the change rules of its phase and electric vector. The rotated $\boldsymbol{J}$ matrix can be written as $\boldsymbol{J}^R = \boldsymbol{R}(\varphi)\boldsymbol{J}\boldsymbol{R}(-\varphi)$. Substituting Eq. (4.27) and Eq. (4.29b) into this yields:

$$\begin{aligned}
\boldsymbol{J}^R = \boldsymbol{R}(\varphi)\boldsymbol{J}\boldsymbol{R}(-\varphi) &= \pm\frac{1}{2}i\boldsymbol{R}(\varphi)\begin{bmatrix} \mp i & 1 \\ 1 & \pm i \end{bmatrix}\begin{bmatrix} \cos\varphi & \sin\varphi \\ -\sin\varphi & \cos\varphi \end{bmatrix} \\
&= \pm\frac{1}{2}i\boldsymbol{R}(\varphi)\begin{bmatrix} \mp i\cos\varphi - \sin\varphi & \mp i\sin\varphi + \cos\varphi \\ \cos\varphi \mp i\sin\varphi & \sin\varphi \pm i\cos\varphi \end{bmatrix} \\
&= \pm\frac{1}{2}i\boldsymbol{R}(\varphi)\begin{bmatrix} \cos\varphi & -\sin\varphi \\ \sin\varphi & \cos\varphi \end{bmatrix}\begin{bmatrix} \mp i & 1 \\ 1 & \pm i \end{bmatrix} \\
&= \pm\frac{1}{2}i\boldsymbol{R}(\varphi)\boldsymbol{R}(\varphi)\begin{bmatrix} \mp i & 1 \\ 1 & \pm i \end{bmatrix} = \boldsymbol{R}(2\varphi)\boldsymbol{J}
\end{aligned} \tag{4.34}$$

The transmitted wave at this point is:

$$\boldsymbol{E}^{t^*} = \boldsymbol{J}^R\boldsymbol{E}^i = \boldsymbol{R}(2\varphi)\boldsymbol{J}\boldsymbol{E}^i = \boldsymbol{R}(2\varphi)\boldsymbol{E}^t. \tag{4.35}$$

From the equation, it can be seen that the new transmitted electric vector $\boldsymbol{E}^{t^*}$ is equivalent to the original transmitted electric vector $\boldsymbol{E}^t$ (before rotating $\boldsymbol{J}$) rotated by 2φ. Furthermore, substituting Eq. (4.29b) and Eq. (4.33) into Eq. (4.35) gives:

$$\begin{aligned}
&\boldsymbol{R}(2\varphi)\boldsymbol{E}_{RCP}^t = \boldsymbol{R}(2\varphi)\boldsymbol{\Lambda}\hat{\boldsymbol{E}}^t \\
&= \frac{\sqrt{2}}{2}\begin{bmatrix} e^{-i2\varphi} & e^{i2\varphi} \\ ie^{-i2\varphi} & -ie^{i2\varphi} \end{bmatrix}\begin{bmatrix} E_R^t \\ 0 \end{bmatrix} = \frac{\sqrt{2}}{2}e^{-i2\varphi}\begin{bmatrix} E_R^t \\ iE_R^t \end{bmatrix} = e^{-i2\varphi}\boldsymbol{E}_{RCP}^t,
\end{aligned} \tag{4.36a}$$

$$R(2\varphi)E_{LCP}^{t} = R(2\varphi)\Lambda\hat{E}^{t}$$
$$= \frac{\sqrt{2}}{2}\begin{bmatrix} e^{-i2\varphi} & e^{i2\varphi} \\ ie^{-i2\varphi} & -ie^{i2\varphi} \end{bmatrix}\begin{bmatrix} 0 \\ E_L^t \end{bmatrix} = \frac{\sqrt{2}}{2}e^{i2\varphi}\begin{bmatrix} E_L^t \\ -iE_L^t \end{bmatrix} = e^{i2\varphi}E_{LCP}^{t} \quad (4.36b)$$

From the above equations, it is known that the amplitude of the new transmitted wave remains unchanged. The phase delays for the RCP and LCP transmitted waves are -2φ and 2φ, respectively. This is a typical characteristic of the PB phase. From Eq. (4.33) to Eq. (4.36), it proves that rotating the meta-molecule represented by J can yield a rich PB phase without additional structures, and without altering the handedness feature or amplitude of the transmitted wave. This provides the feasibility of manipulating the wavefront of transmitted circularly polarized waves using only one type of meta-molecule.

4.2.2 Physical Realization of the Device

All numerical simulations were performed using the time-domain solver in the commercial software CST Microwave Studio. Sample fabrication employed standard plasma etching on silicon (Si) wafer technology (using high-resistivity silicon with a relative permittivity of 11.9 and a thickness of 500 μm). The design of the all-silicon dielectric meta-atoms is shown in Fig. 4.2(a). The lattice constant for the three square meta-atoms is $P = 160$ μm. The electromagnetic wave is incident from the $-z$ direction, and the operating frequency is

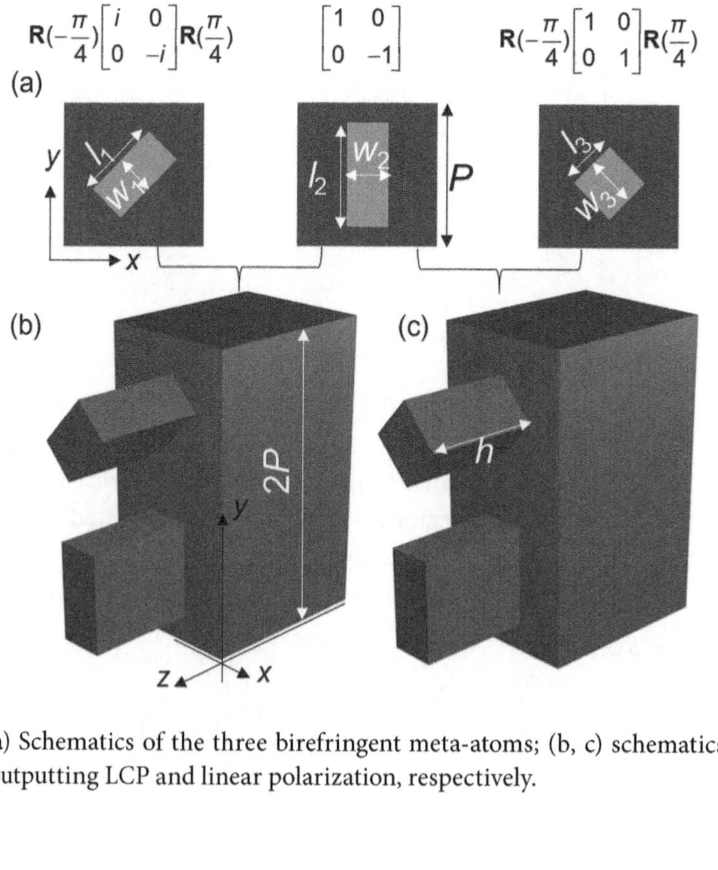

FIGURE 4.2 (a) Schematics of the three birefringent meta-atoms; (b, c) schematics of the meta-molecules for outputting LCP and linear polarization, respectively.

designed to be 1 THz. The final structural parameters determined through a series of parameter scans are: $l_1 = 74$ μm, $w_1 = 47$ μm, $l_2 = 104$ μm, $w_2 = 50$ μm, $l_3 = w_3 = 58.5$ μm, and $h = 200$ μm. A meta-molecule (a complete unit cell) consists of two meta-atoms, as shown in Fig. 4.2(b, c). The designed output waves for the two meta-molecules are LCP (Fig. 4.2(b)) and x-polarization (Fig. 4.2(c)). In Fig. 4.2(b), one meta-atom has an angle of −45° relative to the y-direction, and the transmitted waves along its fast and slow axes are imparted phase delays of $\pi/2$ and $-\pi/2$, respectively. The other meta-atom has an angle of zero relative to the y-direction, and the transmitted waves along its fast and slow axes are imparted phase delays of 0 and $-\pi$, respectively. In Fig. 4.2(c), one meta-atom also has an angle of −45° relative to the y-direction, but the phase delays for the transmitted waves along its fast and slow axes are both zero. The other meta-atom is the same as in Fig. 4.2(b). The simulation results are shown in Fig. 4.3, with the three meta-atoms labelled A, B, and C. As shown in Fig. 4.3(a, b), meta-atom A has a length of 74 μm, a width of 47 μm, and an angle of −45° relative to the y-direction. When unrotated, the transmitted waves along its fast axis (x-axis) and slow axis (y-axis) are imparted phase shifts of $\pi/2$ and $-\pi/2$, respectively. As shown in Fig. 4.3(c, d), meta-atom B has a length of 104 μm, a width of 50 μm, and an angle of zero relative to the y-direction. The transmitted waves along its fast axis (x-axis) and slow axis (y-axis) are imparted phase shifts of 0 and $-\pi$, respectively. As shown in Fig. 4.3(e, f), meta-atom C has equal length and width of 58.5 μm and an angle of −45° relative to the y-direction. The transmissions along its fast and slow axes are equivalent, thus having the same amplitude and phase shift, and the phase shift is zero or 2π.

For the circular polarization meta-polarizer, let the right-handed and left-handed basis vectors in the circular basis be $\vec{\mathbf{e}}_{\mathrm{R}} = \frac{1}{\sqrt{2}}\begin{bmatrix} 1 \\ i \end{bmatrix}$ and $\vec{\mathbf{e}}_{\mathrm{L}} = \frac{1}{\sqrt{2}}\begin{bmatrix} 1 \\ -i \end{bmatrix}$, respectively. Their relationship with the Cartesian basis vectors $\vec{\mathbf{e}}_{\mathrm{x}} = \begin{bmatrix} 1 \\ 0 \end{bmatrix}$ and $\vec{\mathbf{e}}_{\mathrm{y}} = \begin{bmatrix} 0 \\ 1 \end{bmatrix}$ is:

$$\vec{\mathbf{e}}_{\mathrm{R}} = \frac{1}{\sqrt{2}}(\vec{\mathbf{e}}_{\mathrm{x}} + i\vec{\mathbf{e}}_{\mathrm{y}}), \vec{\mathbf{e}}_{\mathrm{L}} = \frac{1}{\sqrt{2}}(\vec{\mathbf{e}}_{\mathrm{x}} - i\vec{\mathbf{e}}_{\mathrm{y}}). \tag{4.37}$$

Through basis vector transformation, the electric amplitude vector of arbitrary polarization in the Cartesian basis is expressed in the circular basis as follows:

$$\begin{aligned} \begin{bmatrix} E_x \\ E_y \end{bmatrix} &= E_x\vec{\mathbf{e}}_{\mathrm{x}} + E_y\vec{\mathbf{e}}_{\mathrm{y}} = E_R\vec{\mathbf{e}}_{\mathrm{R}} + E_L\vec{\mathbf{e}}_{\mathrm{L}} \\ &= \frac{1}{\sqrt{2}}(E_x - iE_y)\vec{\mathbf{e}}_{\mathrm{R}} + \frac{1}{\sqrt{2}}(E_x + iE_y)\vec{\mathbf{e}}_{\mathrm{L}} \end{aligned}, \tag{4.38}$$

where $E_R = \frac{1}{\sqrt{2}}(E_x - iE_y)$ and $E_L = \frac{1}{\sqrt{2}}(E_x + iE_y)$ denote the electric field amplitudes of the right-handed and left-handed circular polarization components, respectively. Therefore, if the x and y component electric fields of the transmitted wave are obtained, the RCP and LCP component electric fields of the transmitted wave can be acquired according to

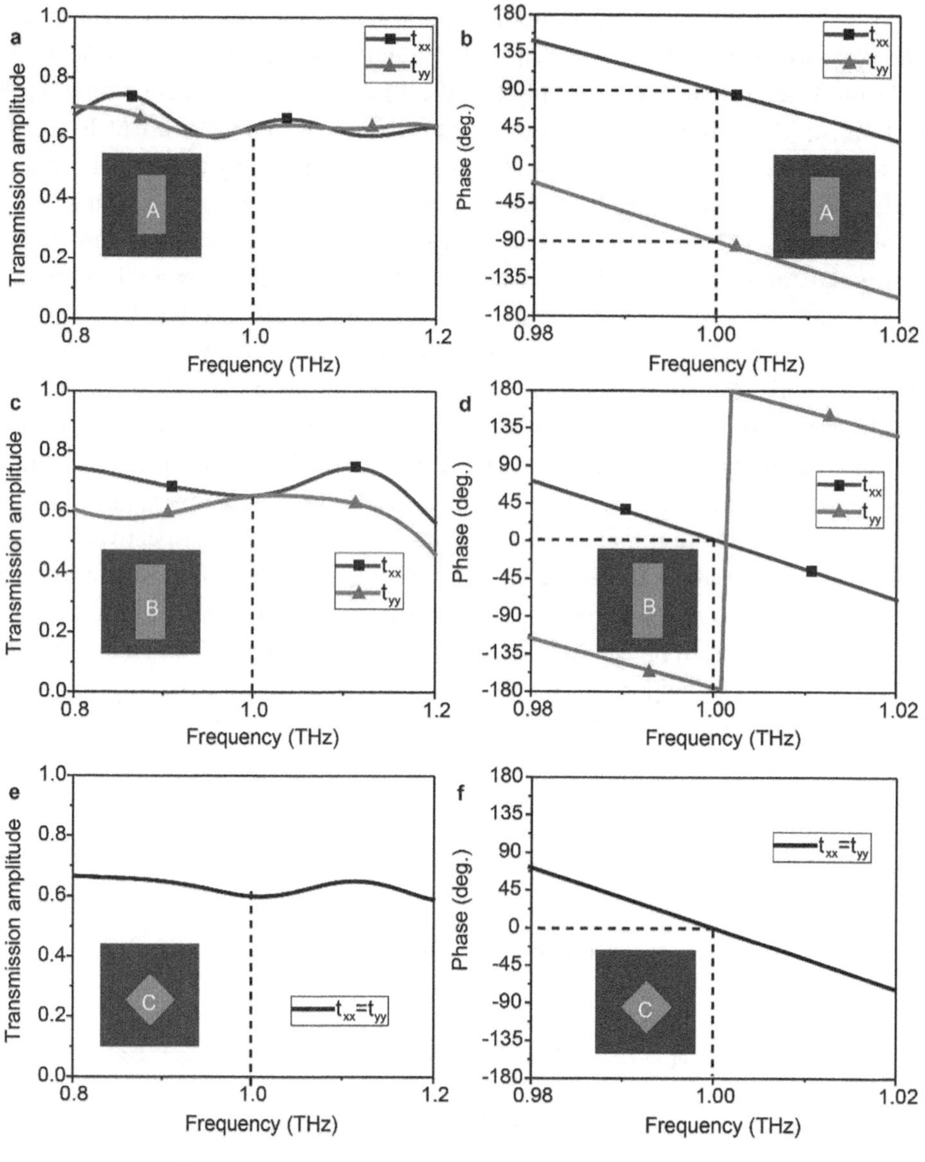

FIGURE 4.3 (a, b) Normalized co-polarized transmission amplitude and phase of meta-atom A; (c, d) normalized co-polarized transmission amplitude and phase of meta-atom B; (e, f) normalized co-polarized transmission amplitude and phase of meta-atom C.

Eq. (4.38). Figure 4.4(a, b) depict the amplitudes $A_L = |E_L|$ and $A_R = |E_R|$ of the RCP and LCP in the transmitted wave under RCP and LCP incidence, respectively. Figure 4.4(c) depicts the normalized amplitudes $A_x = |E_x|$ and $A_y = |E_y|$ of the background transmission x and y components through the substrate under circular polarization incidence. Clearly, the normalized transmittance $|A_x|^2 + |A_y|^2$ of the substrate background is the same for any circular polarization incidence. For the linear polarization meta-polarizer, Fig. 4.4(d, e) depicts the normalized amplitudes of the x and y components in the transmitted wave under x- and

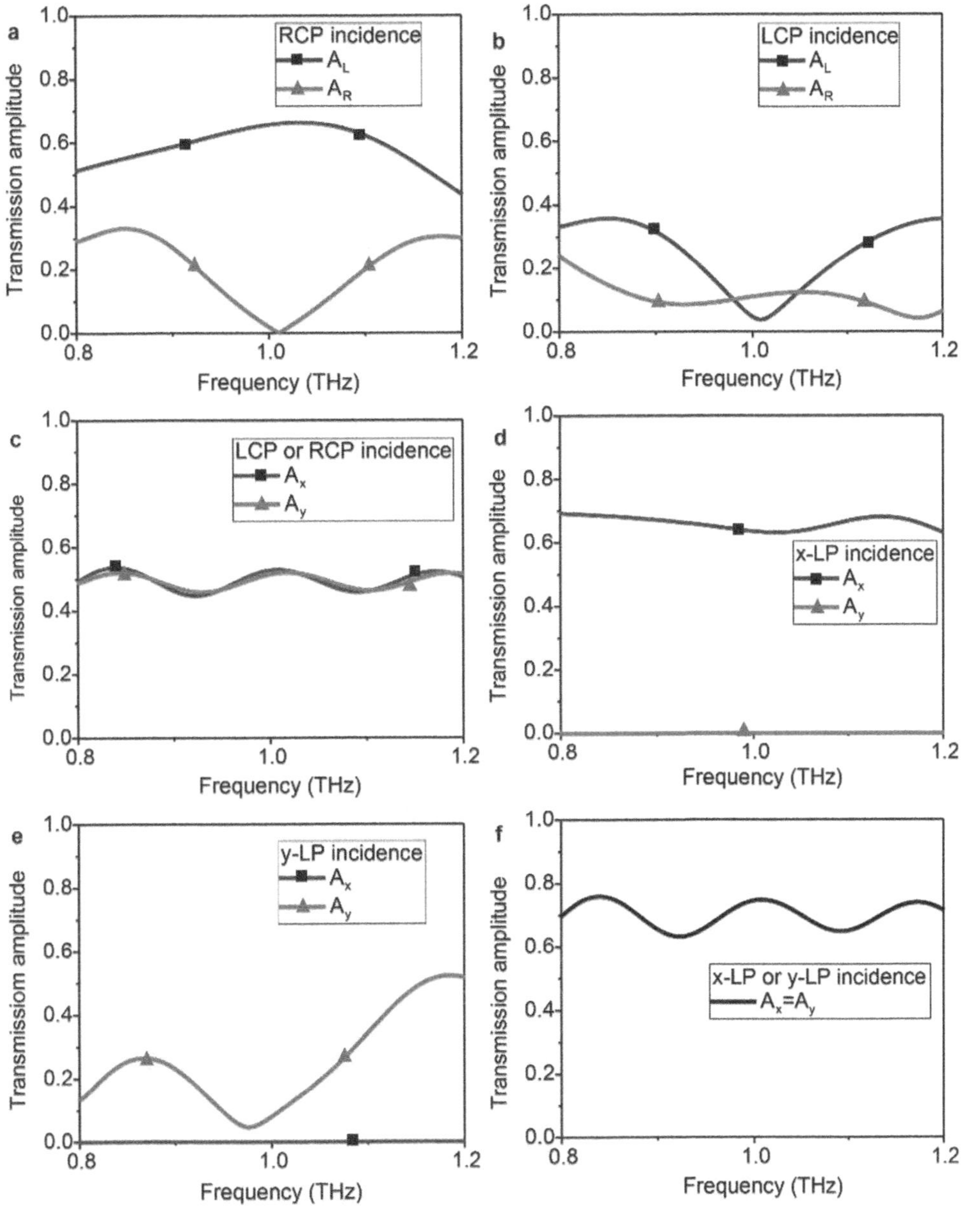

FIGURE 4.4 For the circular polarization meta-polarizer: (a, b) normalized amplitude of the circularly polarized output wave from the meta-molecule under RCP and LCP incidence, respectively. Here, A_L and A_R mean the amplitudes of the LCP and RCP waves, respectively; (c) normalized x and y transmission amplitudes through the substrate background under circularly polarized wave incidence. For the linear polarization meta-polarizer: (d, e) normalized amplitudes of the x- and y-polarized output waves from the meta-molecule under x- and y-polarization incidence, respectively; (f) normalized x and y transmission amplitudes through the substrate background under linearly polarized wave incidence.

y-polarized incidence, respectively. Figure 4.4(f) depicts the normalized amplitudes of the background transmission x and y components through the substrate under linear polarization incidence. Clearly, the normalized transmission amplitude through the substrate background is the same for any linear polarization incidence.

For the meta-molecule in Fig. 4.2(b), it is essentially designed to allow only the handedness-flipped state of the RCP component from the incident wave to transmit at 1 THz. The polarization conversion efficiency is calculated and shown in Fig. 4.5(a). It is defined as $T_{L/R}/T_{Sub} = |A_{L/R}|^2/(|A_x|^2 + |A_y|^2)$, where $T_{L/R}$ means the transmittance of the LCP or RCP output wave, with amplitude A_L or A_R. The T_{Sub} is the background transmittance of the incident light through the substrate, with amplitudes of its x and y components being A_x and A_y, respectively. From Fig. 4.5(a), when RCP is incident, the LCP conversion efficiency

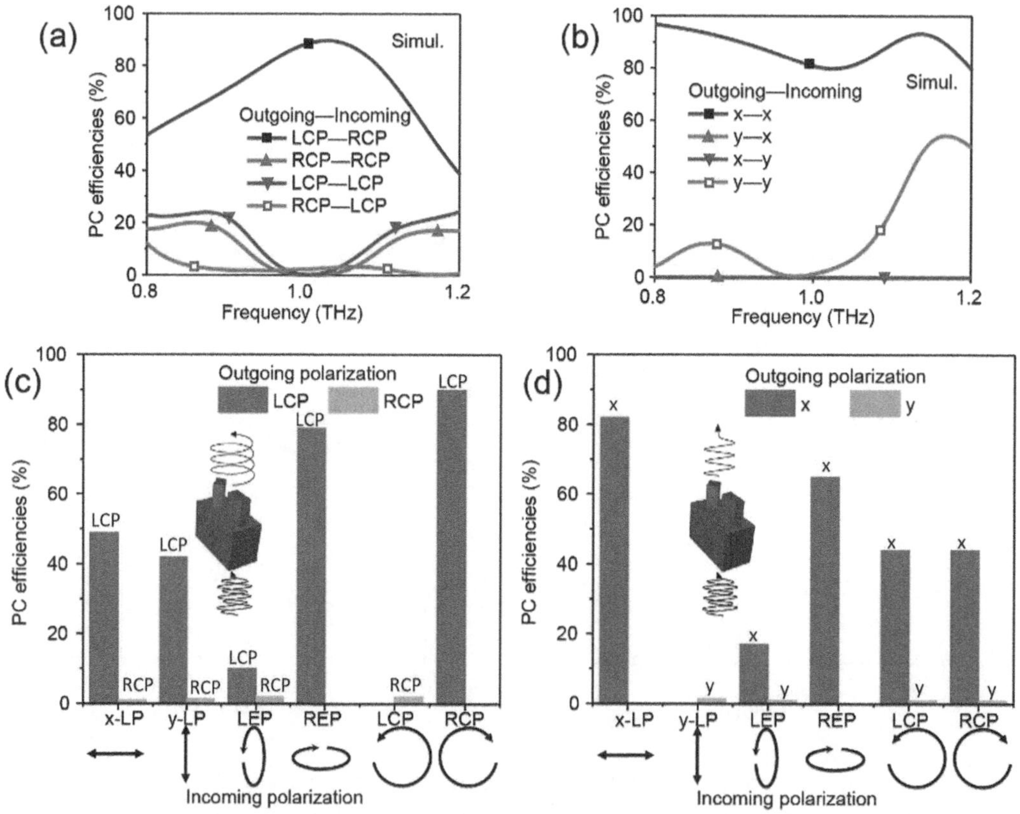

FIGURE 4.5 (a) Polarization conversion efficiency for LCP and RCP when circularly polarized light is incident on the meta-molecule shown in Fig. 4.2(b), and (b) polarization conversion efficiency for x and y components when linearly polarized light is incident on the meta-molecule shown in Fig. 4.2(c), where the curves y-x and x-y are almost zero; (c) polarization conversion efficiency for LCP and RCP when arbitrary polarized light is incident on the meta-molecule shown in Fig. 4.2(b), and (d) polarization conversion efficiency for x and y components when arbitrary polarized light is incident on the meta-molecule shown in Fig. 4.2(c). The ratio of the major to minor axes of the elliptical polarization in (c, d) is 2:1.

at 1 THz exceeds 85%, while the incident LCP wave is almost completely blocked. Since any polarization can be decomposed into RCP and LCP components, an LCP transmitted wave will be generated when linearly polarized, elliptically polarized, or RCP waves are incident. The output efficiencies of LCP and RCP for arbitrary polarized light incidence are summarized in Fig. 4.5(c). Except for LCP, incident light of any polarization on the meta-molecule in Fig. 4.2(b) primarily outputs an LCP wave. Therefore, when unpolarized light is incident, such a meta-molecule will inevitably output an LCP wave, which is consistent with the theoretical derivation described above. Additionally, due to design errors of the meta-molecule, a small amount of RCP output (less than 2% as shown in Fig. 4.5(c)) is inevitable but negligible.

For the meta-molecule in Fig. 4.2(c), it is essentially designed to allow only the x-component of the incident wave to transmit at 1 THz. The polarization conversion efficiency for x and y components is shown in Fig. 4.5(b), defined as $T_{x/y}/T_{Sub} = |A''_{x/y}|^2/(|A_x|^2 + |A_y|^2)$, where $T_{x/y}$ means the transmittance of the x- or y-polarized output wave, with amplitude A''_x or A''_y. The output efficiency of the co-polarized x component exceeds 80%, with almost no cross-polarization conversion, and the co-polarized y component output is completely blocked. Since any polarization can be decomposed into x and y components, incident light of any polarization except y-polarization will only produce an x-polarized transmitted wave. Figure 4.5(d) summarizes the output efficiencies of x and y polarization after arbitrary polarized light is incident on the meta-molecule in Fig. 4.2(c). Almost no y component is detected (less than 1% as shown in Fig. 4.5(d)). Therefore, when unpolarized light is incident, such a meta-molecule will inevitably output an x-polarized wave, which is consistent with the theoretical derivation described above.

To facilitate the design of wavefront-controllable meta-polarizers, the actual unit cell consists of four identical meta-atoms arranged in pairs (as shown in Fig. 4.6(a, c)), forming a square meta-molecule with C_2 symmetry. Such a square meta-molecule is equivalent to the meta-molecules in Fig. 4.2(b, c), which can be proven by Fig. 4.6(b, d). In Fig. 4.6(b, d), the experimental results of polarization conversion efficiency are basically consistent with the simulation results in Fig. 4.5(a, b). Manipulating the wavefront of the output wave must rely on a phase delay gradient covering 0–2π, but the method of obtaining the phase gradient differs for linear and circular polarization. When manipulating the wavefront of the output linearly polarized wave, it is necessary to obtain a dynamic phase by changing the size parameters of the structure. Eight meta-molecules were designed, with structures shown in Fig. 4.7(a) and size parameters listed in Table 4.1. According to Eq. (4.32), the phase delays of the two meta-atoms in each meta-molecule must be synchronized. All eight meta-molecules (U1–U8) output x-polarized waves with roughly consistent amplitude (Fig. 4.7(b)), and the phase delay perfectly covers 0–2π (Fig. 4.7(c)). Based on these eight meta-molecules, simple wavefront manipulation of the x-polarized wave is demonstrated. As shown in Fig. 4.7(d), if the eight meta-molecules are arranged according to a phase profile from 0 to 2π, it not only ensures the generation of an x-polarized output wave from the incident unpolarized terahertz wave but also causes the equiphase surface of the x-polarized wave to tilt, indicating beam deflection.

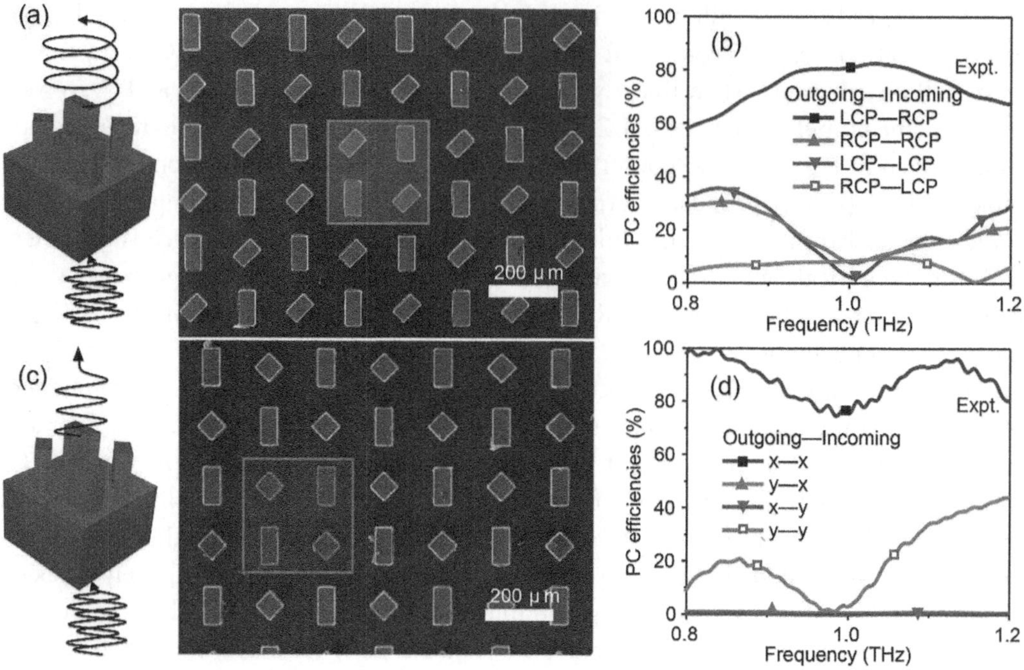

FIGURE 4.6 (a, b) SEM image of the circular polarization meta-polarizer and experimental test results of its conversion efficiency; (c, d) SEM image of the linear polarization meta-polarizer and experimental test results of its conversion efficiency.

The dynamic phase always relies on a tedious process of selecting structural parameters, but this is not necessary for wavefront manipulation of circular polarization. The theoretical derivation above has shown that simply rotating the meta-molecule can yield a rich PB phase without changing the polarization or amplitude of the output wave. There are two available rotation methods: one is to rotate the entire meta-molecule as a whole, and the other is to rotate each meta-atom individually. According to Eq. (4.29), we have:

$$\boldsymbol{J}^R = \boldsymbol{R}(\varphi)\boldsymbol{J}\boldsymbol{R}(-\varphi) = \boldsymbol{R}(\varphi)\boldsymbol{J}_A\boldsymbol{R}(-\varphi) + \boldsymbol{R}(\varphi)\boldsymbol{J}_B\boldsymbol{R}(-\varphi). \tag{4.39}$$

This indicates that the result of rotating the entire meta-molecule is theoretically equivalent to rotating each meta-atom within the meta-molecule. When adopting the overall rotation approach to obtain the PB phase (Fig. 4.7(e)), slight adjustments to the position of each meta-atom are necessary to ensure the meta-molecule's dimensions do not overflow the lattice during overall rotation. Specifically, taking the centre of the meta-molecule as the coordinate origin, the centre coordinates of the two A atoms are (−85.5 μm, 85.5 μm) and (85.5 μm, −85.5 μm), and the centre coordinates of the two B atoms are (74.5 μm, 74.5 μm) and (−74.5 μm, −74.5 μm). The results show that as the meta-molecule rotates from 0 to π, the amplitude of the output LCP remains basically unchanged (Fig. 4.7(f)), and the phase delay can perfectly cover 0–2π (Fig. 4.7(g)). The abundant phase allows for wavefront manipulation of the output LCP wave, such as deflection, beam splitting,

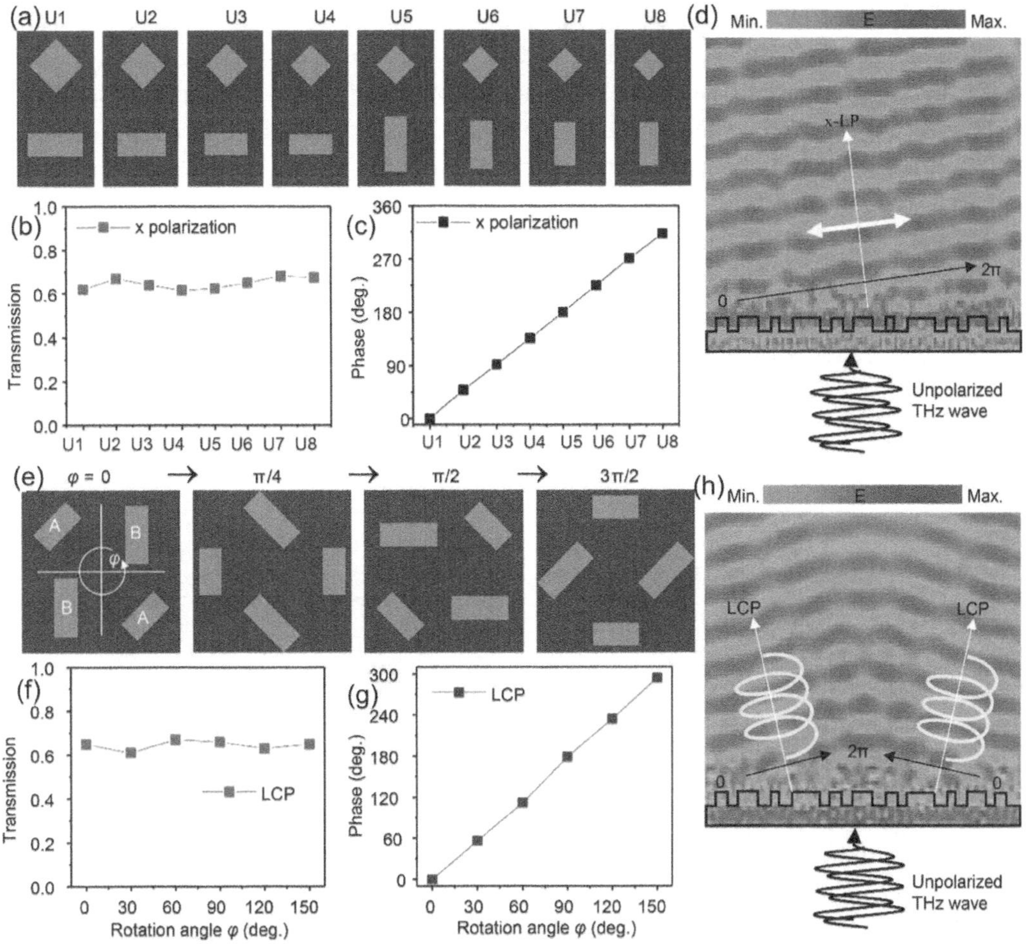

FIGURE 4.7 (a) Schematic of eight meta-molecules composed of meta-atoms of different sizes; (b, c) trend of amplitude and phase change of the x-polarized output wave from these meta-molecules at 1 THz, and (d) electric field phase distribution when these meta-molecules are used for beam deflection of the x-polarized wave; (e) schematic of rotating the entire square meta-molecule by different angles; (f, g) trend of amplitude and phase change of the LCP output wave from this meta-molecule at 1 THz with rotation angle, and (h) electric field phase distribution when the meta-molecule is used for beam splitting of the LCP wave.

TABLE 4.1 Structural Dimensions of the Eight Meta-Molecules in Fig. 4.7(a) (unit: μm)

Atoms	Parameters	U1	U2	U3	U4	U5	U6	U7	U8
A	$l = w$	81	69	64	61	58.5	56	52.5	45.5
B	l(x-axis)	104	81	74	74	50	50	47	40
B	w(y-axis)	50	50	47	40	104	81	74	74

and focusing. A simple wavefront manipulation schematic is shown in Fig. 4.7(h). When meta-molecules with different rotation angles are arranged according to a phase profile of 0–2π–0, the equiphase surface of the LCP output wave easily tilts in different directions to achieve beam splitting. When rotating each meta-atom individually (Fig. 4.8(a)), a rich PB phase can also be obtained by rotating each atom within the meta-molecule, which also allows wavefront manipulation of the output LCP wave, such as deflection (Fig. 4.8(b)), beam splitting (Fig. 4.8(c)), and focusing (Fig. 4.8(d)).

Focusing of the circularly polarized output wave is presented as an important demonstration of wavefront manipulation, as shown in Fig. 4.9. A metalens with circular polarization conversion functionality can replace traditional optical circular polarizers and focusing lenses. Furthermore, ordinary metalenses with circular polarization conversion rely on a specific polarized wave input, which requires an additional polarizer to generate

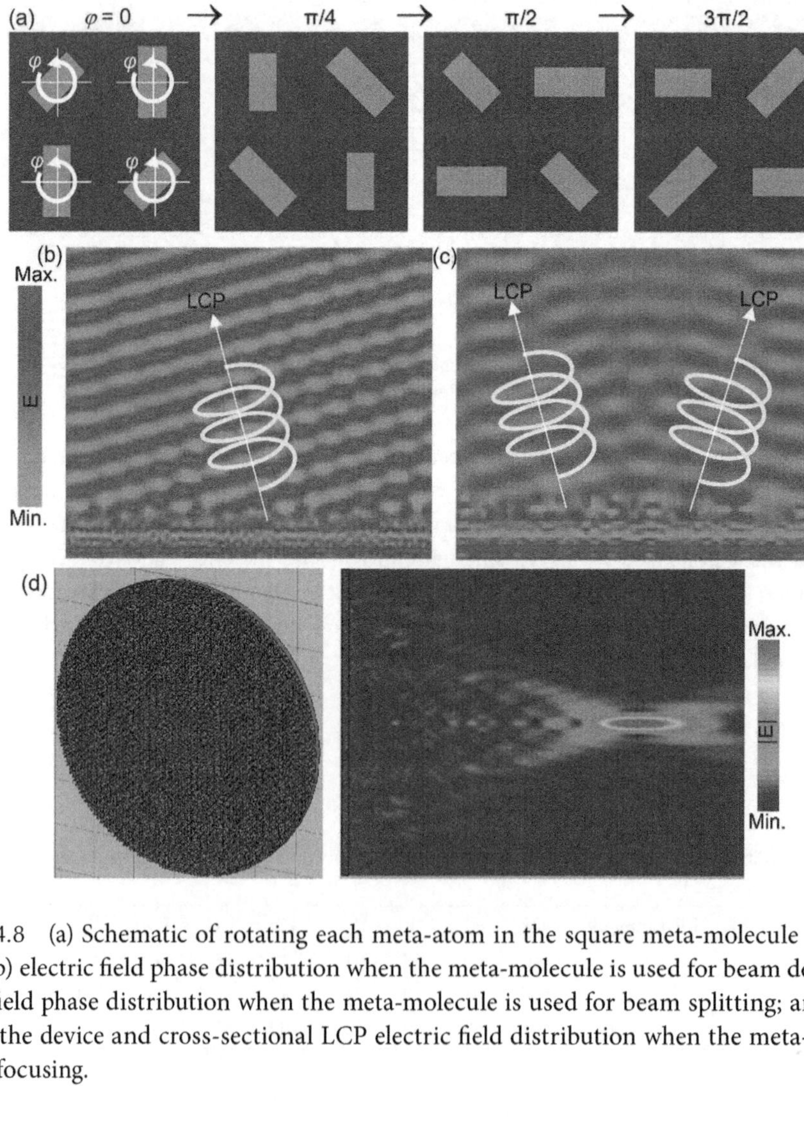

FIGURE 4.8 (a) Schematic of rotating each meta-atom in the square meta-molecule by different angles; (b) electric field phase distribution when the meta-molecule is used for beam deflection; (c) electric field phase distribution when the meta-molecule is used for beam splitting; and (d) schematic of the device and cross-sectional LCP electric field distribution when the meta-molecule is used for focusing.

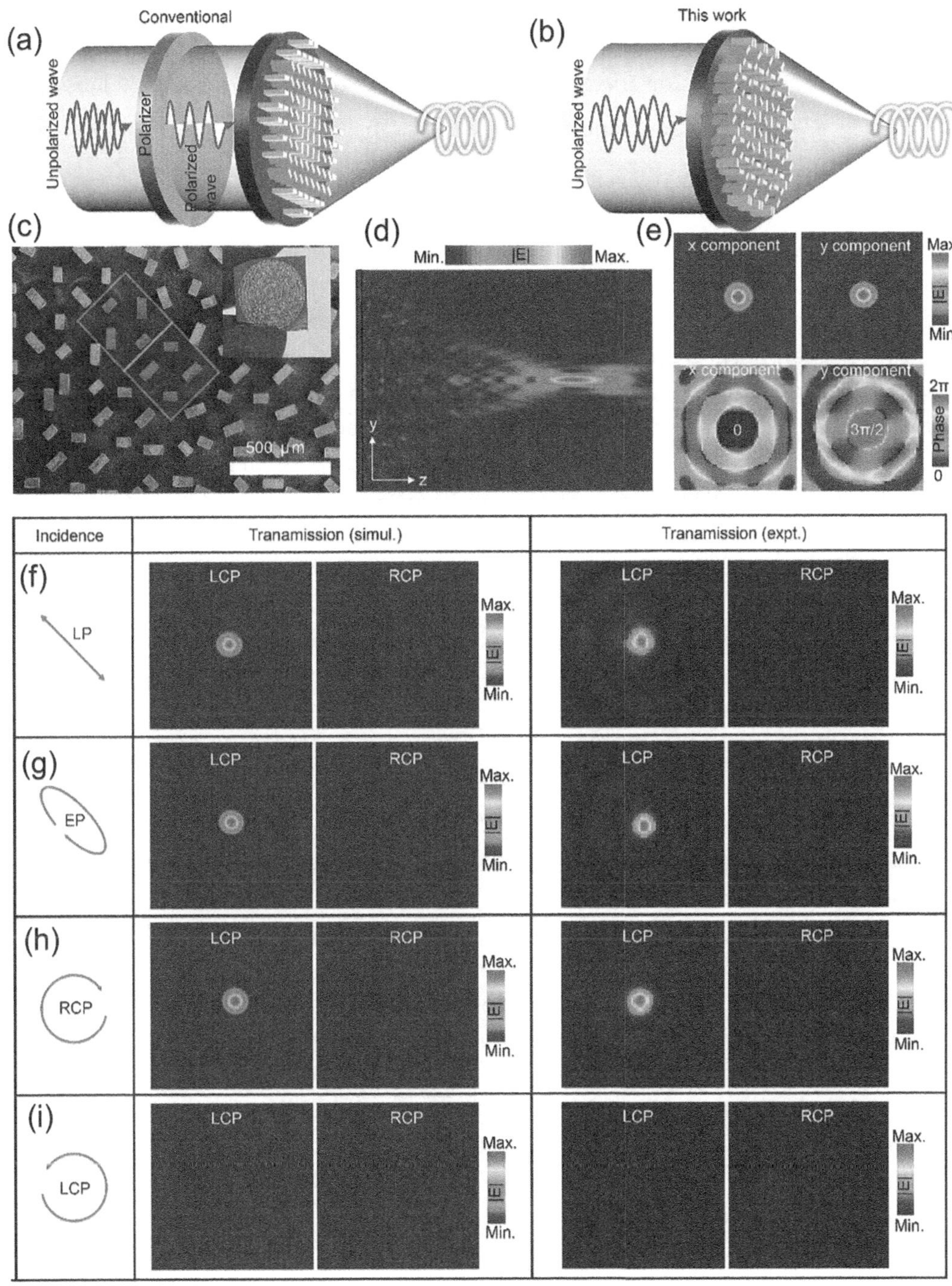

FIGURE 4.9 (a) To produce circular polarization, traditional metalenses require an additional polarizer to generate firstly a specific input polarization; (b) the metalens proposed in this work can directly obtain circular polarization from unpolarized light; (c) SEM image of the sample; (d) the *y-z* cross-sectional LCP electric field distribution map; the electric field clearly converges to the focal point; (e) amplitude and phase distribution of the *x* and *y* components on the focal plane; (f–i) simulation and experimental results of the LCP and RCP transmitted electric field intensity distribution on the focal plane under different polarization incidences.

the specific input polarization (Fig. 4.9(a)). However, this metalens directly obtains a circularly polarized wave from an unpolarized wave (Fig. 4.9(b)), thereby omitting the additional polarizer, thus offering integration advantages over conventional metalenses. As shown in Fig. 4.9(c), the phase profile of this metalens satisfies $\varphi = -\frac{2\pi}{\lambda}\left(\sqrt{x^2+y^2+f^2}-f\right)$, where the operating wavelength $\lambda = 300$ μm, and (x, y) are the coordinates of each point on the metasurface with the centre of the metasurface as the origin, and the focal length $f = 16$ mm. As shown in Figs. 4.9(d, e), the electric field focusing intensities of the transmitted x and y components are almost equal. At the focal point, the phase of the y component leads that of the x component by $3\pi/2$; in other words, the phase of the y component lags behind that of the x component by $\pi/2$; therefore, the output wave is an LCP wave. Simulation and experimental results (Fig. 4.9(f–i)) show that when incident with any polarization of a terahertz wave except LCP, an LCP wave is output and focused, indicating that this metalens can directly obtain a circularly polarized wave from an unpolarized wave and focus it. In Fig. 4.9(f–h), the polarization conversion efficiencies for the LCP wave on the focal plane are approximately 47%, 44%, and 85%, respectively. The definition of polarization conversion efficiency on the focal plane is similar to that described in Fig. 4.5, i.e. T_L/T_{Sub}. However, due to the non-uniform distribution of the electric field on the focal plane, the actual T_L and T_{Sub} should be calculated as the sum of the transmittance at each point on the focal plane, i.e. $T_L = \int\int t_L \mathrm{d}x\mathrm{d}y$ and $T_{Sub} = \int\int t_{Sub} \mathrm{d}x\mathrm{d}y$, where t_L and t_{Sub} represent the LCP transmittance and substrate background transmittance, respectively, which vary with the coordinates on the focal plane.

4.3 CHAPTER SUMMARY

This chapter first introduced the general principle of an arbitrary polarization generation based on the interference of two meta-atoms. Based on this, the principles and design methods of meta-polarizers for circularly polarized waves and linearly polarized waves were introduced. Furthermore, to enhance the functionality and integration of the meta-polarizers, phase gradient design was incorporated, enabling the devices to perform wavefront manipulation simultaneously with polarization generation. A meta-molecule (a complete unit cell) of the device consists of two meta-atoms. When the desired output wave is circularly polarized, the function of the meta-molecule is designed to allow the handedness flipping and transmission of a specific spin circularly polarized wave while blocking the circularly polarized wave with the opposite spin. When the desired output wave is linearly polarized, the function of the meta-molecule is designed to allow only co-polarized linear polarization to transmit while blocking the orthogonal linear polarization. This chapter used incident waves of arbitrary polarization instead of unpolarized waves to demonstrate the effect of such meta-polarizers. Additionally, for linear polarization, meta-molecules of different sizes were designed to obtain a dynamic phase delay covering $0–2\pi$, thereby achieving wavefront manipulation of the transmitted linearly polarized wave. For circular polarization, simply rotating the meta-molecule yields a rich PB phase, enabling wavefront manipulation of the transmitted circularly polarized wave. Various wavefront manipulations were demonstrated, including wave beam deflection, beam splitting, and focusing. It is worth emphasizing again, as stated at the beginning of this chapter, that besides using the beam interference of two meta-atoms to achieve polarization generation, polarization generation can also be achieved based on special resonance effects, such as BICs, which will be disclosed in Chapter 8.

REFERENCES

1. Tamayama, Y., Nakanishi, T., Sugiyama, K., et al. (2006) 'Observation of Brewster's effect for transverse-electric electromagnetic waves in metamaterials: experiment and theory', Physical Review B, 73, 193104.
2. Watanabe, R., Iwanaga, M. and Ishihara, T. (2008) 's-polarization Brewster's angle of stratified metal–dielectric metamaterial in optical regime', Physica Status Solidi B, 245, 2696–2701.
3. Alu, A., D'Aguanno, G., Mattiucci, N., et al. (2011) 'Plasmonic Brewster angle: broadband extraordinary transmission through optical gratings', Physical Review Letters, 106, 123902.
4. Wang, C., Zhu, Z., Cui, W., et al. (2019) 'All-angle Brewster effect observed on a terahertz metasurface', Applied Physics Letters, 114, 191902.
5. Shelby, R.A., Smith, D.R. and Schultz, S. (2001) 'Experimental verification of a negative index of refraction', Science, 292, 77–79.
6. Fan, H., Chu, H., Luo, H., et al. (2022) 'Brewster metasurfaces for ultrabroadband reflectionless absorption at grazing incidence', Optica, 9, 1138–1148.
7. Fan, H.Y. and Luo, J. (2023) 'Research progress of reflectionless electromagnetic metasurfaces', Opto-Electronic Engineering, 50(9), 230147.
8. Luo, H., Luo, J., Zhang, Z., et al. (2024) 'Dielectric metamaterials with effective self-duality and full-polarization omnidirectional Brewster effect', Light: Science & Applications, 13, 262.
9. Wang, S., Deng, Z.-L., Wang, Y., et al. (2021) 'Arbitrary polarization conversion dichroism metasurfaces for all-in-one full Poincaré sphere polarizers', Light: Science & Applications, 10, 24.
10. Wu, P.C., Tsai, W.-Y., Chen, W.T., et al. (2017) 'Versatile polarization generation with an aluminum plasmonic metasurface', Nano Letters, 17, 445–452.
11. Ding, F., Chang, B., Wei, Q., et al. (2020) 'Versatile polarization generation and manipulation using dielectric metasurfaces', Laser & Photonics Reviews, 14, 2000116.
12. Li, J.T., Wang, G.C., Yue, Z., et al. (2022) 'Dynamic phase assembled terahertz metalens for reversible conversion between linear polarization and arbitrary circular polarization', Opto-Electronic Advances, 5, 210062.
13. Zhang, F., Pu, M., Li, X., et al. (2017) 'All-dielectric metasurfaces for simultaneous giant circular asymmetric transmission and wavefront shaping based on asymmetric photonic spin–orbit interactions', Advanced Functional Materials, 27, 1704295.
14. Fan, Q., Liu, M., Zhang, C., et al. (2020) 'Independent amplitude control of arbitrary orthogonal states of polarization via dielectric metasurfaces', Physical Review Letters, 125, 267402.
15. Yue, Z., Li, J.T., Li, J. et al. (2022) 'Terahertz metasurface zone plates with arbitrary polarizations to a fixed polarization conversion', Opto-Electronic Science, 1, 210014.
16. Zheng, C., Li, J., Yue, Z. et al. (2022) 'All-dielectric trifunctional metasurface capable of independent amplitude and phase modulation', Laser & Photonics Reviews, 16, 2200051.
17. Kim, T.-T., Oh, S.S., Kim, H.-D., et al. (2017) 'Electrical access to critical coupling of circularly polarized waves in graphene chiral metamaterials', Science Advances, 3, e1701377.
18. Lin, B.-Q., Lv, L.-T., Guo, J.-X., et al. (2020) 'Ultra-wideband linear-to-circular polarization conversion metasurface', Chinese Physics B, 29, 104205.
19. Wang, Q., Kong, X., Yan, X., et al. (2019) 'Flexible broadband polarization converter based on metasurface at microwave band', Chinese Physics B, 28, 074205.
20. Li, J., Li, J., Zheng, C., et al. (2021) 'Active controllable spin-selective terahertz asymmetric transmission based on all-silicon metasurfaces', Applied Physics Letters, 118, 221110.

21. Akram, M.R., Mehmood, M.Q., Bai, X., et al. (2019) 'High efficiency ultrathin transmissive metasurfaces', Advanced Optical Materials, 7, 1801628.
22. Rubin, N.A., D'Aversa, G., Chevalier, P., et al. (2019) 'Matrix Fourier optics enables a compact full-Stokes polarization camera', Science, 365, eaax1839.
23. Cao, G., Xu, H.-X., Zhou, L.-M. et al. (2021) 'Infrared metasurface-enabled compact polarization nanodevices', Materials Today, 50, 499–515.
24. Menzel, C., Rockstuhl, C. and Lederer, F. (2010) 'Advanced Jones calculus for the classification of periodic metamaterials', Physical Review A, 82, 053811.
25. Li, J., Yue, Z., Li, J., et al. (2023) 'Wavefront-controllable all-silicon terahertz meta-polarizer', Science China Materials, 66(1), 300–308.

CHAPTER 5

Chiral Photonic Manipulation

Abstract

Chiral photonic devices can control the selective propagation of circularly polarized states with different spin directions, and this extraordinary capability makes them highly valuable for both practical applications and scientific research. Previous Chapter 4 mentioned that selective transmission of a pair of orthogonal circularly polarized states can be achieved through the interference of multiple meta-atoms, indicating that the device gains chiral transmission properties. In fact, research on chiral devices based on electromagnetic resonance for selective transmission of circularly polarized waves is more prevalent, as their physical essence differs entirely from the birefringence-like effect relied upon by the interference of multiple meta-atoms. This chapter will introduce the general principles of chiral optical responses using Jones matrices and demonstrate the construction and simple applications of chiral meta-devices. It should be noted that the physical essence of chiral responses described in this chapter is electromagnetic eigen-resonance, and the demonstrated chiral devices are based on conventional chiral eigen-resonances. Some special electromagnetic eigen-resonances, such as bound states in continuum, can also achieve chiral propagation, which will be discussed in Chapter 8.

5.1 CHIRAL RESPONSE PRINCIPLE OF DEVICES

From the perspective of device geometry, chiral devices are defined as: in the same reference plane, the device structure cannot be overlapped with its own mirror body through an appropriate rotation operation [1]. Here, emphasizing the "same reference plane" is necessary, and its reason will be explained later. This is like the relationship between a person's left and right hands, which can be seen as a pair of mirror images, and we cannot rotate one hand to overlap with the other. In recent years, researchers have achieved selective manipulation of circularly polarized light with different spin directions by designing chiral metasurfaces with specific structures. The transmission and reflection characteristics of incident light from the back of metasurfaces are different from those from the front, and this characteristic is called

DOI: 10.1201/9781003784821-5

spin-selective asymmetric propagation [2,3]. They can be used to make asymmetric filters and optical anti-counterfeiting devices [4,5] and can also be used in many fields such as chiral imaging, optical communication, quantum optics, and optical sensing.

Chiral metasurfaces can be classified into transmissive and reflective types based on their electromagnetic wave propagation ports. A metasurface that can selectively output circularly polarized waves in a certain spin direction at the transmission port is called a transmissive chiral metasurface, and its spin-selective transmission can be achieved by constructing chiral structures using single-layer metal materials [6]. However, in order to achieve strong circular dichroism (CD), further reliance on multi-layer metal design is generally required. In addition, high refractive index and low loss all-dielectric (such as Si-SiO_2 system or SiO_2-TiO_2 system) micro-nano structures are also a good choice for achieving strong CD [7,8]. Reflective metasurfaces completely close the transmission port and improve reflection efficiency by blocking any transmitted light. Their chiral properties can be understood as the difference in reflectivity between different spin circularly polarized waves due to the structural chirality of the metasurface. Reflective chiral metasurfaces can strongly reflect one circularly polarized light and suppress another circularly polarized light. In addition to static manipulation of circular polarization, researchers also hope to prepare tunable chiral metasurfaces to expand their applications [9]. The methods reported involve mechanical drive, magneto-optical control, electrical gating, optical carrier excitation, etc. [10,11]. This chapter will demonstrate the design of transmissive and reflective chiral devices with tunable CD, as well as their simple applications. Before demonstrating these designs and applications, this chapter will elaborate on the general principles of chiral devices.

5.1.1 Jones Matrix Description of Circular Polarization Response

When studying chiral response, it is most important to qualitatively assess the overall optical response of the device, rather than focusing on the specific values of material parameters. The Jones matrix can phenomenologically express the overall optical response of a device using complex reflection and transmission coefficients. Therefore, using the Jones matrix to study the chiral response of a device is a very effective tool [12].

We still need to start from the linear polarization response of the device to derive its circular polarization response characteristics. The schematic diagram of the device is shown in Fig. 5.1(a), which defines the forward port and backward port of the device. Assuming linearly polarized light is incident from the forward port, its incident wave E_x^i, E_y^i and transmitted wave E_x^t, E_y^t can be related by the transmission matrix $\boldsymbol{T}^f$, expressed as:

$$\begin{pmatrix} E_x^t \\ E_y^t \end{pmatrix} = \begin{pmatrix} t_{xx} & t_{xy} \\ t_{yx} & t_{yy} \end{pmatrix} \begin{pmatrix} E_x^i \\ E_y^i \end{pmatrix} = \begin{pmatrix} A & B \\ C & D \end{pmatrix} \begin{pmatrix} E_x^i \\ E_y^i \end{pmatrix} = \boldsymbol{T}^f \begin{pmatrix} E_x^i \\ E_y^i \end{pmatrix}, \tag{5.1}$$

where t_{jk} ($j, k \in \{x, y\}$) represents the j component transmission coefficients with the k component incidence. For the convenience of discussion, the transmission coefficients in the $\boldsymbol{T}^f$ matrix are replaced by A, B, C, and D. When light is incident from the backward port of the device, considering optical reciprocity (time-reversal symmetry), there is

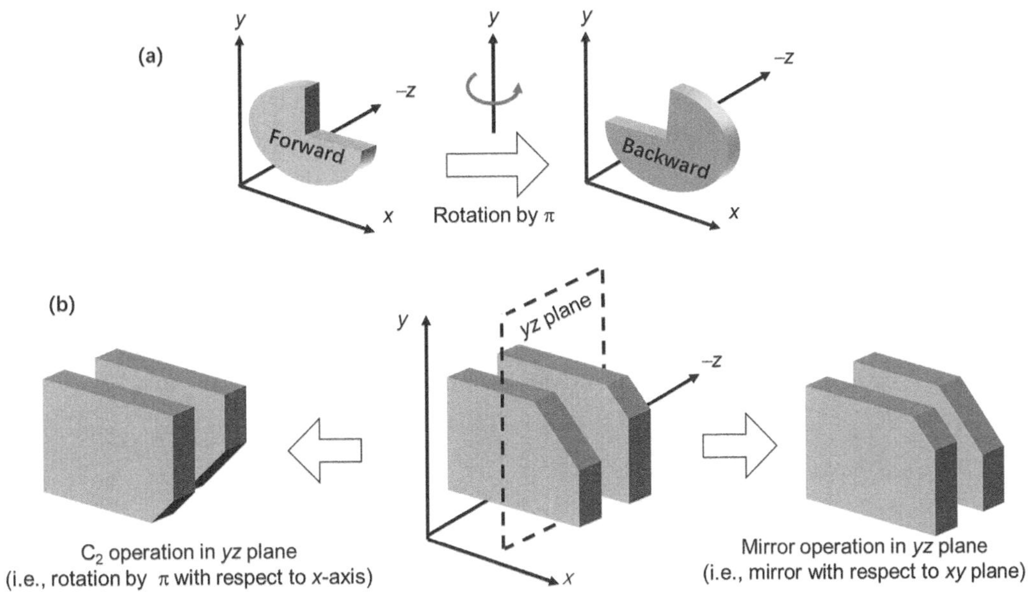

FIGURE 5.1 (a) Schematic diagram of the optical response system from the forward and backward; (b) schematic diagram of rotation and mirror operation of the structure within the same plane (e.g. *yz* plane).

$E_t = TE_i \rightarrow E_i = T^{-1}E_t \rightarrow E_i^* = \left(T^{-1}\right)^* E_t^*$ [13]. Also, the T should be unitary (satisfy $T^{+}T = \mathbf{I}$) due to non-loss and reciprocity of the system. Hence, there is $T^{-1} = T^{+} = \left(T^{T}\right)^* \rightarrow \left(T^{-1}\right)^* = T^{T}$, where T^T is the transpose matrix of T, and T^+ is the conjugate transpose of T. Therefore, the off-diagonal elements of the transfer matrix T^f are swapped. Meanwhile, considering coordinate transformation, this is equivalent to rotating the device by 180° with respect to the *y*-axis (meaning $x \rightarrow -x$). At this point, off-diagonal elements will be added with negative signs, and the final transfer matrix T^b is:

$$T^b = \begin{pmatrix} A & -C \\ -B & D \end{pmatrix}. \tag{5.2}$$

In order to investigate the circularly polarized response, it is necessary to transform the transfer matrix in the linear polarization basis into the transfer matrix in the circular polarization basis. The transformation from linear to circular polarization basis is completed through a basis matrix $\Lambda = \frac{1}{\sqrt{2}}\begin{pmatrix} 1 & 1 \\ -\mathrm{i} & i \end{pmatrix}$, and there is $E_{x,y} = \Lambda E_{+,-}$, where $E_{+,-}$ represents the electric vector of the circularly polarized wave, and the subscripts "+" and "–" represent the electric vectors rotating clockwise and counterclockwise viewed along –z direction, respectively. Therefore, we have:

$$\begin{aligned} &E_{x,y}^{t} = T_{x,y}\, E_{x,y}^{i} \rightarrow \Lambda E_{+,-}^{t} = T_{x,y}\Lambda E_{+,-}^{i} \rightarrow \Lambda^{-1}\Lambda E_{+,-}^{r} = \Lambda^{-1}T_{x,y}\Lambda E_{+,-}^{r} \\ &\Rightarrow E_{+,-}^{r} = T_{+,-}E_{+,-}^{r} \rightarrow T_{+,-} = \Lambda^{-1}T_{x,y}\Lambda \end{aligned} \tag{5.3}$$

By substituting the transfer matrices T^f and T^b in the linear polarization basis into the above equation, the transmission response in the circular polarization basis can be obtained. When circularly polarized light is incident from the forward port of the device, the transmission transfer matrix is as follows:

$$T^f_{+,-} = \Lambda^{-1} T^f \Lambda = \begin{pmatrix} t_{++} & t_{+-} \\ t_{-+} & t_{--} \end{pmatrix} = \frac{1}{2} \begin{pmatrix} A+D+i(B-C) & A-D-i(B+C) \\ A-D+i(B+C) & A+D-i(B-C) \end{pmatrix}. \quad (5.4)$$

Using Eqs. (5.2) and (5.4), when circularly polarized light is incident from the backward port of the device, the transmission transfer matrix is:

$$T^b_{+,-} = \Lambda^{-1} T^b \Lambda = \begin{pmatrix} t_{++} & -t_{-+} \\ -t_{+-} & t_{--} \end{pmatrix}. \quad (5.5)$$

It should be noted that the transfer matrix in an arbitrary basis is not simply given by Eq. (5.2), which imitates the form of Eq. (5.2) to exchange negative off-diagonal elements, but rather by applying the corresponding changes in the basis matrix Λ to T^f and T^b in the linear basis. The performance indicator for evaluating asymmetric transmission of circularly polarized waves is CD Δ_{cir}, which is generally defined as the difference in response of circularly polarized light incident from the forward and backward ports of the device, expressed as:

$$\Delta_{cir} = \left|T_1^f\right|^2 - \left|T_1^b\right|^2 = |t_{-+}|^2 - |t_{+-}|^2 \quad or \quad \left|T_2^f\right|^2 - \left|T_2^b\right|^2 = |t_{+-}|^2 - |t_{-+}|^2. \quad (5.6)$$

There are also reports that define CD as the difference in response of different input circular polarization states at the same port. For example, for the device described in Eq. (5.4), CD can be defined as:

$$\Delta_{cir} = \left|T_1^f\right|^2 - \left|T_2^f\right|^2 = \left(|t_{++}|^2 + |t_{-+}|^2\right) - \left(|t_{--}|^2 + |t_{+-}|^2\right). \quad (5.7)$$

For two-port devices, both definitions mentioned above are applicable, but for single-port devices such as pure reflective devices, only the response differences of different input polarization states of the reflective port can be used to describe CD. In this case, the transmission coefficient in Eq. (5.7) only needs to be replaced with the reflection coefficient.

5.1.2 Discussion on Geometric Symmetry

For a structure without any symmetry, the elements in its T matrix are different, making it easy to form chiral responses. However, most artificially designed structures have one or more symmetries, and we need to discuss the impact of these symmetries on the chiral response of the devices. Several main geometric symmetries and polarization responses are summarized in Table 5.1.

TABLE 5.1 Schematic Diagram of the Example Structures with Different Geometric Symmetries, and Their Corresponding Linear T Matrix and Eigenpolarizations

Symmetry	**Examples**	**Linear *T* Matrix**	**Eigenpolarizations**
No symmetry at all		$T=\begin{pmatrix} A & B \\ C & D \end{pmatrix}$	Elliptic
Mirror symmetry with respect to the yz or xz plane	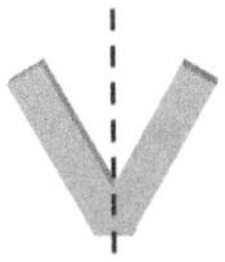	$T=\begin{pmatrix} A & 0 \\ 0 & D \end{pmatrix}$	Linear
Planar structure (mirror symmetry with respect to the xy plane)	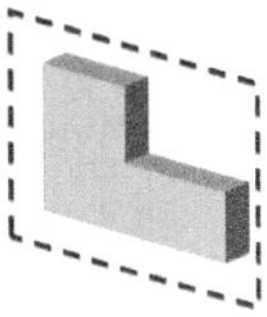	$T=\begin{pmatrix} A & B \\ B & D \end{pmatrix}$	Elliptic
$C_{2,z}$ symmetry	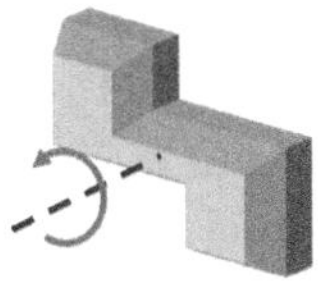	$T=\begin{pmatrix} A & B \\ C & D \end{pmatrix}$	Elliptic
Planar structure with $C_{2,z}$ symmetry		$T=\begin{pmatrix} A & B \\ B & D \end{pmatrix}$	Elliptic
$C_{4,z}$ ($C_{3,z}$) symmetry		$T=\begin{pmatrix} A & B \\ -B & A \end{pmatrix}$	Circular
Planar structure with $C_{4,z}$ ($C_{3,z}$) symmetry	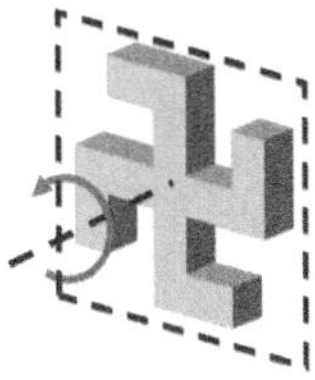	$T=\begin{pmatrix} A & 0 \\ 0 & A \end{pmatrix}$	Linear
Non-planar structure with $C_{2,x}$ ($C_{2,y}$) symmetry	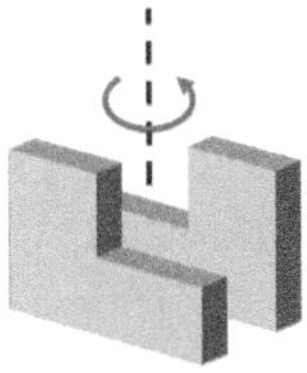	$T=\begin{pmatrix} A & B \\ -B & D \end{pmatrix}$	Elliptic

If a structure has mirror symmetry with respect to the *yz* plane, which means that the Jones matrices of the device before and after coordinate mirror transformation are equal, we have:

$$M_y = \begin{pmatrix} -1 & 0 \\ 0 & 1 \end{pmatrix}, M_y \boldsymbol{T}^f M_y^{-1} = \begin{pmatrix} \mathrm{A} & -B \\ -C & \mathrm{D} \end{pmatrix} = \begin{pmatrix} \mathrm{A} & \mathrm{B} \\ \mathrm{C} & \mathrm{D} \end{pmatrix} \rightarrow \boldsymbol{T}^f = \begin{pmatrix} \mathrm{A} & 0 \\ 0 & \mathrm{D} \end{pmatrix}, \quad (5.8)$$

where M_y is the coordinate mirror transformation matrix. When the structure is mirror symmetric with respect to the *yz* plane, its coordinate transformation follows $x \rightarrow -x$ and $y \rightarrow y$. Therefore, the first element in the matrix M_y has a negative sign. It is easy to understand that when the structure is mirror symmetric with respect to the *xz* plane, its coordinate transformation follows $x \rightarrow x$ and $y \rightarrow -y$. At this point, the fourth element in the M matrix becomes negative, but the final result of $\boldsymbol{T}^f$ remains unchanged, i.e. $\mathrm{B} = \mathrm{C} = 0$ (the mathematical operations process will not be repeated here). It should be noted that since the coordinate transformation we described is for the Cartesian coordinate system, the coordinate transformation is applied to the linearly polarized $\boldsymbol{T}$ matrix instead of the circularly polarized $\boldsymbol{T}$ matrix. It is easy to understand that such mirror symmetric structures inevitably lack chiral response. By substituting Eq. (5.8) into Eq. (5.4), this conclusion can be proven (obtaining $t_{++} = t_{--} = 0.5(A + D)$, $t_{-+} = t_{+-} = 0.5(A - D)$, $\Delta_{cir} = 0$).

When the structure has C_2 symmetry with respect to the *z*-axis (180° rotational symmetry in the *xy* plane), we have:

$$R_\pi = \begin{pmatrix} -1 & 0 \\ 0 & -1 \end{pmatrix}, R_\pi \boldsymbol{T}^f R_\pi^{-1} = \begin{pmatrix} A & B \\ C & D \end{pmatrix} = \boldsymbol{T}^f, \quad (5.9)$$

where $R = \begin{pmatrix} \cos\theta & -\sin\theta \\ \sin\theta & \cos\theta \end{pmatrix}$ is the rotation matrix. It can be seen that rotating the C_2 symmetry structure by 180° with respect to the *z*-axis will not change any response parameters. Whether such structures can exhibit chiral response requires further confirmation by combining additional symmetry. When the structure has C_3 symmetry with respect to the *z*-axis (120° rotational symmetry in the *xy* plane), we have:

$$R_{2\pi/3} = \begin{pmatrix} -1/2 & -\sqrt{3}/2 \\ \sqrt{3}/2 & -1/2 \end{pmatrix} \rightarrow \boldsymbol{T}^f = \begin{pmatrix} A & B \\ -B & A \end{pmatrix}. \quad (5.10)$$

When the structure has C_4 symmetry with respect to the *z*-axis (90° rotational symmetry in the *xy* plane), we have:

$$R_{\pi/2} = \begin{pmatrix} 0 & -1 \\ 1 & 0 \end{pmatrix}, R_{\pi/2} \boldsymbol{T}^f R_{\pi/2}^{-1} = \begin{pmatrix} \mathrm{D} & -\mathrm{C} \\ -B & \mathrm{A} \end{pmatrix} = \begin{pmatrix} \mathrm{A} & \mathrm{B} \\ \mathrm{C} & \mathrm{D} \end{pmatrix} \rightarrow \boldsymbol{T}^f = \begin{pmatrix} A & B \\ -B & A \end{pmatrix}. \quad (5.11)$$

It can be seen that the device is insensitive to any linear polarization state at this time. A structure that satisfies C_4 symmetry must also satisfy C_2 symmetry. If there is no additional symmetry, the structure is chiral. By substituting Eq. (5.11) into Eq. (5.4), this conclusion can be proved (obtaining $t_{++} = A + iB$, $t_{--} = A - iB$, $t_{-+} = t_{+-} = 0$, $\Delta_{cir} \neq 0$). If this structure has an additional mirror symmetry with respect to the *yz* or *xz* plane (meaning that Eqs. (5.11) and (5.8) hold simultaneously), all off-diagonal elements become zero ($B = 0$) so that the structure loses its chiral characteristics.

For an arbitrary planar structure, a plane perpendicular to the *z*-axis can be found, making the structure mirror symmetric with respect to this plane. When light is incident from the backward port of the device, if we consider coordinate transformation, it is equivalent to mirror operation of the structure with respect to the *yz* plane (Fig. 5.1(a)), that is, coordinate transformation follows $x \rightarrow x$ and $y \rightarrow -y$, and its transformation basis matrix is the same as M_y in Eq. (5.8). We have:

$$M_y \boldsymbol{T}^f M_y^{-1} = \begin{pmatrix} A & -B \\ -C & D \end{pmatrix} = \boldsymbol{T}^b = \begin{pmatrix} A & -C \\ -B & D \end{pmatrix} \rightarrow \boldsymbol{T}^f = \begin{pmatrix} A & B \\ B & D \end{pmatrix}. \tag{5.12}$$

If there is no additional symmetry, then B in the above equation is not zero, making the device chiral. Substituting Eq. (5.12) into Eq. (5.4) can prove this conclusion (obtaining $t_{++} = t_{--} = 0.5(A + D)$, $t_{-+} = 0.5(A - D) + iB$, $t_{+-} = 0.5(A - D) - iB$, $\Delta_{cir} \neq 0$). However, if this structure has an additional C_4 symmetry with respect to the *z*-axis (meaning that Eqs. (5.12) and (5.11) hold simultaneously), all off-diagonal elements will also disappear ($B = 0$) so that the device will lose its chiral characteristics.

Based on the above discussion, we now can understand why the definition of chiral devices in the first paragraph of this section specifically emphasizes "in the same reference plane". We use an example here to illustrate the consequences without the prerequisite of "in the same reference plane" in the definition of chiral structures. For example (seeing Fig. 5.1(a)), it is easy to know that for any planar structure without additional symmetry, it must be chiral because Eq. (5.12) holds and B is not zero. However, if such a planar structure is rotated by 180° with respect to the *x* or *y* axis (rotation operation in the *yz* or *xz* plane), it will inevitably coincide with its mirror body in the *xy* plane, which violates the definition of chiral structures. Obviously, the rotation operation and mirror operation here are not in the same reference plane, but they can still cause structural overlap. If we do not emphasize "in the same reference plane", we will draw an erroneous conclusion that the structure is achiral.

The above example has demonstrated the importance of making the prerequisite of "in the same reference plane" in defining chiral structures. However, it should be noted that "the same reference plane" does not necessarily mean it is the *xy* plane. For example (seeing Fig. 5.1(b)), for a non-planar structure without mirror symmetry in the *xy* plane, not only does the rotation operation in the *xy* plane not coincide with its mirror body in the *xy* plane, but the rotation operation in the *yz* or *xz* plane also cannot coincide with its mirror body in the *yz* or *xz* plane.

For non-planar structures that may have out of plane C_2 symmetry, the impact of this symmetry on the chiral response also needs to be considered. If a structure can coincide with itself after rotating 180° with respect to the x- or y-axis, it has out of plane C_2 symmetry. For this structure, it appears the same when viewed from the forward and backward. Therefore, we can obtain:

$$\boldsymbol{T}^f = \begin{pmatrix} A & B \\ C & D \end{pmatrix} = \boldsymbol{T}^b = \begin{pmatrix} A & -C \\ -B & D \end{pmatrix} \rightarrow \boldsymbol{T}^f = \begin{pmatrix} A & B \\ -B & D \end{pmatrix}. \tag{5.13}$$

If there is no additional symmetry, the B in the above equation is not zero, and the described structure must be chiral. By substituting Eq. (5.13) into Eq. (5.4), this conclusion can be proved (obtaining $t_{++} = 0.5(A + D) + iB$, $t_{--} = 0.5(A + D) - iB$, $t_{-+} = t_{+-} = 0$, $\Delta_{cir} \neq 0$). If the structure has an additional mirror symmetry about the yz or xz plane, the B in the above equation becomes zero (meaning that Eqs. (5.13) and (5.8) hold simultaneously) so that the structure loses its chiral properties.

5.1.3 Eigenstate Vectors

The eigenstates of the transfer $\boldsymbol{T}$ matrix are directly related to structural symmetry, and the macroscopic chiral response of the device is inevitably related to elliptical and circularly polarized electromagnetic eigenstates. Solving the eigenvectors of the $\boldsymbol{T}$ matrix can provide an essential understanding of chiral response from the perspective of electromagnetic resonance, which requires solving the following eigenvalue problems:

$$\boldsymbol{Ti} = \kappa \boldsymbol{i} \rightarrow \begin{pmatrix} A & B \\ C & D \end{pmatrix} \begin{pmatrix} i_x \\ i_y \end{pmatrix} = \kappa \begin{pmatrix} i_x \\ i_y \end{pmatrix}, \tag{5.14}$$

where k is the eigenvalue of the $\boldsymbol{T}$ matrix, and $\boldsymbol{i}$ is the corresponding eigenvector. By examining the general scenario where all elements in the $\boldsymbol{T}$ matrix are non-zero, the eigenvalues can be obtained through simple mathematical calculations as follows:

$$\kappa_{\pm} = \frac{1}{2}\left[(A+D) \pm \sqrt{(A-D)^2 + 4BC}\right]. \tag{5.15}$$

By substituting Eq. (5.15) into Eq. (5.14), we can obtain the eigenvectors. We have:

$$\begin{gathered} (\boldsymbol{T} - \kappa \boldsymbol{E})\boldsymbol{i} = \begin{pmatrix} A-\kappa & B \\ C & D-\kappa \end{pmatrix} \begin{pmatrix} i_x \\ i_y \end{pmatrix} = 0 \\ \Rightarrow \boldsymbol{i}_1 = \begin{pmatrix} 1 \\ \dfrac{\kappa_+ - A}{B} \end{pmatrix} or \begin{pmatrix} \dfrac{\kappa_+ - D}{C} \\ 1 \end{pmatrix}, \boldsymbol{i}_2 = \begin{pmatrix} 1 \\ \dfrac{\kappa_- - A}{B} \end{pmatrix} or \begin{pmatrix} \dfrac{\kappa_- - D}{C} \\ 1 \end{pmatrix}. \end{gathered} \tag{5.16}$$

Note that every vector has two solutions, which are essentially the same, but expressed in different mathematical forms. Two vectors can form a pair of vector bases. For the convenience of future discussions, we will take:

$$\begin{pmatrix} \boldsymbol{i}_1 & \boldsymbol{i}_2 \end{pmatrix} = \begin{pmatrix} 1 & \dfrac{\kappa_+ - D}{C} \\ \dfrac{\kappa_+ - A}{B} & 1 \end{pmatrix} = \begin{pmatrix} 1 & 1 \\ \dfrac{\mathrm{K}}{2B} & -\left(\dfrac{\mathrm{K}}{2C}\right)^{-1} \end{pmatrix}, \tag{5.17}$$

where $\mathrm{K} = D - A + \sqrt{(A-D)^2 + 4BC}$. The two fractions in the above equation are complex numbers, so they can be written in the complex form of $ye^{i\varphi}$, where y is the real amplitude value and φ is the phase. The vector bases can be further written as:

$$\begin{pmatrix} \boldsymbol{i}_1 & \boldsymbol{i}_2 \end{pmatrix} = \begin{pmatrix} 1 & 1 \\ y_1 e^{i\varphi_1} & -\dfrac{1}{y_2} e^{-i\varphi_2} \end{pmatrix}. \tag{5.18}$$

The above equation indicates that in general (A, B, C, D are non-zero), the eigenvectors of the $\boldsymbol{T}$ matrix are elliptically polarized.

Next, we will further investigate the eigenvectors of eigenstates when the structure exhibits certain symmetries. When the structure has mirror symmetry with respect to the yz or xz plane, there is $B = C = 0$, and the eigenvalues $k_+ = \mathrm{A}$ and $k_- = \mathrm{D}$ can be easily obtained, as well as the eigenvectors $\boldsymbol{i}_1 = [1\ 0]^{\mathrm{T}}$ and $\boldsymbol{i}_2 = [0\ 1]^{\mathrm{T}}$, which are clearly linearly polarized and therefore achiral. For planar structures with C_4 symmetry (without additional symmetry), there is also $B = C = 0$, and it is also easy to obtain that the eigenvector is a linearly polarized state, which will not be repeated here. In addition, for other symmetry scenarios discussed in Section 5.1.2, they can be classified into two categories: $B = -C \neq 0$ or $B = C \neq 0$. When $B = -C \neq 0$, there are $y_1 = -y_2$ and $\varphi_1 = \varphi_2$ in Eq. (5.18), and the two elliptical polarization vectors are orthogonal ($\boldsymbol{i}_1 \cdot \boldsymbol{i}_2^* = 0$), where $\boldsymbol{i}_2^*$ represents the conjugate of $\boldsymbol{i}_2$; specifically, when the structure has a C_4 symmetry with respect to the z-axis but no other symmetry, in addition to satisfying $B = -C \neq 0$, it also satisfies $A = D$. By substituting this into Eq. (5.17), it is easy to calculate $\boldsymbol{i}_1 = (1\ i)^{\mathrm{T}}$ and $\boldsymbol{i}_2 = (1\ -i)^{\mathrm{T}}$, which means obtaining a pair of orthogonal circularly polarized eigenstates. When $B = C \neq 0$, there are $y_1 = y_2$ and $\varphi_1 = \varphi_2$ in Eq. (5.18), and the two elliptical polarization vectors are nonorthogonal ($\boldsymbol{i}_1 \cdot \boldsymbol{i}_2^* \neq 0$).

5.2 DESIGN AND ANALYSIS OF TRANSMISSIVE CHIRAL DEVICES

This section uses all-silicon materials to demonstrate the design of transmissive chiral devices [14]. High-resistance silicon wafers not only have the advantages of high refractive index and low loss but also have abundant sources and mature processing technologies. Compared with other all-dielectric metasurfaces, the process of all-silicon metasurfaces is simpler and more cost-effective. Compared to metasurfaces with multi-layer metal frameworks, all-silicon metasurfaces do not require cumbersome processes. In order to enhance the control ability of metasurfaces over electromagnetic waves, dynamically tunable metasurfaces are a further pursuit, which happens to be the capability that silicon media may possess. Silicon materials have a good response to infrared photons and can generate non-equilibrium charge carriers at appropriate infrared excitation frequencies, thereby affecting the resonance of metasurfaces. This means that for all-silicon metasurfaces, the principle of optical excitation can be introduced to dynamically tune the electromagnetic wave transmission of metasurfaces.

The RCP and LCP waves are defined here as circularly polarized waves with spin directions clockwise and counterclockwise, respectively, when observed along the direction of wave propagation. The relationship between the incident RCP and LCP electric fields and the transmitted RCP and LCP electric fields can be expressed as:

$$\begin{pmatrix} E_R^t \\ E_L^t \end{pmatrix} = \begin{pmatrix} T_{RR} & T_{RL} \\ T_{LR} & T_{LL} \end{pmatrix} \begin{pmatrix} E_R^i \\ E_L^i \end{pmatrix} = \boldsymbol{T} \begin{pmatrix} E_R^i \\ E_L^i \end{pmatrix}, \tag{5.19}$$

where $\boldsymbol{T}$ is a transmission Jones matrix composed of four transmission coefficients T_{jk} (j, $k \in \{R, L\}$), and T_{jk} represents the transmission coefficient of j-circularly polarized waves under the incidence of k-circularly polarized waves. If the light is directed along −z and incident on the metasurface, which is the same as the observation direction defined by the subscripts "+" and "−" in Eq. (5.4), then the transfer matrix of its circularly polarized transmission can directly use Eq. (5.4), as:

$$\boldsymbol{T} = \begin{pmatrix} t_{++} & t_{+-} \\ t_{-+} & t_{--} \end{pmatrix} = \begin{pmatrix} T_{RR} & T_{RL} \\ T_{LR} & T_{LL} \end{pmatrix} = \frac{1}{2} \begin{pmatrix} t_{xx} + t_{yy} + i(t_{xy} - t_{yx}) & t_{xx} - t_{yy} - i(t_{xy} + t_{yx}) \\ t_{xx} - t_{yy} + i(t_{xy} + t_{yx}) & t_{xx} + t_{yy} - i(t_{xy} - t_{yx}) \end{pmatrix}. \tag{5.20}$$

Using commercial CST microwave studio software and its time-domain solver, an all-silicon metasurface structure (high-resistance silicon, relative permittivity $\varepsilon = 11.9$) operating in the terahertz band was designed and simulated. As shown in Fig. 5.2(a, b), starting from a symmetrical structure (a pair of elliptical bars) without chirality (zero CD), weak chirality (non-zero CD) is obtained by breaking the C_2 rotational symmetry in-plane. Then, the major axis length of the two elliptical bars is kept the same, and the minor axis length of one of the elliptical bars is changed. Finally, in order to further enhance the interaction between electromagnetic waves and unit cells, the elliptical bars with smaller minor axes are divided into a pair and their centre positions are adjusted appropriately to obtain maximum CD. Here, CD is defined as $\text{CD} = |T_{\text{RR}}|^2 + |T_{\text{LR}}|^2 - |T_{\text{LL}}|^2 - |T_{\text{RL}}|^2$. As shown in Fig. 5.2(c), spin-selective asymmetric transmission is characterized using a polarization controllable terahertz time-domain spectrometer (THz-TDS) system. Meanwhile, in order to achieve active control of CD, a continuous wave laser with a wavelength of 1064 nm is used as the pump source.

As shown in Fig. 5.3(a), the unit cell of all-silicon metasurface is composed of three elliptical bars. The total thickness of the sample is 500 μm, where the height of the silicon bar is $h_1 = 200$ μm, the thickness of the silicon substrate is $h_2 = 300$ μm, the unit cell period is

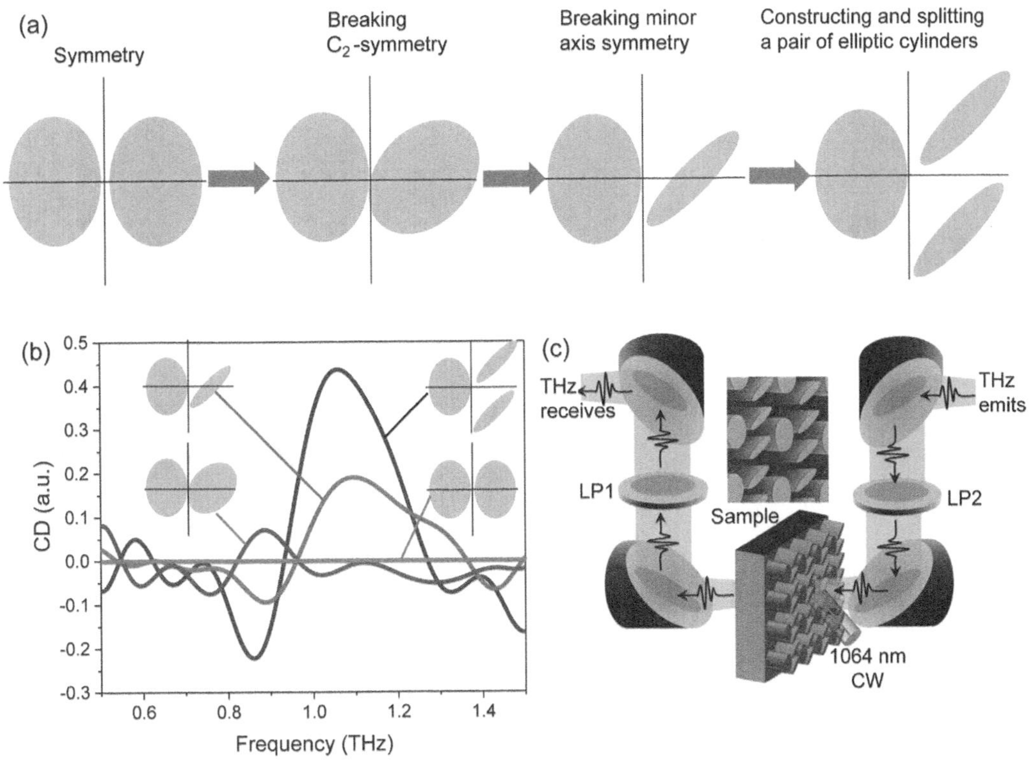

FIGURE 5.2 (a) The evolution process of chiral unit cell design based on a pair of elliptical strips corresponds to (b) their CD changes. (c) Schematic diagram of polarization controllable terahertz time-domain spectroscopy testing; rotating two linear polarizers, LP1 and LP2, can obtain the emission and reception of x-polarized and y-polarized waves, and then obtain circular polarization information through data post-processing.

$P = 160$ μm, and the major axis of all elliptical bars is $L = 100$ μm, with the minor axes being $W_1 = 60$ μm and $W_2 = 20$ μm. The centre coordinate of the left silicon bar is marked as (−40 μm, 0), and the centre coordinates of the two right silicon bars are (40 μm, ±40 μm). It has a tilt angle of 45° with respect to the x-axis. The spin-selective asymmetric transmission of metasurfaces can be described as follows. At 1.06 THz as the centre, for terahertz waves incident from the front of the metasurface, the incident RCP wave is converted into LCP wave and transmitted (transmission coefficient expressed as T_{LR}), but the incident LCP wave is reflected. For waves incident from the back of the metasurface, the incident RCP wave is reflected, but the incident LCP wave is converted into RCP wave and transmitted (transmission coefficient expressed as T_{RL}). The transmission spectra of T_{LR} and T_{RL} on the front and back of the metasurface are shown in Fig. 5.3(b, c). Significant spin-selection and CD can be observed in the range of 0.9–1.3 THz, with the incident wave at 1.06 THz exhibiting the strongest asymmetric transmission.

We studied the x, y, and z component electric and magnetic fields at 1.06 THz to qualitatively analyse the main resonance modes. Here, we choose to monitor the surface electric and magnetic field distribution of the structure when electromagnetic waves are incident

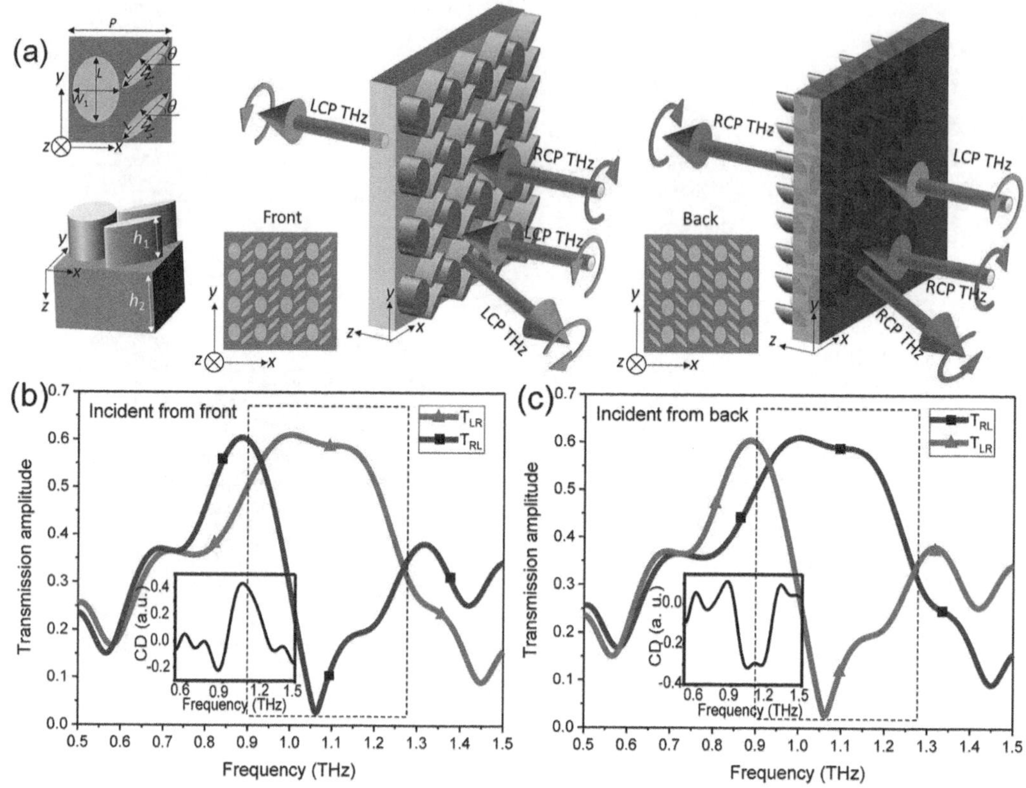

FIGURE 5.3 (a) Schematic diagram of the three-dimensional structure and spin-selective asymmetric transmission of all-silicon metasurface. (b) The cross-polarized transmission spectra of RCP and LCP waves when waves incident from the front of the metasurface are illustrated with their corresponding CD curves. (c) The cross-polarized transmission spectra of RCP and LCP waves when waves incident from the back of the metasurface are illustrated with their corresponding CD curves.

from the back of the metasurface. For the scenario of electromagnetic waves incident from the back, it is known from Fig. 5.3(a, c) that LCP waves are converted into RCP and transmitted, but RCP incident waves are reflected. Figure 5.4(a, b) shows the electric and magnetic field components in the *x*, *y*, and *z* directions when LCP waves are incident. In the *x* direction, the electric field amplitudes of the left and right elliptical bars are opposite, which excites the magnetic field in the *z* direction. In the *z* direction, the electric field amplitudes of the left and right elliptical bars are also opposite, which excites strong magnetic fields in the *x* and *y* directions. As the monitoring phase changes, all electric field amplitude distributions rotate around the *z*-axis, and these features include magnetic dipole and toroidal dipole resonances, which are the main electromagnetic resonance modes that allow LCP wave transmission. Figure 5.4(c, d) shows the electric and magnetic field components in the *x*, *y*, and *z* directions when RCP waves are incident from the back. Observation shows that a weak *x* direction electric field excites a weak *z* direction magnetic field, and a weak *z* direction electric field excites weak *x* and *y* direction magnetic field components, indicating that compared with the LCP incident wave, the resonance modes of magnetic dipole

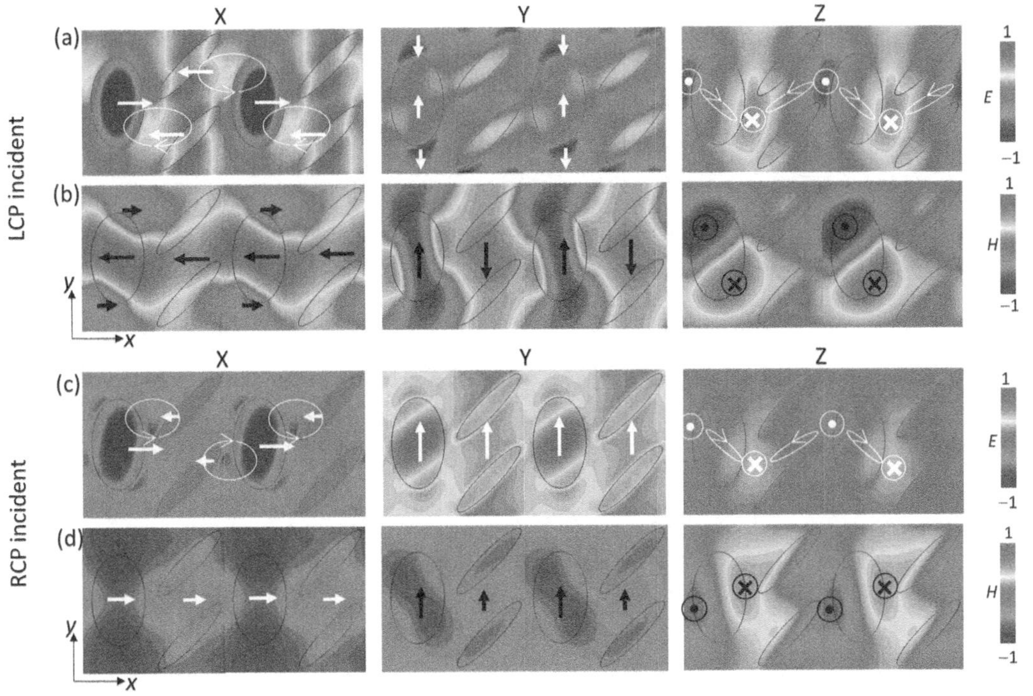

FIGURE 5.4 (a, b) When LCP waves are incident from the back of the metasurface, the (a) electric field and (b) magnetic field distribution of the *x*, *y*, and *z* components. (c, d) When RCP waves are incident from the back of the metasurface, the (c) electric field and (d) magnetic field distribution of the *x*, *y*, and *z* components.

and toroidal dipole excited by the RCP incident wave are significantly weakened. However, compared with the LCP incident wave, the *y* direction electric field excited by the RCP incident wave becomes stronger, exhibiting in-plane up-down oscillations, which is a characteristic of electric dipole resonance, indicating that the electric dipole resonance mode dominates the reflection of the incident RCP wave.

Figure 5.5(a, b) show scanning electron microscope images of all-silicon metasurface samples. The experiment effectively verified the spin-selective asymmetric transmission characteristics of all-silicon metasurfaces. Figure 5.5(c, d) shows the co-polarized and cross-polarized transmission spectra of incident RCP and LCP waves from the front of the metasurface. From experimental and simulation results, it can be seen that there is a significant amplitude difference within the operating frequency band of 0.9–1.3 THz. This metasurface causes cross-polarization of incident waves and allows for RCP (cross-polarization to LCP) wave transmission, but prevents LCP (cross-polarization to RCP) wave transmission. When terahertz waves are incident from the back of the metasurface, Fig. 5.5(e, f) shows that the device allows LCP (cross-polarization to RCP) wave transmission, but RCP (cross-polarization to LCP) wave transmission is prevented.

Further exploration is conducted on the active control of the transmission amplitude of the metasurface when light waves are incident from the front. As shown in Fig. 5.6(a),

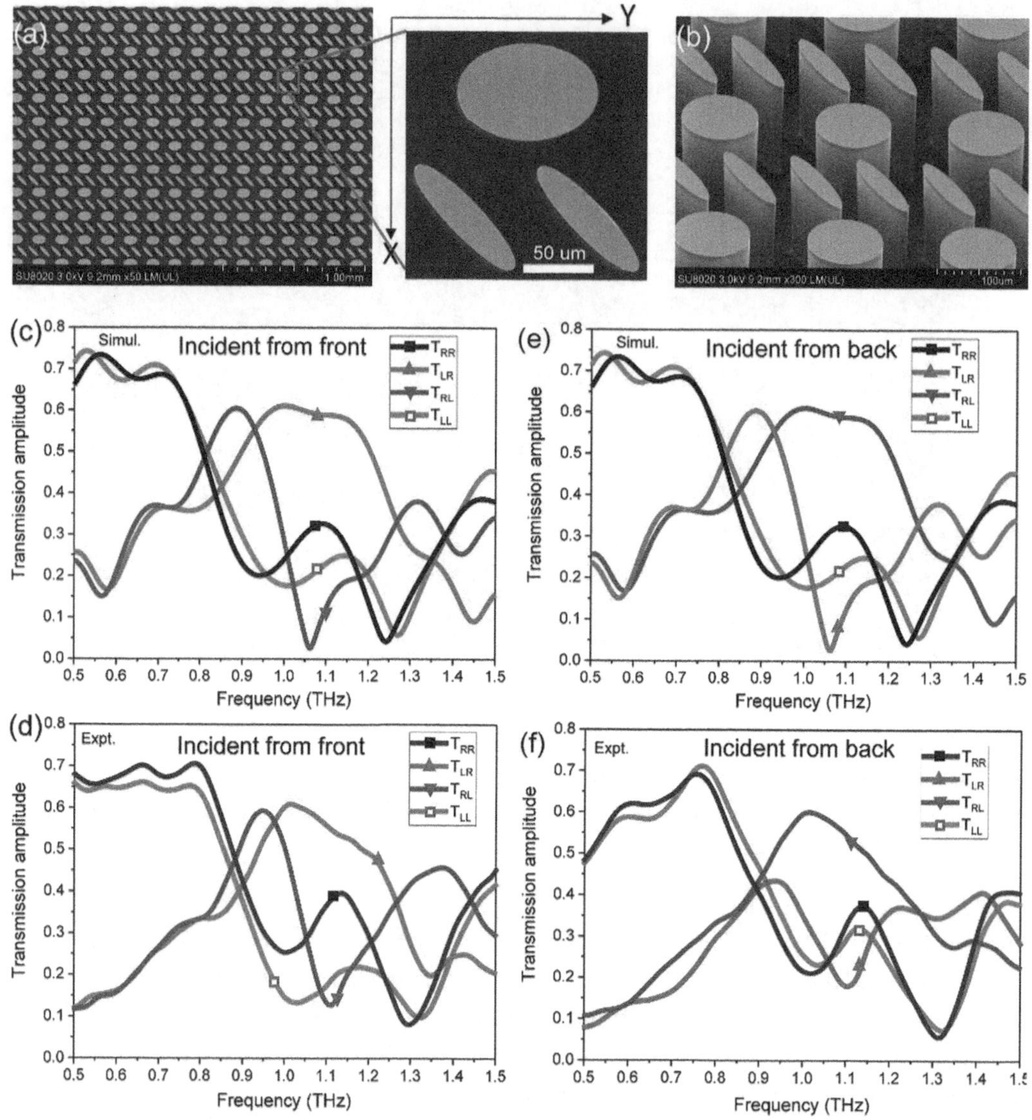

FIGURE 5.5 (a, b) SEM images of metasurface samples. (c, d) Simulation and experimental results of co-polarization and cross-polarization transmission amplitudes in the scenario of waves incident from the front of the metasurface. (e, f) Simulation and experimental results of co-polarization and cross-polarization transmission amplitudes in the scenario of waves incident from the back of a metasurface.

due to the photoelectric effect of semiconductors, the carrier concentration in silicon can be changed under light excitation with photon energy greater than the silicon bandgap. Therefore, the transmission amplitude of the all-silicon metasurface can be modulated by a 1064 nm continuous wave excitation. The penetration depth of 1064 nm continuous light is greater than the thickness of the silicon sample, which causes the change in conductivity to occur throughout the entire silicon sample, not limited to the sample

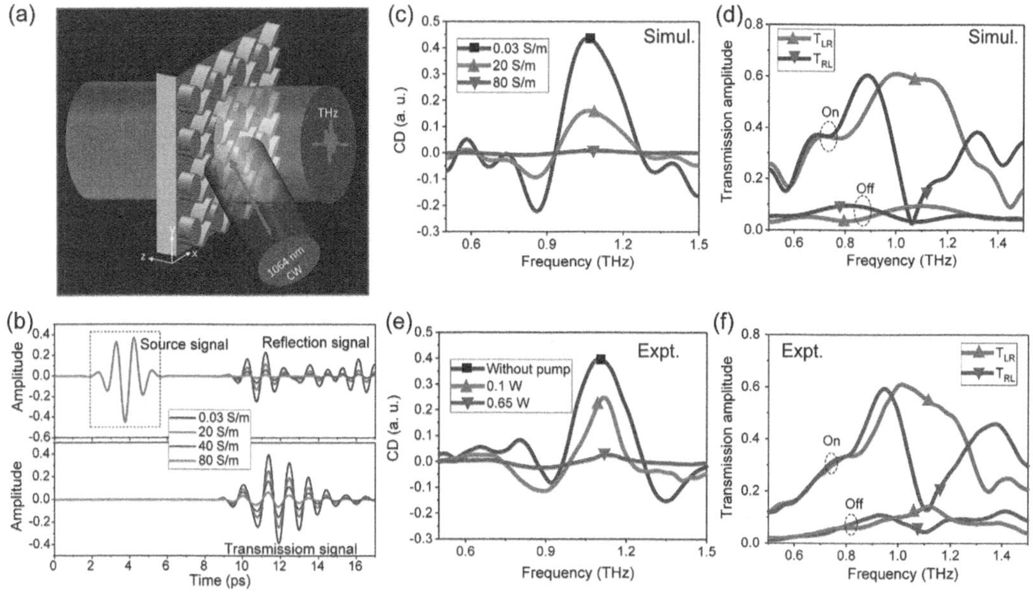

FIGURE 5.6 (a) Schematic diagram of active controlled metasurface transmission amplitude. (b) Time-domain terahertz reflection and transmission simulation signals under different silicon conductivities. (c, e) The variation of CD values with silicon conductivity in simulations and experiments. (d, f) Simulation and experimental transmission spectra of cross-polarization of incident RCP and LCP waves in chiral "on" and "off" states.

surface. Here, by changing the silicon conductivity parameters to simulate optical pump excitation reasonably, the pump power that varies from 0 to 0.65 W can be matched to the silicon conductivity that varies from 0.03 to 80 S/m. As shown in Fig. 5.6(b), the changes in reflection and transmission time-domain signals with increasing silicon conductivity are monitored to study the effects of carrier concentration changes. As the optical pump power increased and the carrier concentration increased, both reflection and transmission signals gradually decreased. The increase in carrier concentration will lead to the absorption of terahertz energy by silicon materials, further resulting in a decrease and near disappearance of CD, as shown in Fig. 5.6(c, e). Strong CD can be considered as the chiral "on" state, while almost disappearing CD can be considered as the chiral "off" state. Correspondingly, as shown in Fig. 5.6(d, f), the transmission amplitudes of cross-polarized T_{LR} and T_{RL} decrease synchronously, and their transmission amplitude difference approaches disappearance.

5.3 DESIGN AND ANALYSIS OF REFLECTIVE CHIRAL DEVICES

This section demonstrates the design of reflective chiral devices based on graphene and metal materials [15], with the terahertz wave band still chosen as the operating band. Chiral metasurfaces can selectively reflect circularly polarized waves with a specific spin direction, while maintaining the same spin direction as the incident wave, which is different from traditional mirrors that cause handedness flipping of circularly polarized waves. The process of deriving the circularly polarized Jones matrix for reflective metasurfaces and transmissive metasurfaces is essentially the same. Assuming that electromagnetic waves are incident from the front of a metasurface, the relationship between the incident linear component E_x^i, E_y^i and the reflected linear component E_x^r, E_y^r can be expressed as:

$$\begin{pmatrix} E_x^r \\ E_y^r \end{pmatrix} = \begin{pmatrix} r_{xx} & r_{xy} \\ r_{yx} & r_{yy} \end{pmatrix} \begin{pmatrix} E_x^i \\ E_y^i \end{pmatrix} = \boldsymbol{r} \begin{pmatrix} E_x^i \\ E_y^i \end{pmatrix}, \tag{5.21}$$

where r_{jk} (j, $k \in \{R, L\}$) represents the reflection coefficient of a j-polarized wave under the incidence of a k-polarized wave. Since two linearly polarized waves can be combined into a circularly polarized wave, the reflection Jones matrix $\boldsymbol{r}'$ of circularly polarized waves can also be obtained from the reflection Jones matrix $\boldsymbol{r}$ of the linear component. The relationship between the two can be transformed by the basis matrix Λ into:

$$\boldsymbol{r}' = \begin{pmatrix} r_{++} & r_{+-} \\ r_{-+} & r_{--} \end{pmatrix} = \boldsymbol{\Lambda}^{-1} \boldsymbol{r} \boldsymbol{\Lambda} = \frac{1}{2} \begin{pmatrix} r_{xx} + r_{yy} + i\left(r_{xy} - r_{yx}\right) & r_{xx} - r_{yy} - i\left(r_{xy} + r_{yx}\right) \\ r_{xx} - r_{yy} + i\left(r_{xy} + r_{yx}\right) & r_{xx} + r_{yy} - i\left(r_{xy} - r_{yx}\right) \end{pmatrix}, \tag{5.22}$$

where the original definitions of subscripts "+" and "–" are (electric vector) clockwise and counterclockwise circularly polarized waves viewed along the –z direction, respectively, without considering the wave propagation direction. If circularly polarized waves are incident on the metasurface along the –z direction, the reflected waves propagate along the +z direction. At this point, in the incident direction, the defined symbols "+" and "–" still correspond to RCP and LCP waves, respectively. But in the reflection direction, the defined symbols "+" and "–" should correspond to LCP and RCP waves, respectively, due to the fact that the propagation direction of the reflected wave is opposite to the view direction on which the symbols "+" and "–" depend. Therefore, it is necessary to adjust the transfer matrix $\boldsymbol{r}'$ to adapt to the direction of wave propagation, and the corresponding relationship after adjustment is as follows:

$$\begin{pmatrix} r_{++} & r_{+-} \\ r_{-+} & r_{--} \end{pmatrix} = \begin{pmatrix} R_{LR} & R_{LL} \\ R_{RR} & R_{RL} \end{pmatrix}. \tag{5.23}$$

Therefore, the reflection Jones matrix that meets the circular polarization definition in this chapter is:

$$\boldsymbol{R} = \begin{pmatrix} R_{RR} & R_{RL} \\ R_{LR} & R_{LL} \end{pmatrix} = \begin{pmatrix} r_{-+} & r_{--} \\ r_{++} & r_{+-} \end{pmatrix} = \frac{1}{2} \begin{pmatrix} r_{xx} - r_{yy} + i\left(r_{xy} + r_{yx}\right) & r_{xx} + r_{yy} - i\left(r_{xy} - r_{yx}\right) \\ r_{xx} + r_{yy} + i\left(r_{xy} - r_{yx}\right) & r_{xx} - r_{yy} - i\left(r_{xy} + r_{yx}\right) \end{pmatrix}. \tag{5.24}$$

Furthermore, considering the characteristics of graphene, single-layer graphene (with a thickness less than 1 nm) as a "sheet-like" structure, its main physical parameter is conductivity. According to the Kubo model, the electronic transitions in graphene include both inter-band and intra-band transitions, i.e. $\delta = \delta_{\text{inter}} + \delta_{\text{intra}}$. Due to the low photon energy of terahertz, it hardly causes inter-band electron transitions, so we only consider intra-band electron transitions [16]. We have:

$$\delta \approx \delta_{\text{intra}} = \frac{ie^2 K_B T}{\pi\hbar^2\left(\omega + i\tau^{-1}\right)}\left[\frac{E_F}{K_B T} + 2\ln\left(\exp\left(-\frac{E_F}{K_B T}\right) + 1\right)\right], \tag{5.25}$$

where e is the elementary charge, ω is the incident wave angular frequency, the carrier relaxation time τ is 0.3 ps here, K_B is the Boltzmann constant, and the Kelvin temperature T is 293 K. The Fermi level (or chemical potential) E_F varies from 0.01 to 1 eV. For the low concentration doped graphene, which is $|E_F| \gg K_B T$ and $\hbar \ll 2|E_F|$, the Drude model is usually used to further simplify Eq. (5.25) as follows:

$$\delta \approx \frac{ie^2 E_F}{\pi\hbar^2\left(\omega + i\tau^{-1}\right)} = \frac{ie^2 V_F \hbar\left(\pi|n|\right)^{1/2}}{\pi\hbar^2\left(\omega + i\tau^{-1}\right)}, \tag{5.26}$$

where $E_F = V_F\hbar(\pi|n|)^{1/2}$ is related to the Fermi velocity $V_F = 1 \times 10^6$ m/s and electron concentration n. By adjusting E_F, the conductivity of graphene can be tuned within a wide range.

The unit cell design of the reflective chiral metasurface is shown in Fig. 5.7(a). Single-layer graphene is located on the surface of a low resistivity silicon substrate (relative permittivity $\varepsilon = 11.9$, conductivity $\delta > 10$ S/m). Gold (Au) cylinders with a thickness of 200 nm are deposited on graphene. A 30 μm thick polyimide (PI, $\mu = 2.93 + 0.13i$) is located between

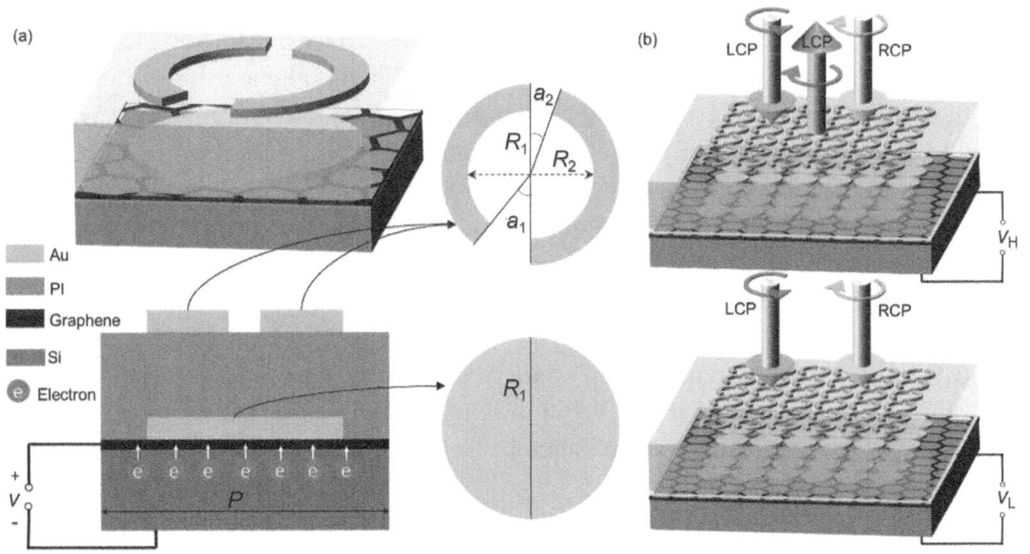

FIGURE 5.7 (a) Schematic unit cell of graphene-metal hybrid chiral metasurface. (b) Functional diagram of chiral metasurface.

the bottom metal cylinder and the top metal circular split ring (CSR). The outer diameter R_1 of the top metal CSR with a thickness of 200 nm is 90 μm, and the inner diameter R_2 is 60 μm. The two splitting angles of metal CSR are $a_1 = 40°$ and $a_2 = 20°$, respectively. Adjusting the Fermi level of graphene can obtain different reflection and absorption intensities, thereby affecting the chiral properties of metasurfaces. However, the reflective ability of single-layer graphene is weak, so the presence of bottom gold cylinders is necessary to enhance the reflection of electromagnetic waves, which ensures good chiral properties of the metasurface. It should be noted that there is a competitive relationship between the diameter of the bottom gold cylinders and the control ability of graphene. Bottom gold cylinders with larger diameters can more effectively reflect electromagnetic waves to obtain better chirality, but will correspondingly weaken the dynamic control ability of graphene on the chiral properties of metasurfaces. The diameter of the bottom gold cylinders can be determined using the same parameters as the outer diameter of the top metal CSR, i.e. $R_1 =$ 90 μm, which ensures both the great chirality of the metasurface and the dynamic control capability of graphene.

The fundamental principle behind the Fermi level shift in graphene relies on changes in carrier concentration. When a positive voltage is applied from graphene to the silicon substrate, some electrons from the silicon are injected into graphene to raise its Fermi level, which is a phenomenon that has been experimentally demonstrated as feasible. Conversely, when the voltage is reversed, some holes from the silicon are injected into graphene, reducing the free electrons in graphene and thereby lowering its Fermi level. The purpose of using a low-resistance silicon substrate is to leverage its abundant free carriers, and the carrier concentration in silicon can be further increased through additional excitation (e.g. optical excitation). Additionally, the silicon square side length of the metasurface unit is $P = 100$ μm, and the silicon thickness is also 100 μm, making the sample easy to fabricate in practice. The general fabrication steps are as follows. Grow monolayer graphene on copper and transfer it onto the surface of a low-resistance silicon substrate; then deposit a 200 nm thick gold layer on the graphene using common thin film deposition techniques such as thermal evaporation and magnetron sputtering, followed by photolithography to form an array of gold cylinders. Subsequently, spin coat a 30 μm thick PI layer onto the gold cylinders, which is then dried and solidified. Finally, deposit another 200 nm-thick gold layer on the PI and further form a CSR array after photolithography. The chiral metasurface is illustrated in Fig. 5.7(b), with its functional description as follows: when a sufficiently high positive voltage is applied, the Fermi level of graphene increases to enable chirality, primarily reflecting LCP waves and absorbing RCP waves (referred to as the "on" state of chirality). When a sufficiently low negative voltage is applied, the Fermi level of graphene decreases to deactivate chirality, at which point both LCP and RCP waves exhibit equal reflection, known as the "off" state of chirality.

When graphene $E_F = 1$ eV, the metasurface achieved significant selective reflection of LCP and RCP waves at 0.96 THz, as shown in Fig. 5.8. For the structure shown in the illustration of Fig. 5.8(a), the LCP incident from the front is strongly reflected (labelled LCP-LCP), but the co-polarized reflection of RCP (labelled RCP-RCP) is suppressed.

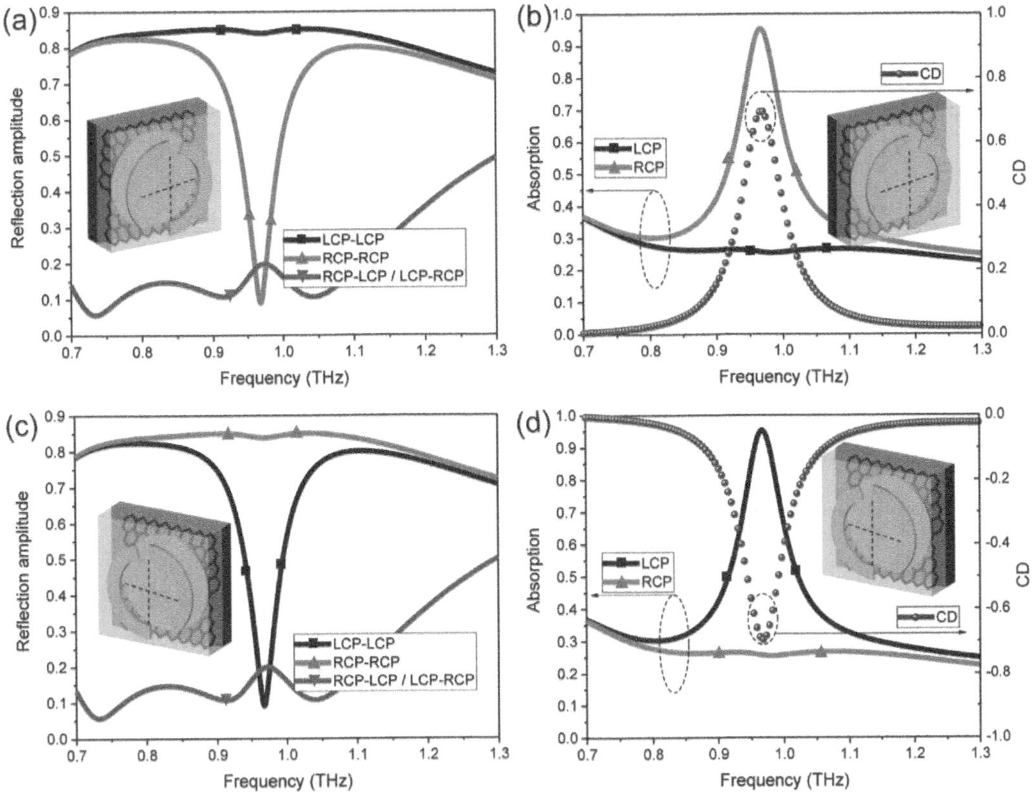

FIGURE 5.8 (a, b) Normalized reflection amplitude, normalized absorption, and CD of the device. (c, d) Normalized reflection amplitude, normalized absorption rate, and CD of its mirror structure.

For RCP and LCP incident light, the cross-polarized reflection components are the same (labelled RCP-LCP and LCP-RCP). The CD of the metasurface is as high as 0.7 at 0.96 THz, as shown in Fig. 5.8(b). Here, CD is defined as CD = $A_{RCP} - A_{LCP}$, where A_{RCP} and A_{LCP} represent the absorption rates of RCP and LCP waves in the metasurface, respectively. In addition, $A_{RCP} = 1 - |R_{RR}|^2 - |R_{LR}|^2$ and $A_{LCP} = 1 - |R_{RL}|^2 - |R_{LL}|^2$. Of course, the mirror structure of the metasurface will exhibit the opposite trend, where RCP incident from the front can be greatly reflected, but the incident LCP will be significantly absorbed, as shown in Fig. 5.8(c, d).

Further, we discuss the dynamic control of graphene on the CD of metasurfaces. As shown in Fig. 5.9(a, b), the co-polarized reflection of incident LCP and RCP waves increases with the increase of graphene E_F. Correspondingly, Fig. 5.9(c) shows that the cross-polarized reflection of the incident LCP and RCP waves is suppressed. At 0.96 THz, the co-polarized reflection of the incident RCP wave remains at a low value. Figure 5.9(d, e) shows the co-polarized reflection and absorption of LCP and RCP at 0.96 THz. It can be clearly seen that with the increase of graphene E_F, the selective reflection and absorption of LCP and RCP waves are very obvious. Figure 5.9(f) shows the variation of CD at 0.96 THz. When E_F drops to 0.2 eV, CD decreases to 0.35, which is about half of the highest value. At

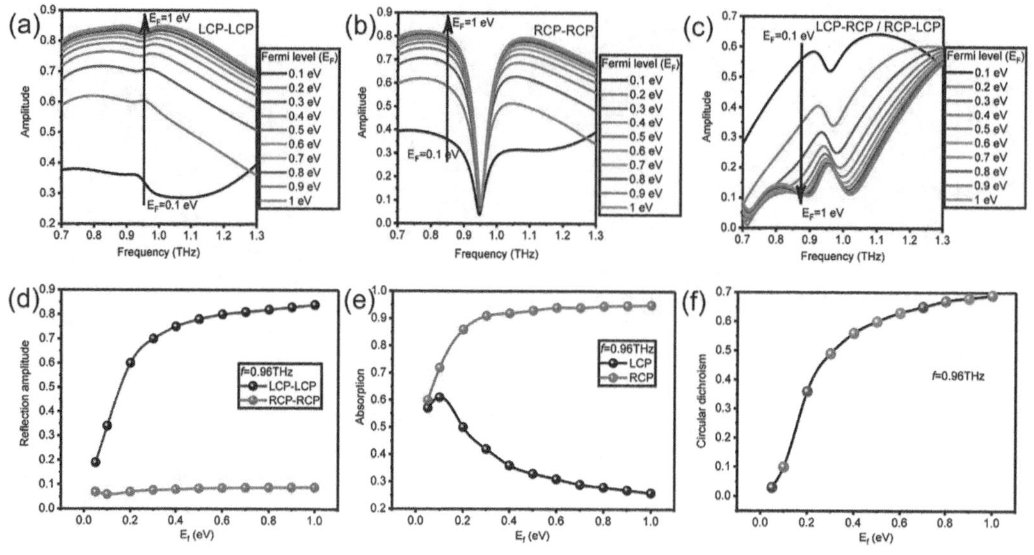

FIGURE 5.9 (a, c) The co-polarized and cross-polarized reflectance spectra of the device accompanied by changes in graphene Fermi level, and (d–f) represent the co-polarized reflection amplitude, incident wave absorption, and CD spectra of the device accompanied by changes in graphene Fermi level at 0.96 THz, respectively.

this point, chirality can be considered to be in a sub-closed state. When graphene $E_F = 0.05$ eV, CD decreases to near zero and chirality is completely turned off.

To understand the physical mechanism of the variation of CD, the electric field changes under different states are explored. Here, two extreme scenarios are discussed, corresponding to the typical chiral "on" and "off" states shown in Fig. 5.10, namely $E_F = 1$ eV and $E_F = 0.05$ eV. In addition, the characteristic changes of the weak chiral frequency region and the strong chiral frequency region are compared, and 0.8 and 0.96 THz are selected as representative frequency points for these two regions, respectively. The resonance frequency of the metasurface is determined by metal CSR. However, the reflection and absorption changes of metasurfaces are controlled by the complex electromagnetic interactions between light fields and graphene electrons. Generally, for high E_F, the increase in conductivity in graphene enhances wave reflection and reduces wave absorption. On the contrary, for low E_F, the decrease in conductivity in graphene weakens wave reflection and promotes wave absorption. Therefore, in Fig. 5.10(a), when the E_F increases from 0.05 to 1 eV, the decrease in the absorption of metasurface LCP in the entire 0.7–1.3 THz range is mainly attributed to the decrease in wave absorption of graphene, rather than resonance in metal CSR. As shown in Fig. 5.10(b), the monitored electric field strengths of graphene are from strong to weak, corresponding to absorption enhancement at E_F = 0.05 eV and absorption attenuation at $E_F = 1$ eV. As shown in Fig. 5.10(c), comparing the electric field intensity distribution in metal CSR with $E_F = 0.05$ eV and $E_F = 1$ eV, it can be found that the resonance intensity of LCP waves in metal CSR changes very little. For the RCP wave absorption on the metasurface in Fig. 5.10(d), when the E_F reaches 1 eV, the

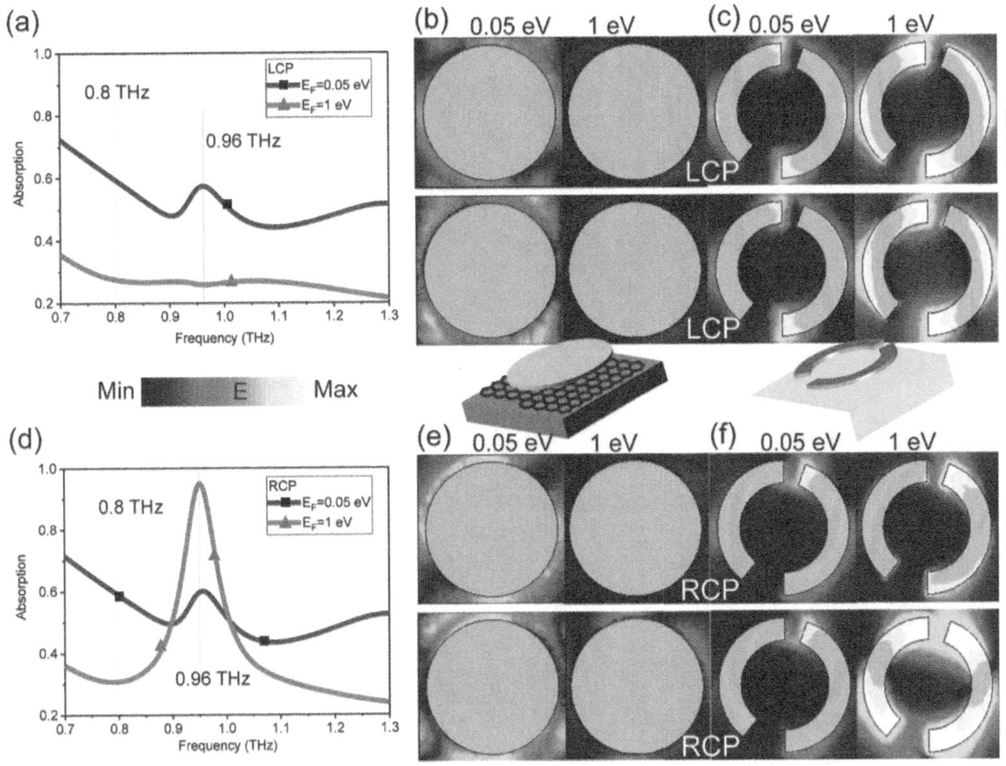

FIGURE 5.10 (a) Comparison of absorption of LCP waves in chiral "on" and "off" states, corresponding to (b, c) the electric field changes monitored on the graphene surface and metal CSR surface. (d) Comparison of absorption of RCP waves in chiral "on" and "off" states, corresponding to (e, f) the electric field changes monitored on the graphene surface and metal CSR surface.

variation in the weak chiral frequency region (e.g. 0.8 THz) still depends on the absorption attenuation of graphene on terahertz waves, as shown in Fig. 5.10(e). However, the decrease in absorption means a relative increase in reflection from graphene to metal CSR. Therefore, in the strong chiral frequency region (e.g. 0.96 THz), more electromagnetic energy will participate in the resonance of metal CSR, resulting in strong absorption. This can be demonstrated by Fig. 5.10(f), which shows that when $E_F = 1$ eV, the resonance at 0.96 THz in metal CSR is significantly enhanced compared to that with $E_F = 0.05$ eV.

Furthermore, taking the proposed reflective dynamic chiral metasurfaces as an example, their rich applications will be demonstrated. We take vortex focusing applications as an example to demonstrate the advantages of this metasurface. The designed array consists of 50 × 50 unit cells, as shown in Fig. 5.11(a). Rotating metal CSR at an angle of θ can result in a PB phase delay of 2θ for the LCP reflected wave, and the amplitude of the LCP reflected wave remains relatively stable, as shown in Fig. 5.11(b). In addition, when the E_F of graphene decreases from 1 eV to 0.05 eV, the amplitude of LCP reflection waves decreases from 0.85 to around 0.2. This feature can be used to develop tunable multifunctional

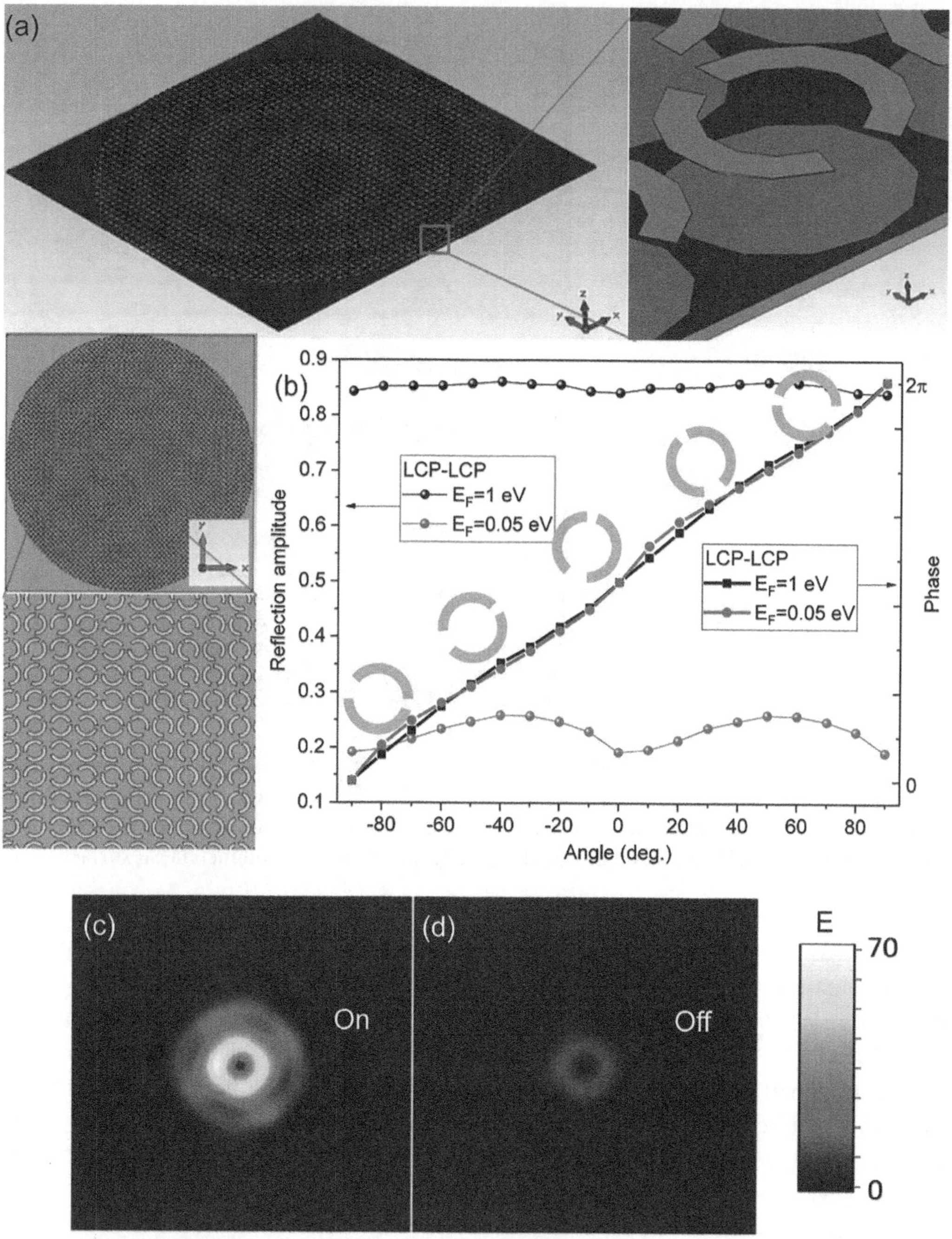

FIGURE 5.11 (a) Schematic diagram of vortex focusing application device based on chiral LCP reflection. (b) The amplitude and phase variations of LCP reflected waves at different graphene Fermi levels and rotation angles. (c, d) Distribution of LCP reflection electric field intensity at 0.96 THz on the focal plane for $E_F = 1$ and 0.05 eV, respectively.

applications such as perfect anomalous reflection, beam splitting, diffuse scattering, focusing, vortex beam generation, and holographic image generation. To achieve vortex focusing, the phase distribution of the metasurface should satisfy the following conditions $\varphi = -\frac{2\pi}{\lambda}\left(\sqrt{(x-x_0)^2+(y-y_0)^2+f^2}-f\right)+l\times\delta$, where λ corresponds to a wavelength of 0.96 THz, $f = 3$ mm is the focal length, and the geometric centre of the metasurface is taken as the coordinate origin. The coordinates of any point on the metasurface are represented as (x, y), and the projection coordinates of the focal point on the metasurface are (x_0, y_0); δ represents the azimuth angle, which is used as a phase factor to control the vortex of the wave; l represents the topological charge, which is a factor that controls the orbital angular momentum of the vortex beam. From Fig. 5.11(c, d), it can be seen that adjusting the E_F of graphene to chiral "on" and "off" states, respectively, can achieve a decrease in the electric field intensity of vortex focusing from bright to dark.

The above application scenarios are mainly based on the consideration of the PB phase and the strong reflection component (LCP). In fact, considering only the electromagnetic wave reflection intensity without considering the PB phase can also achieve interesting applications. The original metasurface reflects LCP waves but absorbs RCP waves, while its mirror structure absorbs LCP waves but reflects RCP waves. Therefore, utilizing these characteristics, near-field display applications can be developed by arranging these metasurfaces and their mirror structures into a pattern and then detecting the difference in reflected electric fields of LCP and RCP waves to display the pattern. This display performs well when graphene $E_F = 1$ eV, but is difficult to distinguish when graphene E_F decreases to 0.05 eV, which is the chiral "off" state (Fig. 5.12(a, b)). The displayed image is shown in Fig. 5.12(c). The original structure forms a T-shaped array, while the mirrored structure is arranged around the T-shaped array. The number of unit cells in the entire array is 21×24. For a LCP wave incident from the front, the reflection field of the co-polarized LCP wave is shown in Fig. 5.12(d, f), where the bright "T" is displayed at the "on" state and disappears at the "off" state. For a RCP wave incident from the front, the reflection field of the co-polarized RCP wave is shown in Fig. 5.12(e, g), where the bright area around the dark state "T" is displayed at the "on" state and disappears in the "off" state, and the RCP wave imaging and LCP wave imaging complement each other.

5.4 CHAPTER SUMMARY

This chapter first elaborates on the general principle of chiral response. Based on the Jones matrix, the circular polarization response characteristics of the device are demonstrated, and the scattering response and eigenstate polarizations corresponding to structures with different symmetries are analysed. Then, this chapter demonstrated the design and application of dynamic chiral devices. Specifically, a transmissive chiral metasurface based on all-silicon dielectric is demonstrated, with structural units composed of three elliptical bars of different minor axis lengths. Metasurfaces allow for the transmission of circularly polarized electromagnetic waves in specific spin directions, while preventing circularly polarized waves in opposite spin directions, due to the different electromagnetic resonance modes excited in all-silicon metasurfaces. At the same time, 1064 nm continuous

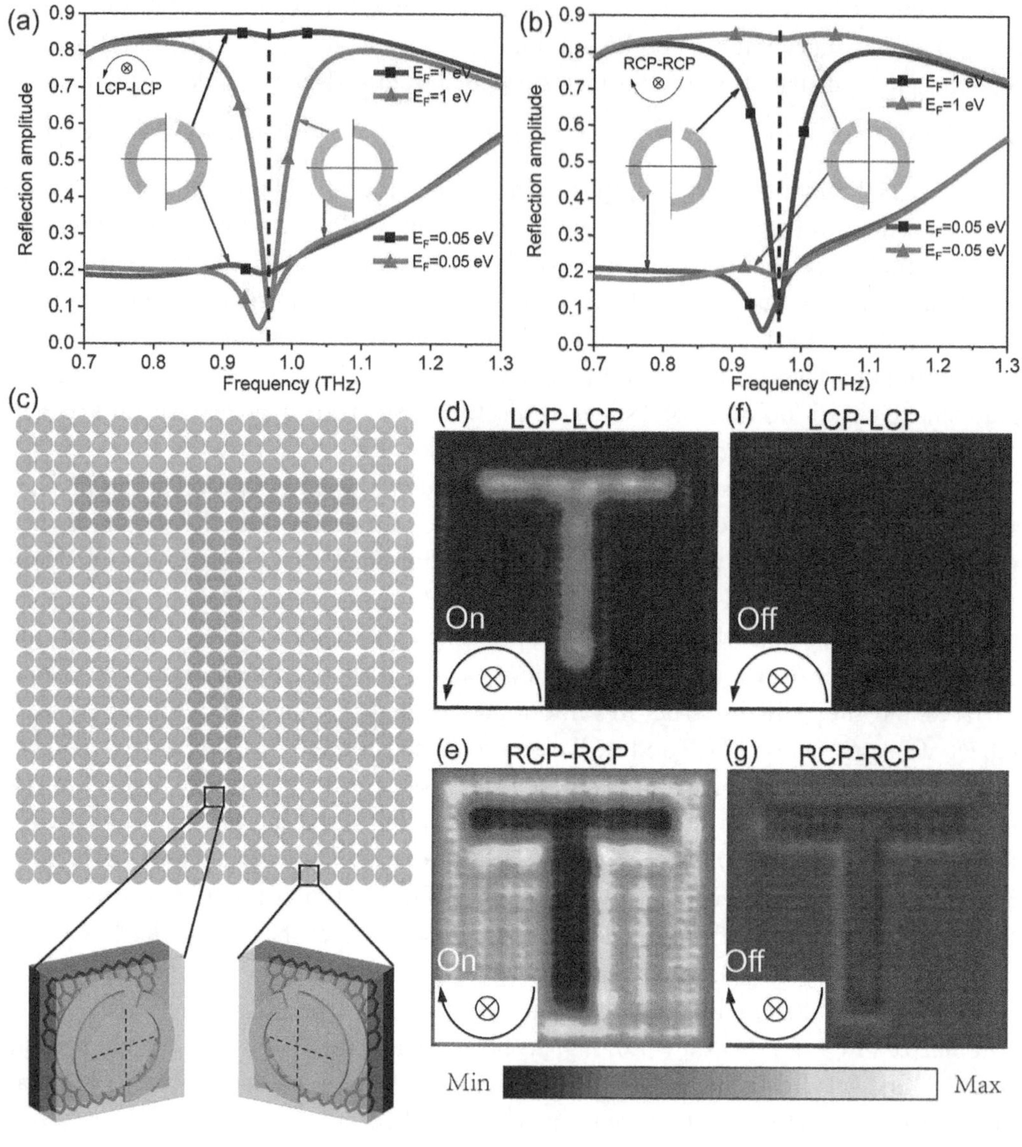

FIGURE 5.12 (a) The reflection amplitude of co-polarized LCP waves of the original structure and its mirror structure in chiral "on" and "off" states. (b) The reflection amplitude of co-polarized RCP waves of the original structure and its mirror structure in chiral "on" and "off" states. (c) Schematic pattern array for near-field display. (d, f) Imaging of co-polarized wave LCP reflection field in chiral "on" and "off" states. (e, g) Imaging of co-polarized RCP wave reflection field, with an imaging plane distance of 60 μm from the metal CSR surface.

optical pumping was introduced to excite more charge carriers in silicon, modulating the transmission amplitude of the metasurface. This can strongly weaken the transmission amplitude difference between different circularly polarized waves, causing the CD state to switch from "on" to "off". Meanwhile, a reflective chiral metasurface based on graphene-metal material was demonstrated. It uses a single layer of graphene on a low-resistance silicon substrate as the electromagnetic wave reflection medium. Applying a voltage between graphene and silicon injects electrons from silicon into graphene, thereby modulating the Fermi level of graphene. As the Fermi level of graphene decreases from high to low, the weakening (enhancement) of electromagnetic reflection (absorption) in graphene can dynamically control the chiral properties of metasurfaces from "on" to "off". These dynamic controls are also effective for various further applications, such as vortex focusing and near-field display. It should be reiterated that the physical essence of chiral response described in this chapter is the conventional electromagnetic resonance effect. The use of special electromagnetic eigen-resonances (such as BICs) to achieve chiral propagation will be introduced in Chapter 8.

REFERENCES

1. Pendry, J.B. (2004) 'A chiral route to negative refraction', Science, 306, 1353–1355.
2. Fedotov, V.A., Mladyonov, P.L., Prosvirnin, S.L., et al. (2006) 'Asymmetric propagation of electromagnetic waves through a planar chiral structure', Physical Review Letters, 97(16), 167401.
3. Li, W., Coppens, Z.J., Besteiro, L.V., et al. (2015) 'Circularly polarized light detection with hot electrons in chiral plasmonic metamaterials', Nature Communications, 6, 8379.
4. Zhang, Y., Li, Z., Liu, W., et al. (2019) 'Spin-selective and wavelength-selective demultiplexing based on waveguide-integrated all-dielectric metasurfaces', Advanced Optical Materials, 7, 1801273.
5. Khorasaninejad, M., Ambrosio, A., Kanhaiya, P., et al. (2016) 'Broadband and chiral binary dielectric meta-holograms', Science Advances, 2(5), e150125.
6. Zhang, Y., Liu, H., Cheng, H., et al. (2020) 'Multidimensional manipulation of wave fields based on artificial microstructures', Opto-Electronic Advances, 3(11), 200002.
7. Ma, D., Li, Z., Zhang, Y., et al. (2019) 'Giant spin-selective asymmetric transmission in multipolar-modulated metasurfaces', Optics Letters, 44(15), 3805–3808.
8. Zhu, A.Y., Chen, W.T., Zaidi, A., et al. (2018) 'Giant intrinsic chiro-optical activity in planar dielectric nanostructures', Light: Science & Applications, 7, 17158.
9. Li, J., Li, J.T., Yang, Y., et al. (2020) 'Metal-graphene hybrid active chiral metasurfaces for dynamic terahertz wavefront modulation and near field imaging', Carbon, 163, 34–42.
10. Xiao, S., Wang, T., Liu, T., et al. (2020) 'Active metamaterials and metadevices: a review', Journal of Physics D: Applied Physics, 53, 503002.
11. Cui, T., Bai, B. and Sun, H. (2019) 'Tunable metasurfaces based on active materials', Advanced Functional Materials, 29(10), 1806692.
12. Menzel, C., Rockstuhl, C. and Lederer, F. (2010) 'Advanced Jones calculus for the classification of periodic metamaterials', Physical Review A, 82, 053811.
13. Potton, R.J. (2004) 'Reciprocity in optics', Reports on Progress in Physics, 67, 717–754.
14. Li, J., Li, J., Zheng, C., et al. (2021) 'Active controllable spin-selective terahertz asymmetric transmission based on all-silicon metasurfaces', Applied Physics Letters, 118, 221110.
15. Li, J., Li, J., Zheng, C., et al. (2021) 'Dynamic control of reflective chiral terahertz metasurface with a new application developing in full grayscale near field imaging', Carbon, 172, 189–199.
16. Falkovsky, L.A. and Pershoguba, S.S. (2007) 'Optical far-infrared properties of a graphene monolayer and multilayer', Physical Review B, 76, 153410.

CHAPTER 6

Vector Light Field Generation

Abstract

The meta-photonic devices presented in the preceding chapters widely employ spatially uniform polarization states of light fields, i.e. scalar light fields, which play a central role in fields such as information encoding, optical sensing, and measurement. In contrast to scalar light fields, there exists a novel class of light field structures whose polarization states exhibit spatially non-uniform distributions and are tightly coupled with the spatial modes of propagation. Such light fields are defined as vector light fields, whose characteristics extend beyond the conventional concept of uniform polarization states. In scalar light fields, the distribution of polarization states within the cross section (in the x–y plane) of light beam is uniform and does not vary with the coordinate positions (x, y). Vector light fields, distinct from scalar light fields, possess non-uniform spatial polarization distributions at the wavefront. Their wavefront shapes and polarizations can be tailored into arbitrary desired states, offering higher degrees of freedom for manipulation to meet diverse demands. After continuous optimization and development, the theory of vector light fields has developed into a relatively comprehensive theoretical framework. This chapter will subsequently focus on introducing the principles and device design methods for vector light fields with spatially varying polarization states. It should be noted that the vector light fields discussed here pertain to the far field, as the near field is inherently vector in nature.

6.1 THEORETICAL DESCRIPTION OF VECTOR LIGHT FIELD

In 1980, Pattanayak and his team [1] theoretically and innovatively established a complete descriptive framework for vector light fields. Since then, this field has sparked a widespread and in-depth research boom globally, with significant breakthroughs achieved in both related experimental techniques and applied research. Classic vector light fields include radially polarized light and azimuthally polarized light modes in cylindrically symmetric structures. In the 1960s, Snitzer [2] revealed the existence of special polarization modes in cylindrically

DOI: 10.1201/9781003784821-6

symmetric dielectric waveguides, and by the early 1970s, these modes were successfully realized in the laboratory [3,4]. For a long time afterwards, research on vector light fields focused on relatively simple types such as cylindrical symmetry and did not receive sufficient attention. A critical turning point occurred in 2000 when Youngworth and Brown [5] applied the vector diffraction theory of Richards and Wolf [6] to analyse the focusing characteristics of radially and azimuthally polarized light. They discovered that such light fields possess unique advantages in super-resolution imaging and extended depth of focus, and this theoretical prediction was subsequently verified experimentally. Gradually, researchers were no longer satisfied with only classic cylindrically symmetric vector light fields. More novel and complex light fields, such as hybrid vector light fields, full Poincaré light fields [7,8], and array vector light fields [9,10], were proposed and studied. The unique distribution properties of these vector light fields enable them to play important roles in fields such as super-resolution, optical storage, micro-nano fabrication, and optical communication.

6.1.1 Classic Cylindrical Symmetric Vector Light Field

Cylindrical vector beams are vector beam solutions to Maxwell's equations, exhibiting axial symmetry in both amplitude and phase. In free space, typical paraxial beam-like solutions with harmonic time dependence can be obtained by solving the scalar Helmholtz equation [11,12]:

$$(\nabla^2 + k^2)E = 0, \tag{6.1}$$

where $k = 2\pi/\lambda$ is the wavenumber. In Cartesian coordinates (x, y, z) and cylindrical coordinates (r, ϕ, z), the general forms of the electric field propagating along the z-axis with angular frequency ω are:

$$\begin{aligned} E(x,y,z,t) &= u(x,y,z)\exp[i(kz-\omega t)] \\ E(r,\phi,z,t) &= u(r,\phi,z)\exp[i(kz-\omega t)] \end{aligned}. \tag{6.2}$$

In Cartesian coordinates, the paraxial beam-like solution to Eq. (6.1) is the Hermite-Gaussian (HG) mode, with the general form:

$$u(x,y,z) = E_0 H_m\left(\frac{\sqrt{2}x}{w(z)}\right) H_n\left(\frac{\sqrt{2}y}{w(z)}\right)\frac{w_0}{w(z)}\exp[-i\varphi_{mn}(z)]\exp\left[i\frac{k(x^2+y^2)}{2q(z)}\right], \tag{6.3}$$

where $H_m(\cdot)$ and $H_n(\cdot)$ are Hermite polynomials, and m and n represent the nodal numbers of the beam in the x and y directions, respectively. E_0 is the electric field amplitude constant, $w(z)$ is the Gaussian beam radius, and w_0 is the beam radius at the beam waist; $q(z) = z - iz_0$ is the complex beam parameter, where $z_0 = \pi w_0^2/\lambda$ is the Rayleigh range of the Gaussian mode; $\varphi_{mn}(z) = (m+n+1)\varphi(z)$, where $\varphi(z) = \arctan(z/z_0)$ is the Gouy phase shift of the fundamental Gaussian beam. When $m = n = 0$, the fundamental Gaussian beam solution is obtained:

$$u(r,z) = E_0\frac{w_0}{w(z)}\exp[-i\varphi(z)]\exp\left[i\frac{k}{2q(z)}r^2\right]. \tag{6.4}$$

In cylindrical coordinates, the paraxial beam-like solution to Eq. (6.1) is the Laguerre-Gaussian (LG) mode, with the general form:

$$u(r,\phi,z)=E_0\left(\frac{\sqrt{2}r}{w(z)}\right)^l L_p^l\left(\frac{2r^2}{w^2(z)}\right)\frac{w_0}{w(z)}\exp[-i\varphi_{pl}(z)]\exp\left[i\frac{kr^2}{2q(z)}\right]\exp(il\phi), \tag{6.5}$$

where $L_p^l(\cdot)$ is the Laguerre polynomial, l and p are the radial and azimuthal indices, respectively, and $\varphi_{pl}(z) = (2p + l + 1)\varphi(z)$. It is easy to observe that when $l \neq 0$, this LG solution exhibits a vortex phase $e^{il\phi}$, while when $p = l = 0$, the LG solution also reduces to the fundamental Gaussian beam solution.

The above solutions correspond to the case where the polarization is uniformly distributed across the beam cross-section (scalar beams). If the full vector wave scenario is considered, the vector Helmholtz equation becomes:

$$\nabla\times(\nabla\times\vec{E})-k^2\vec{E}=0. \tag{6.6}$$

In this case, the cylindrically symmetric beam-like vector solution for the azimuthal electric field takes the following form:

$$\vec{E}(r,z)=U(r,z)\exp[i(kz-\omega t)]\vec{e}_\phi. \tag{6.7}$$

A trial solution that satisfies azimuthal polarization symmetry exists:

$$U(r,z)=E_0J_1\left(\frac{\beta r}{1+iz/z_0}\right)\exp\left[-\frac{i\beta^2 z/(2k)}{1+iz/z_0}\right]u(r,z), \tag{6.8}$$

where β is a constant scaling parameter, $u(r, z)$ is the fundamental Gaussian solution given by Eq. (6.4), and $J_1(\cdot)$ is the first-order Bessel function of the first kind. This solution corresponds to the azimuthal vector Bessel-Gaussian beam.

Cylindrically symmetric vector light fields represent the most classic category of vector light fields and currently have the widest range of application scenarios. They originate from the mutual coupling between photonic spin angular momentum (SAM) and orbital angular momentum (OAM), forming a novel spin-orbit entangled state. Figure 6.1(a–e) illustrates HG light fields, while Fig. 6.1(f) illustrates LG light fields. Figure 6.1(a–f) shows that the electric field may have opposite instantaneous directions due to non-uniform phase distributions on the beam, but they are still considered to have spatially uniform polarization distributions. Figure 6.1(g–i) illustrates several vector light fields, with Fig. 6.1(g) and (h) representing radial and azimuthal vector light fields, respectively, and Fig. 6.1(i) depicting a generalized cylindrical vector light field. Radial and azimuthal light fields can be regarded as superpositions of orthogonal HG modes, expressed as:

$$\begin{aligned}\vec{E}_r &= \mathrm{HG}_{10}\vec{e}_x+\mathrm{HG}_{01}\vec{e}_y\\ \vec{E}_\phi &= \mathrm{HG}_{01}\vec{e}_x+\mathrm{HG}_{10}\vec{e}_y\end{aligned}, \tag{6.9}$$

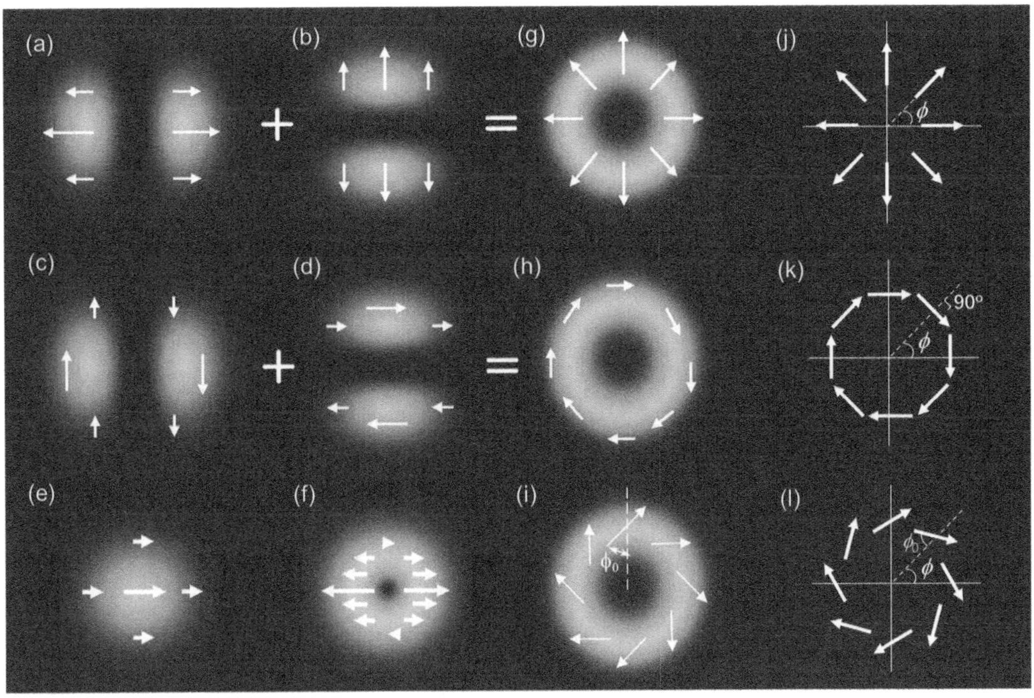

FIGURE 6.1 Electric field distributions of several conventional modes and cylindrically symmetric vector modes. (a) x-polarized HG_{10} mode; (b) y-polarized HG_{01} mode; (c) y-polarized HG_{10} mode; (d) x-polarized HG_{01} mode; (e) x-polarized fundamental Gaussian mode; (f) x-polarized LG mode; (g) radial vector light field; (h) azimuthal vector light field; (i) generalized cylindrical vector light field; (j) geometric structure of radial polarization; (k) geometric structure of azimuthal polarization; and (l) geometric structure of polarization in generalized cylindrical vector light field.

where $\vec{e}_x, \vec{e}_y$ denote unit vectors in the x and y directions. Under the Jones matrix description, radial and azimuthal light fields in the x-y plane can take very simple forms. As shown in Fig. 6.1(j), radial polarization emanates outward from the coordinate centre, with the location coordinate orientation overlapping the vibration direction of the electric field. Through simple geometric analysis, the form of the polarization Jones matrix can be written as:

$$\vec{E}_r = \begin{pmatrix} \cos\phi \\ \sin\phi \end{pmatrix}. \tag{6.10}$$

As shown in Fig. 6.1(k), azimuthal polarization consists of all polarization states connected head-to-tail, forming a closed loop around the coordinate centre. Essentially, it involves rotating the radial polarization state at each spatial location by $\pi/2$, meaning the location coordinate orientation is perpendicular to the vibration direction of the electric field. Therefore, its polarization Jones matrix form is:

$$\vec{E}_\phi = R(\pi/2)\begin{pmatrix} \cos\phi \\ \sin\phi \end{pmatrix} = \begin{pmatrix} 0 & -1 \\ 1 & 0 \end{pmatrix}\begin{pmatrix} \cos\phi \\ \sin\phi \end{pmatrix} = \begin{pmatrix} -\sin\phi \\ \cos\phi \end{pmatrix}, \tag{6.11}$$

where $R(\cdot)$ is the rotation matrix. Radial and azimuthal polarizations are special cases. For more general cases, as shown in Fig. 6.1(l), where the location coordinate orientation is neither perpendicular nor overlapping with the vibration direction of the electric field, it is equivalent to rotating the radial polarization state at each spatial location by an angle of ϕ_0. Thus, we obtain:

$$\vec{E}_{r,\phi} = R(\phi_0)\begin{pmatrix} \cos\phi \\ \sin\phi \end{pmatrix} = \begin{pmatrix} \cos\phi_0 & \sin\phi_0 \\ \sin\phi_0 & \cos\phi_0 \end{pmatrix}\begin{pmatrix} \cos\phi \\ \sin\phi \end{pmatrix} = \begin{pmatrix} \cos(\phi+\phi_0) \\ \sin(\phi+\phi_0) \end{pmatrix}. \tag{6.12}$$

It can be seen that for a generalized cylindrical vector field, it can be considered either as a radial polarization field rotated by an angle ϕ_0 or as a radial light field with an initial phase ϕ_0 added. Both interpretations share the same fundamental meaning. Equations (6.10) to (6.12) correspond to the case of a topological charge number of 1, where the polarization winding once around the coordinate centre corresponds to a phase change of 2π. Considering arbitrary topological charge numbers, the expression for the generalized cylindrical vector field should be extended as:

$$\vec{E}_{r,\phi,l} = \begin{pmatrix} \cos(l\phi+\phi_0) \\ \sin(l\phi+\phi_0) \end{pmatrix}, \tag{6.13}$$

where l is the topological charge number and ϕ_0 is the initial phase. In this case, the polarization winding once around the coordinate centre corresponds to a phase change of $2\pi \cdot l$.

6.1.2 Vector Light Field Described by Poincaré Sphere

Any beam can be expressed as the superposition of a pair of orthogonal circular polarizations [13]:

$$U(r) = u_R(r)e^{i\varphi_R}\begin{bmatrix} 1 \\ i \end{bmatrix} + u_L(r)e^{i\varphi_L}\begin{bmatrix} 1 \\ -i \end{bmatrix}, \tag{6.14}$$

where $u_R(r)$ and $u_L(r)$ represent the amplitudes of the right-handed circularly polarized (RCP) and left-handed circularly polarized (LCP) components, respectively, and φ_R and φ_L are their corresponding phase delays. This indicates that any beam possesses four degrees of freedom: phase, amplitude, polarization ratio, and the relative phase delay between the two components. Using these four degrees of freedom, any vector light field can be perfectly described. Therefore, by adjusting these four parameters, the light field can be controlled and vector light fields of arbitrary properties can be generated. However, to describe a specific type of vector light field, merely understanding vector light fields through these parameters is insufficient; a set of convenient and intuitive characterization methods is also required. Common methods for characterizing the polarization state of a light field include the Jones matrix, polarization ellipse, Stokes parameters, and the Poincaré sphere. Among these, Stokes parameters and some novel types of Poincaré spheres can describe light of any polarization state, while the Jones matrix and polarization ellipse methods can only describe fully polarized light. After confirming the specific characteristics of a vector light field, selecting an appropriate characterization method can greatly facilitate research. In the following, different characterization methods will be used to introduce some common types of vector light fields.

Early research on the properties and generation of vector light fields largely focused on linear vector light fields in cylindrical coordinates. However, linear polarization is only a special case of vector light fields; there has been relatively little research on the broader categories of elliptical and circular polarization states. As research has progressed, researchers have discovered that by methods such as orthogonally superposing the positive and negative first-order diffracted light generated by spatial light modulators [14], vector beams with different polarization characteristics can be produced in cylindrical coordinates. The Poincaré sphere provides a convenient means to characterize such beams. The full Poincaré beam is an extreme type of vector beam, a concept first proposed by Beckley et al. [8] in 2010. They posed and answered the question: Can a light field covering the entire Poincaré sphere be created and conserved under propagation? To this end, they proposed an analytical model based on the superposition of LG modes. They first verified it for monochromatic, fully polarized light, proving that under certain beam conditions, the coverage is indeed complete and conserved. They then performed theoretical derivations and experimental verification for partially polarized light, with results consistent with theory. The family of full Poincaré beams is defined as:

$$E_p(r,\gamma)=\cos\gamma\hat{e}_1U_{00}(r)+\sin\gamma\hat{e}_2U_{01}(r), \tag{6.15}$$

where $\hat{e}_1$ and $\hat{e}_2$ are a pair of orthogonal unit vectors, such as a pair of linear or circular polarization bases. $U_{00}(r)$ is the fundamental Gaussian beam field in cylindrical coordinates, and $U_{01}(r)$ is the lowest-order LG mode with an azimuthal index of 0. They are expressed as:

$$\begin{aligned} U_{00}(r)&=\frac{u_0}{1+iz/z_0}\exp\left[ikz-\frac{x^2+y^2}{w_0^2(1+iz/z_0)}\right], \\ U_{01}(r)&=\frac{\sqrt{2}(x+iy)}{w_0(1+iz/z_0)}U_{00}(r) \end{aligned} \tag{6.16}$$

where u_0 is the amplitude of the beam at the origin, and w_0 and z_0 have the same meanings as in the previous Eq. (6.3). By further transforming the beam family expression, there is:

$$\begin{aligned} E_p(r,\gamma) &=\left(\hat{e}_1+\exp\{i[\varphi-\varphi(z)]\}\frac{\sqrt{2}\rho}{w(z)}\cdot\tan\gamma\hat{e}_2\right) \\ &\cdot\cos\gamma\frac{u_0w_0}{w(z)}\exp\left(-\frac{\rho^2}{w^2(z)}\right), \end{aligned} \tag{6.17}$$

where $\varphi = \arctan(y/x)$ is the azimuthal coordinate, $\varphi(z)$ has the same meaning as in the previous Eq. (6.3), representing the Gouy phase shift, and $\rho^2 = x^2 + y^2$. The first bracket determines the polarization state at a given point, and $(\rho/w)\tan\gamma$ determines the ratio between the two orthogonal polarizations. Near the axis, the polarization tends more towards $\hat{\boldsymbol{e}}_1$, whereas away from this region, it tends more towards $\hat{\boldsymbol{e}}_2$. With z changes, the beam radius also changes, which affects the ratio between the two orthogonal polarizations. Simultaneously, as $\varphi(z)$ changes, the polarization undergoes rotation. It is easy to see that from negative to positive infinity, the rotation ranges from $-\pi/2$ to $\pi/2$. Furthermore, during beam propagation, the polarization state distribution across the beam cross section corresponds to the projection of points on the Poincaré sphere onto the wavefront. As the propagation distance changes, this projection only undergoes scaling and rigid rotation without changing shape. The polarization characteristics of full Poincaré beams enable their application in various fields, including beam shaping, microparticle manipulation, optical communications, and atmospheric turbulence detection.

The higher-order Poincaré sphere is an extension of the standard plane-wave Poincaré sphere. Unlike the basic Poincaré sphere, where each point represents only a scalar light field, each point on the higher-order Poincaré sphere represents a vector light field. For monochromatic, paraxial plane waves, the light field can be expressed as a two-dimensional Jones vector:

$$|\psi_l\rangle = \psi_R^l |R_l\rangle + \psi_L^l |L_l\rangle, \tag{6.18}$$

where ψ_R^l and ψ_L^l can be considered as the complex amplitudes of two basis vectors. Orthogonal left- and right-handed circularly polarized optical vortex beams with opposite topological charges can be expressed as:

$$\begin{aligned} |R_l\rangle &= \exp(-il\varphi)(\hat{\boldsymbol{x}} + i\hat{\boldsymbol{y}})/\sqrt{2} \\ |L_l\rangle &= \exp(il\varphi)(\hat{\boldsymbol{x}} - i\hat{\boldsymbol{y}})/\sqrt{2} \end{aligned} . \tag{6.19}$$

Thus, in the circular polarization basis, by normalizing the higher-order Poincaré sphere, the Stokes parameters can be obtained:

$$\begin{aligned} S_0^l &= |\langle R_l|\psi\rangle|^2 + |\langle L_l|\psi\rangle|^2 = |\psi_R^l|^2 + |\psi_L^l|^2 \\ S_1^l &= 2\,\mathrm{Re}\left(\langle R_l|\psi\rangle^* \langle L_l|\psi\rangle\right) = 2|\psi_R^l||\psi_L^l|\cos\phi \\ S_2^l &= 2\,\mathrm{Im}\left(\langle R_l|\psi\rangle^* \langle L_l|\psi\rangle\right) = 2|\psi_R^l||\psi_L^l|\sin\phi \\ S_3^l &= |\langle R_l|\psi\rangle|^2 - |\langle L_l|\psi\rangle|^2 = |\psi_R^l|^2 - |\psi_L^l|^2 \end{aligned} , \tag{6.20}$$

where $\phi = \arg(\psi_R^l) - \arg(\psi_L^l)$ can be considered as the phase difference between the two complex amplitudes, and l is the OAM quantum number or topological charge number, which also denotes the order of the higher-order Poincaré sphere. The parameter S_0^l represents the total light intensity, while S_1^l and S_2^l contain phase information of the two complex amplitudes. S_3^l indicates the ellipticity of the light field, similar to its definition in the standard Poincaré sphere. When $l = 0$, the higher-order Poincaré sphere reduces to the standard plane-wave Poincaré sphere. For fully polarized light, there is $S_0 = 1, S_0^2 = S_1^2 + S_2^2 + S_3^2$. When $S_0 < 1$, there is $S_0^2 < S_1^2 + S_2^2 + S_3^2$, and the Stokes parameters can describe unpolarized or partially polarized states. For $S_0 = 1$, points on the equator represent linearly polarized vector vortex light fields, while the north and south poles represent circularly polarized vector vortex light fields, i.e. eigenstates of total optical angular momentum. Other points represent elliptical vector vortex fields with continuously varying major axis orientations. It is noteworthy that if the absolute values of the SAM quantum number and OAM quantum number are equal, the two poles of the higher-order Poincaré sphere have the same handedness; otherwise, they have opposite handedness. Figure 6.2 illustrates the representation of a high-order Poincaré light field with an OAM quantum number of +1 and unequal absolute values of SAM quantum numbers [15]. Here, the north and south poles represent orthogonal circularly polarized optical vortex fields with topological charges of +1 and −1, respectively. Points on the equator represent generalized cylindrical vector light fields, while other points represent elliptically polarized cylindrical vector light fields.

Another similar model is the hybrid Poincaré sphere [16], which is formed by superimposing radially and azimuthally polarized vector light fields as basis vectors. The higher-order Poincaré sphere and the hybrid Poincaré sphere differ only by an angle, and both

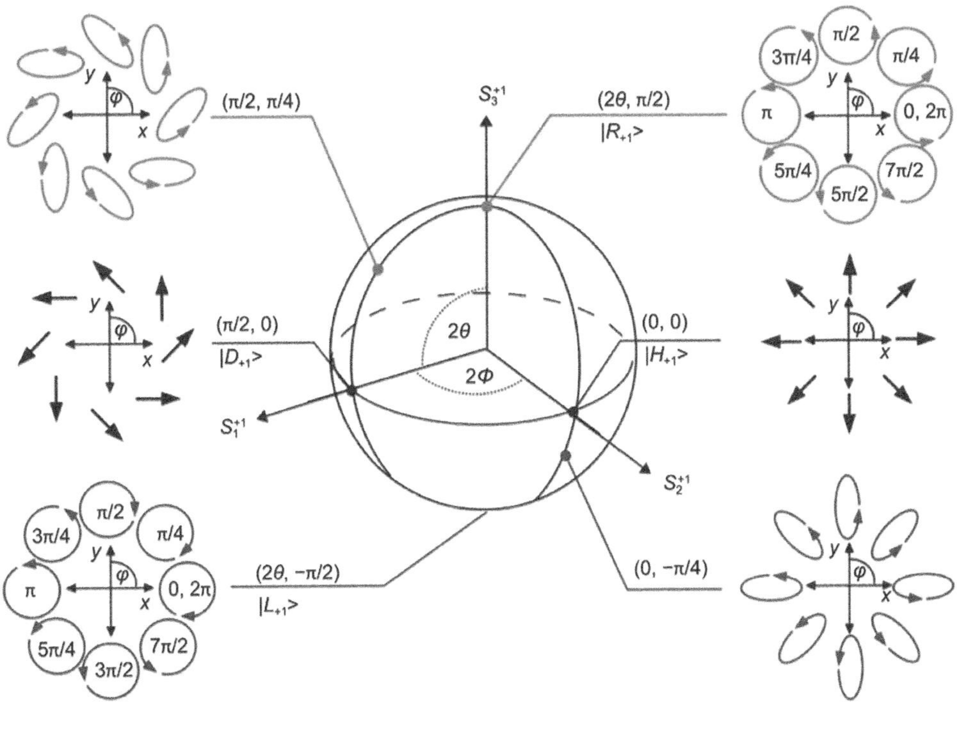

FIGURE 6.2 Schematic of a high-order Poincaré light field with an OAM quantum number of +1, where the absolute values of the OAM and SAM quantum numbers are unequal.

can characterize identical vector light fields. Both the higher-order Poincaré sphere and the hybrid Poincaré sphere are referred to as H-spheres. The higher-order Poincaré sphere method can be used to generate cylindrical vector beams. In 2014, Chen et al. [17] experimentally demonstrated a method to generate arbitrary cylindrical vector beams on the higher-order Poincaré sphere by coaxially superimposing two orthogonally circularly polarized beams with opposite topological charges and adjusting their amplitudes and phases. In the same year, Liu et al. [18] achieved the conversion of uniformly elliptically polarized beams into arbitrary cylindrical vector beams with any polarization distribution on the higher-order Poincaré sphere by locally transforming polarization using inhomogeneous birefringent metasurfaces. Furthermore, the higher-order Poincaré sphere method can be applied in various fields, such as converting Gaussian fundamental modes into HG or LG modes, spatial demultiplexing, optical polarization and singularity resolution, and mode division multiplexing in fibre-optic communications.

However, describing the polarization and phase variations during light field propagation using the higher-order Poincaré sphere has certain limitations, such as when the beam propagates through anisotropic inhomogeneous media. To address this, the concept of the hybrid-order Poincaré sphere was further proposed [19]. Assume that the anisotropic medium is composed of several local waveplates, whose optical axis direction can be expressed as:

$$\alpha(r,\varphi)=q\varphi+\alpha_0, \tag{6.21}$$

where r is the radial coordinate, q is a constant representing the topological charge, ϕ is the azimuthal coordinate, and α_0 is the initial azimuthal angle on the x-axis. Inhomogeneous birefringent elements with specific geometric shapes can be regarded as **q**-plates. When a circularly polarized vortex wave is incident, its expression is:

$$|\psi\rangle=\frac{\sqrt{2}}{2}(\hat{e}_x+i\sigma\hat{e}_y)\exp(il\varphi). \tag{6.22}$$

The SAM and OAM of this vortex wave are σh and lh, respectively. The transmission behaviour of the wave in the **q**-plate can be conveniently described using the Jones matrix. When the thickness of the **q**-plate is much smaller than the Rayleigh range, the diffraction effect can be neglected, and only the evolution of polarization and phase is studied. The evolved light field is obtained as:

$$\begin{aligned}|\Psi_{l,m}\rangle=&\frac{\sqrt{2}}{2}\cos\frac{\delta}{2}(\hat{e}_x+i\sigma\hat{e}_y)\exp(il\varphi)\\&+\frac{\sqrt{2}}{2}\sin\frac{\delta}{2}(\hat{e}_x-i\sigma\hat{e}_y)\exp(im\varphi)-\exp\left[i\left(2\sigma\alpha_0-\frac{\pi}{2}\right)\right],\end{aligned} \tag{6.23}$$

where δ is the phase retardation of the **q**-plate, and $m = l + 2\sigma q$. The above equation can be interpreted as the superposition of two waves: the first wave has the same OAM and OAM as the incident wave, while the second wave has the opposite SAM and an OAM determined by m. The evolved light field has an expression similar to that of a higher-order Poincaré field and can be represented by a two-dimensional Jones vector:

$$
\begin{aligned}
|\psi\rangle &= \psi_N^l |N_L\rangle + \psi_S^m |S_m\rangle \\
N_L &= \frac{\sqrt{2}}{2}(\hat{e}_x + i\sigma\hat{e}_y)\exp(il\varphi) \quad , \\
S_m &= \frac{\sqrt{2}}{2}(\hat{e}_x - i\sigma\hat{e}_y)\exp(im\varphi)
\end{aligned}
\tag{6.24}
$$

where $\boldsymbol{N}_L$ and $\boldsymbol{S}_m$ are orthogonally polarized basis vectors with different topological charges. By varying the coefficients ψ_N^l and ψ_S^m, any polarization state on the hybrid-order Poincaré sphere can be represented. Therefore, Stokes parameters similar to those of the higher-order Poincaré sphere can be redefined to construct a new Poincaré sphere. Since $l \neq m$, it is referred to as the hybrid-order Poincaré sphere. The two poles of the higher-order Poincaré sphere correspond to its basis vectors, which have opposite spin and orbital states, and their OAM magnitudes must be equal. In contrast, the two poles of the hybrid-order Poincaré sphere also represent its basis vectors, but their OAMs are unequal and can be artificially selected, thereby extending the higher-order Poincaré sphere to a more general form. The hybrid-order Poincaré sphere method is primarily used for generating vector vortex beams, tightly focused beams, studying beam propagation in atmospheric turbulence, and other related fields.

While H-spheres provide a convenient and intuitive method for characterizing vector light fields, they can only describe polarization distributions with the same ellipticity but varying major axis orientations at arbitrary positions. They are difficult to generalize to the representation of hybrid vector fields composed of linear, elliptical, and circular polarization states. Additionally, since exchanging the topological charges of the north and south poles results in entirely different vector light field distributions, for topological charges greater than 0, two higher-order Poincaré spheres are required for description. The generalized Poincaré sphere effectively addresses these issues.

Introduce a pair of orthogonal bases with continuously varying ellipticity:

$$
\begin{aligned}
\left|N_{\mathrm{R}}^m\right\rangle &= \frac{1}{\sqrt{2}} e^{-im\varphi}\left(e^{-iR\pi}\hat{e}_x - ie^{iR\pi}\hat{e}_y\right) \\
\left|S_{\mathrm{R}}^m\right\rangle &= \frac{1}{\sqrt{2}} e^{im\varphi}\left(e^{-iR\pi}\hat{e}_x + ie^{iR\pi}\hat{e}_y\right)
\end{aligned}
\quad ,
\tag{6.25}
$$

where m is the topological charge, φ is the azimuthal coordinate, and $R \in [0.5, 1]$ represents the relative phase between the x and y components. When $R = 0.5$, the above orthogonal bases reduce to the orthogonal bases on the H-sphere. When $m = 0$ and $R = 0.5$, they become the left- and right-handed circular polarization bases on the Poincaré sphere. Thus, any azimuthally varying vector field can be expressed as a combination of complex amplitude coefficients ψ_N^m, ψ_S^m, and the orthogonal bases:

$$
\left|\psi^m\right\rangle = \psi_N^m \left|N_{\mathrm{R}}^m\right\rangle + \psi_S^m \left|S_{\mathrm{R}}^m\right\rangle. \tag{6.26}
$$

Using a method similar to Eq. (6.20), the extended higher-order Stokes parameters can be derived. It is noteworthy that when characterizing vector light fields using the Poincaré sphere or H-sphere, the radial parameter is irrelevant for fully polarized light. If this dimension (radial parameter) can be utilized, the sphere's capacity to characterize polarization can be significantly enhanced. Therefore, by introducing R as the radial component, the Stokes parameters for the generalized Poincaré sphere are obtained:

$$\begin{aligned} G_{0R}^{m} &= ES_{0R}^{m}R = R \\ G_{1R}^{m} &= ES_{1R}^{m}R = R\sin 2\beta\cos 2\phi_0 \\ G_{2R}^{m} &= ES_{2R}^{m}R = R\sin 2\beta\sin 2\phi_0 \\ G_{3R}^{m} &= ES_{3R}^{m}R = -R\cos 2\beta \end{aligned}, \tag{6.27}$$

where $\beta \in [0, \pi/2]$ represents the relative weight of the normalized complex amplitude coefficients, and $2\phi_0$ denotes the relative phase of the normalized complex amplitude coefficients. Figure 6.3 illustrates the representation of a generalized Poincaré light field with a topological charge of +1 [20]. The symmetry points on the G_{3R}^{m} axis represent the continuously varying bases. The innermost shell ($R = 0.5$) and the outermost shell ($R = 1$) correspond to H-spheres

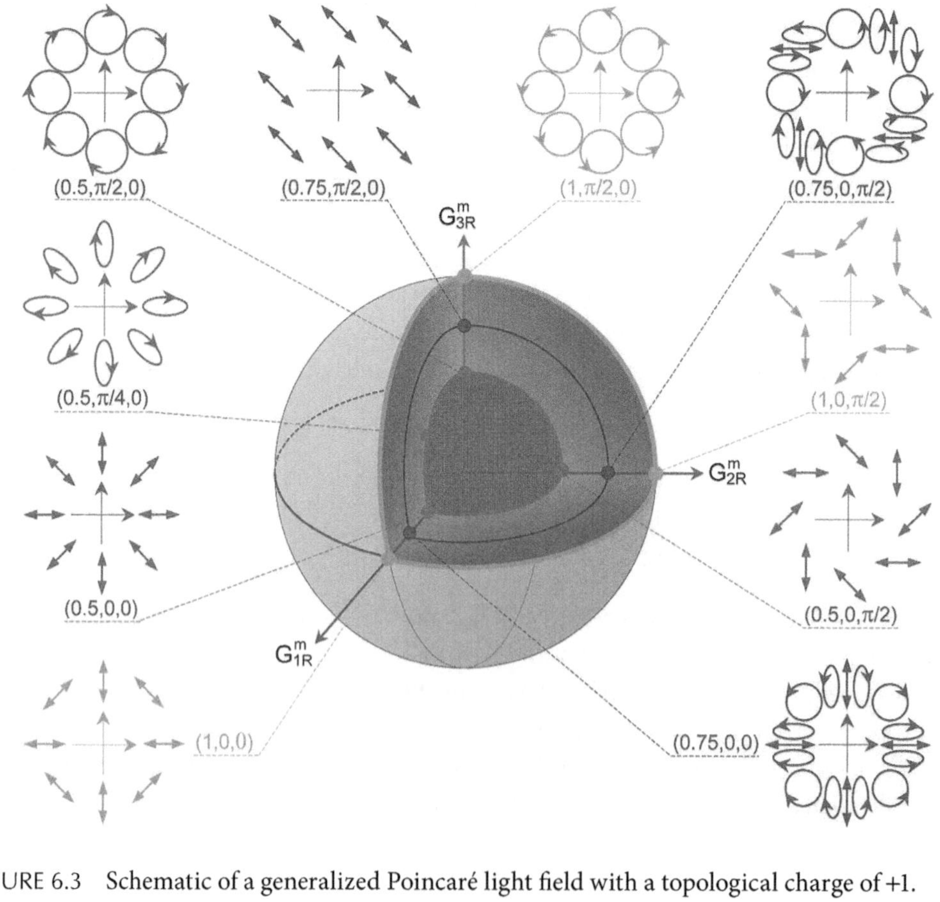

FIGURE 6.3 Schematic of a generalized Poincaré light field with a topological charge of +1.

with $m = 1$ and $m = -1$, respectively. The intermediate shell ($R = 0.75$) represents a vector field with orthogonal linearly polarized bases carrying opposite OAM, where the equator denotes hybrid polarized vector fields. The generalized Poincaré sphere consists of an infinite number of spherical shells with $R \in [0.5, 1]$. The higher-order Poincaré spheres are merely two special spherical surfaces among them. This framework can be used to characterize hybrid vector fields, greatly simplifying other complex polarization problems. Due to its universality, this method is employed to address numerous polarization-related issues.

6.2 DESIGN AND ANALYSIS OF VECTOR LIGHT FIELD GENERATOR

6.2.1 Principle of Spin-Decoupling Technology

Pancharatnam-Berry (PB) phase metasurfaces exhibit excellent capabilities for manipulating circularly polarized waves over a broad bandwidth. However, for the two distinct spin states (left- and right-handed circularly polarized states), the phase modulation provided by PB phase metasurfaces is exactly opposite, resulting in locked and mirrored control for LCP and RCP waves. This limitation restricts the application of metasurfaces in multifunctional devices. Consider the transmitted wave from a birefringent unit cell rotated by an angle θ relative to the x-axis. Its Jones matrix is described in Eq. (3.3) of Chapter 3 and is directly quoted here:

$$\begin{aligned} \boldsymbol{T}_0 &= \begin{pmatrix} \cos\theta & -\sin\theta \\ \sin\theta & \cos\theta \end{pmatrix} \begin{pmatrix} |t_{xx}| e^{i\varphi_{xx}} & 0 \\ 0 & |t_{yy}| e^{i\varphi_{yy}} \end{pmatrix} \begin{pmatrix} \cos\theta & \sin\theta \\ -\sin\theta & \cos\theta \end{pmatrix} \\ &= \begin{pmatrix} \cos^2\theta \cdot |t_{xx}| e^{i\varphi_{xx}} + \sin^2\theta \cdot |t_{yy}| e^{i\varphi_{yy}} & \cos\theta\sin\theta \cdot \left(|t_{xx}| e^{i\varphi_{xx}} - |t_{yy}| e^{i\varphi_{yy}}\right) \\ \cos\theta\sin\theta \cdot \left(|t_{xx}| e^{i\varphi_{xx}} - |t_{yy}| e^{i\varphi_{yy}}\right) & \sin^2\theta \cdot |t_{xx}| e^{i\varphi_{xx}} + \cos^2\theta \cdot |t_{yy}| e^{i\varphi_{yy}} \end{pmatrix} \end{aligned} \quad (6.28)$$

Through a series of mathematical operations, this expression can be further written as:

$$\begin{aligned} \boldsymbol{T}_0 &= \frac{1}{2}\left(|t_{xx}| e^{i\varphi_{xx}} - |t_{yy}| e^{i\varphi_{yy}}\right) \begin{pmatrix} \cos 2\theta & \sin 2\theta \\ \sin 2\theta & -\cos 2\theta \end{pmatrix} + \frac{1}{2}\left(|t_{xx}| e^{i\varphi_{xx}} + |t_{yy}| e^{i\varphi_{yy}}\right) \begin{pmatrix} 1 & 0 \\ 0 & 1 \end{pmatrix} \\ &= \frac{1}{4}\left(|t_{xx}| e^{i\varphi_{xx}} - |t_{yy}| e^{i\varphi_{yy}}\right)\left(e^{i2\theta}\boldsymbol{\sigma}_L + e^{-i2\theta}\boldsymbol{\sigma}_R\right) + \frac{1}{2}\left(|t_{xx}| e^{i\varphi_{xx}} + |t_{yy}| e^{i\varphi_{yy}}\right) \begin{pmatrix} 1 & 0 \\ 0 & 1 \end{pmatrix}, \\ &= \boldsymbol{T}_{cross} + \boldsymbol{T}_{co} \end{aligned} \quad (6.29)$$

where $\boldsymbol{\sigma}_L$ and $\boldsymbol{\sigma}_R$ are circular polarization operators, written as:

$$\boldsymbol{\sigma}_L = \begin{pmatrix} 1 & -i \\ -i & -1 \end{pmatrix}, \boldsymbol{\sigma}_R = \begin{pmatrix} 1 & i \\ i & -1 \end{pmatrix}. \quad (6.30)$$

Note that some literature may represent $\boldsymbol{\sigma}_R$ as the former in this equation and $\boldsymbol{\sigma}_L$ as the latter, because RCP and LCP waves are defined as $[1, -i]^{\mathrm{T}}$ and $[1, i]^{\mathrm{T}}$, respectively, which is opposite to the definition in this book (which uniformly defines RCP as $[1, i]^T$ and LCP as $[1, -i]^T$). For an incident circularly polarized wave (RCP: R, LCP: L), under the action of the circular polarization operators, the handedness of the output circularly polarized wave will be flipped, expressed as:

$$\boldsymbol{\sigma}_L R = L, \boldsymbol{\sigma}_L L = 0, \boldsymbol{\sigma}_R L = R, \boldsymbol{\sigma}_R R = 0. \tag{6.31}$$

Equation (6.29) implies that the output wave consists of two parts. The first term ($\boldsymbol{T}_{cross}$) indicates that a portion of the circularly polarized input wave undergoes handedness flipping under the circular polarization operators and acquires an additional phase delay of $\pm i\theta$. The second term ($\boldsymbol{T}_{co}$) indicates that a portion of the circularly polarized wave maintains the same polarization state upon output. If $|t_{xx}| = |t_{yy}|$ and $|\varphi_{xx} - \varphi_{yy}| = \pi$, i.e. $t_{xx} = -t_{yy}$, the co-polarized output can be eliminated, resulting in $\boldsymbol{T}_0 = \boldsymbol{T}_{cross}$. To simplify the calculation, further assume $|t_{xx}| = |t_{yy}| = 1$, and there is:

$$\begin{aligned} \boldsymbol{E}_o &= \boldsymbol{T}_{cross}\boldsymbol{E}_i \\ &= \frac{1}{4}(e^{i\varphi_{xx}} - e^{i\varphi_{yy}})(e^{i2\theta}\boldsymbol{\sigma}_L + e^{-i2\theta}\boldsymbol{\sigma}_R)\boldsymbol{E}_i \\ &= \frac{1}{4}(1 - e^{i\varphi_{yy}-\varphi_{xx}})(e^{i(2\theta+\varphi_{xx})}\boldsymbol{\sigma}_L + e^{-i(2\theta-\varphi_{xx})}\boldsymbol{\sigma}_R)\boldsymbol{E}_i \\ &= \frac{1}{2}(e^{i(2\theta+\varphi_{xx})}\boldsymbol{\sigma}_L + e^{-i(2\theta-\varphi_{xx})}\boldsymbol{\sigma}_R)\boldsymbol{E}_i \end{aligned} . \tag{6.32}$$

When the incident wave $\boldsymbol{E}_i$ is RCP (where $\boldsymbol{\sigma}_R\boldsymbol{E}_i = 0$), under the action of $\boldsymbol{\sigma}_L$, the output wave $\boldsymbol{E}_o$ will be converted to LCP and acquire a phase delay of $\varphi_L = 2\theta + \varphi_{xx}$. When the incident wave $\boldsymbol{E}_i$ is LCP (where $\boldsymbol{\sigma}_L\boldsymbol{E}_i = 0$), under the action of $\boldsymbol{\sigma}_R$, the output wave $\boldsymbol{E}_o$ will be converted to RCP and acquire a phase delay of $\varphi_R = -(2\theta - \varphi_{xx})$. Therefore, θ, φ_{xx}, and φ_{yy} can all be expressed in terms of φ_L and φ_R:

$$\begin{cases} \theta = (\varphi_L - \varphi_R)/4 \\ \varphi_{xx} = (\varphi_L + \varphi_R)/2 \\ \varphi_{yy} = (\varphi_L + \varphi_R)/2 - \pi \end{cases} . \tag{6.33}$$

This equation shows that for any given φ_L and φ_R, a corresponding set of θ, φ_{xx}, and φ_{yy} can be determined. In other words, the independent control of LCP and RCP (i.e. spin-decoupling) can be designed, and the desired φ_L and φ_R can be achieved through a specific set of θ, φ_{xx}, and φ_{yy}.

6.2.2 Meta-Device Design for Vector Light Field

Here, the design and analysis of focused vector beams are demonstrated. The focusing effect is achieved using the phase profile of a Fresnel zone plate [21]. As shown in Fig. 6.4(a),

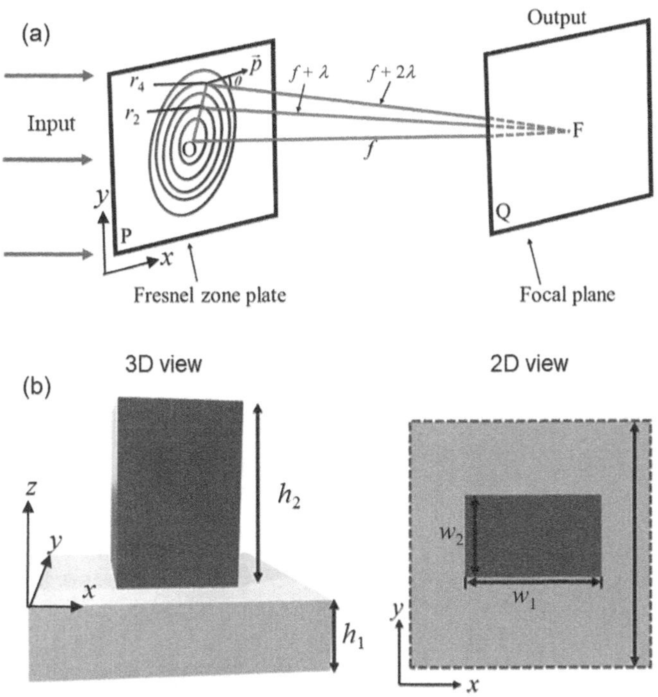

FIGURE 6.4 (a) Geometric schematic illustrating the relationship between radius, focal length, and wavelength for focusing based on a Fresnel zone plate; (b) 3D and 2D schematics of the designed unit cell.

assuming the central ring of the Fresnel zone plate is the first ring with a corresponding radius of r_1, the outer radius r_n of the nth ring can be expressed as [22]:

$$r_n = \sqrt{n\lambda\left(f + \frac{n\lambda}{4}\right)}, \tag{6.34}$$

where f is the focal length, designed as $f = 1.7$ cm, and λ is the operating wavelength, corresponding to a design frequency of 1 THz, i.e. $\lambda = 300$ μm. When a beam is incident on the zone plate, the transmitted beams passing through adjacent zones of the plate will produce a phase difference of π at the focal plane. Specifically, under normal incidence, the phase difference at the focus for beams passing through all even (or odd) zones is an integer multiple of 2π. Several unit cells are introduced here to construct an all-dielectric metasurface zone plate. As shown in Fig. 6.4(b), the unit cell is a rectangular pillar structure made of high-resistivity silicon with a relative permittivity of 11.9. The substrate thickness is $h_1 = 300$ μm, the pillar height is $h_2 = 200$ μm, the lattice constant is $P = 150$ μm, and the lengths along the x and y directions are w_1 and w_2, respectively. To obtain the desired unit cell, the response of the unit cell to linearly polarized waves under different w_1 and w_2 was simulated. By varying the length and width of the rectangular pillar, simulation results were obtained as shown in Fig. 6.5(a–d). Based on this, 15 representative unit cells were

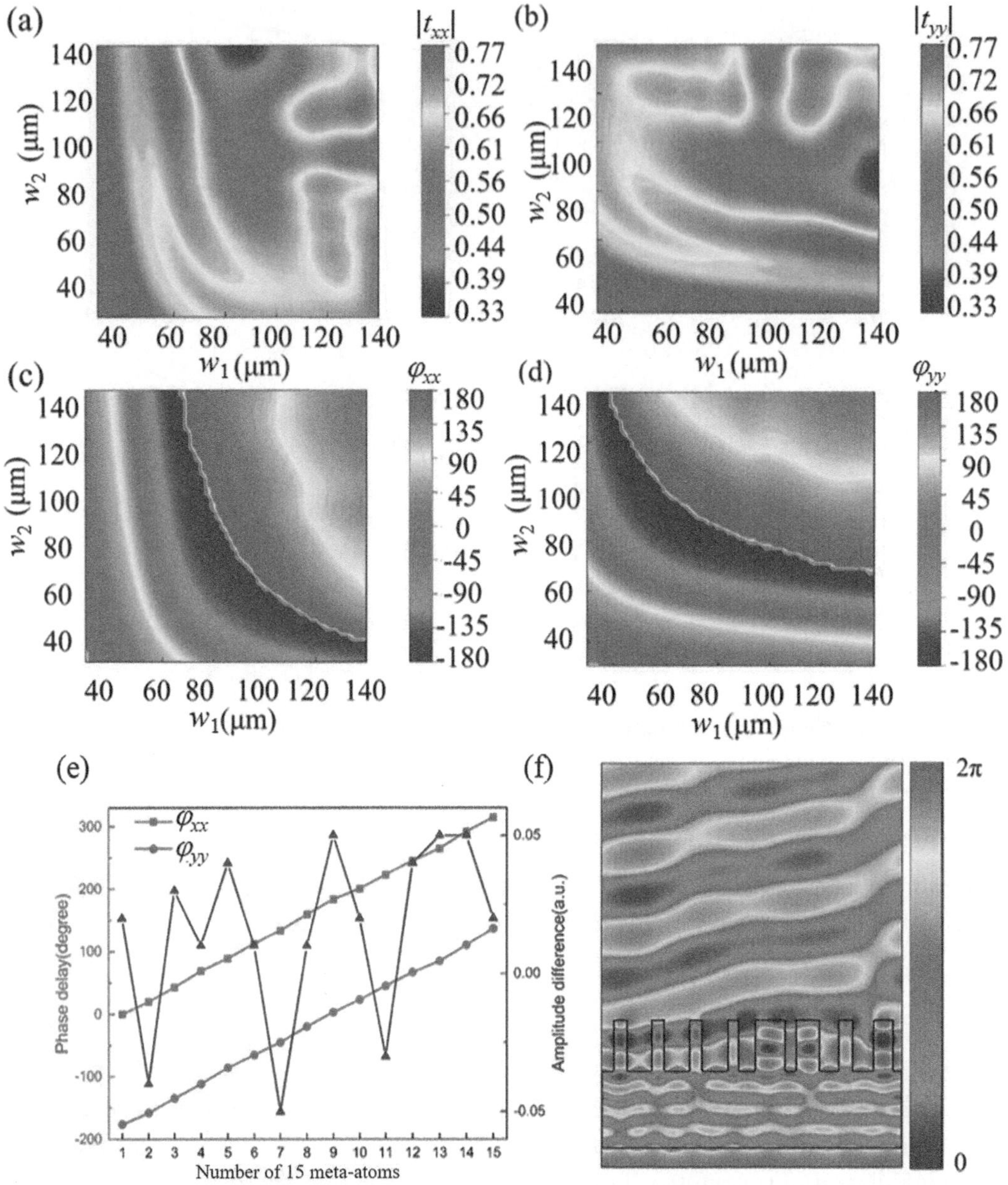

FIGURE 6.5 (a–c) Co-polarized linear transmission amplitude and phase of the unit cell for different w_1 and w_2; (e) phase gradient of co-polarized linear transmission for the selected 15 unit cells; (f) phase distribution of the output wave generated by a periodic structure composed of unit cells arranged in ascending phase order.

selected, with their corresponding simulation results shown in Fig. 6.5(e). The dimensional parameters of the selected 15 unit cells are listed in Table 6.1. The φ_{xx} (φ_{yy}) represents the propagation phase delay for *x*- (*y*-) polarized transmission at the operating frequency of 1 THz. Moreover, the phase delays φ_{xx} and φ_{yy} of these 15 unit cells cover the full range of 0–2π, and the phase difference and amplitude difference for each unit cell are close to π and 0, respectively. Figure 6.5(f) shows the phase distribution of the output wave generated

TABLE 6.1 Size Parameters of the Selected 15 Unit Cells

No.	1	2	3	4	5	6	7	8	9	10	11	12	13	14	15
w_1	50	50	50	49	48	46	42	127	118	108	94	84	80	76	76
w_2	120	106	92	85	80	76	76	51	50	50	50	50	48	46	42

by a periodic structure composed of unit cells arranged in ascending phase order under *x*-polarized incidence, achieving beam deflection, which indicates the desired propagation phase gradient existing on the *xoz* plane.

By introducing the PB phase on top of the aforementioned propagation phase, independent control of two orthogonal circular polarizations can be achieved. Next, based on spin-decoupling, a metasurface zone plate capable of generating focused vector beams will be constructed. Consider superimposing two orthogonal circularly polarized vortex beams to produce the desired vector beam. For this purpose, a metasurface zone plate is designed to generate focused cross-polarized vortex beams carrying OAM of +1 or −1 under circularly polarized incidence. Therefore, the required phase distributions φ_R and φ_L for orthogonal circularly polarized waves can be described as:

$$\begin{aligned} \varphi_R = \varphi_{Rn} &= \begin{cases} l_1\phi, n = 2m-1 \\ l_1\phi + \pi, n = 2m \end{cases}, \\ \varphi_L = \varphi_{Ln} &= \begin{cases} l_2\phi, n = 2m-1 \\ l_2\phi + \pi, n = 2m \end{cases} \end{aligned} \tag{6.35}$$

where m is a positive integer, $\phi = \arctan(y/x)$ represents the azimuthal angle in the x-y plane, and the parameters $l_1 = 1$ and $l_2 = -1$ are the topological charges (OAM) of the vortex beams under LCP and RCP incidence, respectively. For LCP and RCP incidence, the corresponding output waves are:

$$\begin{aligned} \boldsymbol{E}_R &= \begin{cases} e^{i\delta_1} \cdot e^{i\pi(n-1)} \cdot e^{i\phi} \begin{pmatrix} 1 \\ i \end{pmatrix}, n = 2m-1 \\ e^{i\delta_2} \cdot e^{i\pi(n-2)} \cdot e^{i(\phi+\pi)} \begin{pmatrix} 1 \\ i \end{pmatrix}, n = 2m \end{cases}, \\ \boldsymbol{E}_L &= \begin{cases} e^{i\delta_1} \cdot e^{i\pi(n-1)} \cdot e^{-i\phi} \begin{pmatrix} 1 \\ -i \end{pmatrix}, n = 2m-1 \\ e^{i\delta_2} \cdot e^{i\pi(n-2)} \cdot e^{-i(\phi+\pi)} \begin{pmatrix} 1 \\ -i \end{pmatrix}, n = 2m \end{cases} \end{aligned} \tag{6.36}$$

where $\boldsymbol{E}_R$ (or $\boldsymbol{E}_L$) represents the electric field of the output RCP (or LCP) wave under LCP (or RCP) incidence. The δ_1 and δ_2 are the propagation phases at the focal plane for output waves passing through odd and even zones, respectively, with a phase difference of π. An

x-polarized wave can be decomposed into RCP and LCP waves. Therefore, the output wave generated by the metasurface zone plate at the focal plane under x-polarized illumination is:

$$\frac{1}{2}(\boldsymbol{E}_\mathrm{R}+\boldsymbol{E}_\mathrm{L})=e^{i\delta_1}\begin{pmatrix}\cos\phi\\-\sin\phi\end{pmatrix}. \tag{6.37}$$

This expression represents a radially polarized vector field. Note that the y-component of the electric field in this equation is opposite to that in Eq. (6.10), because here the topological charges for LCP and RCP waves are designed as $l_1 = 1$ and $l_2 = -1$, respectively. If the topological charges for LCP and RCP waves are designed as $l_1 = -1$ and $l_2 = 1$, Eq. (6.10) will be obtained. According to Eq. (6.37), the polarization distribution at various positions on the focal plane is inhomogeneous and depends only on the azimuthal angle ϕ, indicating that the output wave on the focal plane is a vector beam. Using a similar derivation approach, it can be concluded that for y-polarized incidence, the transmitted wave of the metasurface zone plate at the focal plane is also a vector beam.

The unit cells were arranged according to the phase defined by Eq. (6.35) to form the metasurface zone plate shown in Fig. 6.6(a). Figure 6.6(b) presents the SEM picture of the fabricated sample. Figure 6.6(c) illustrates the intensity distributions of the co-polarized and cross-polarized components under orthogonal circularly polarized illumination. It is

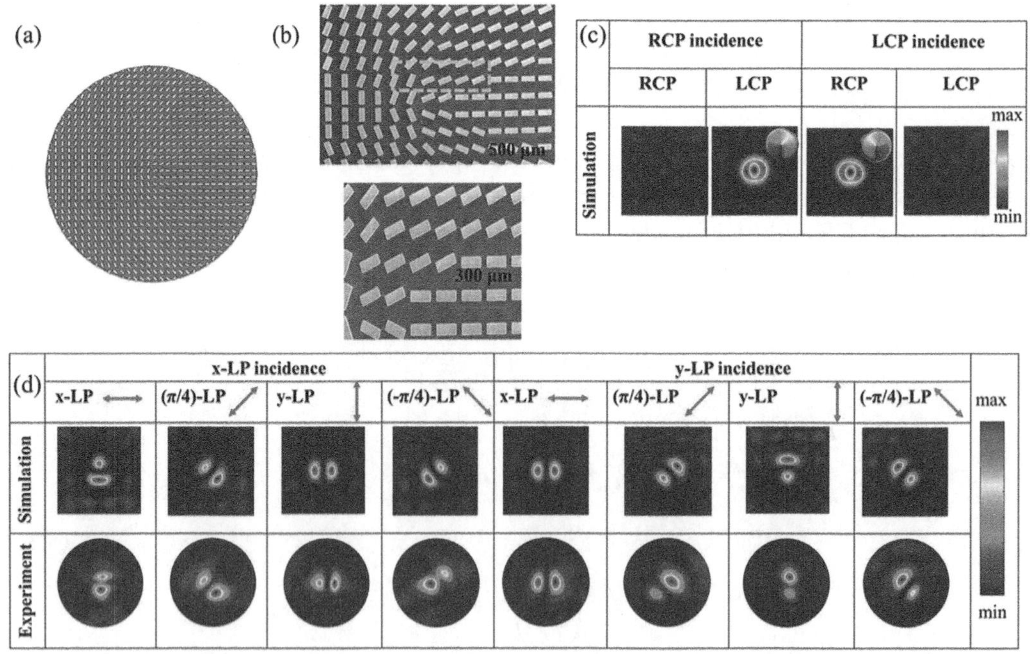

FIGURE 6.6 (a) Schematic of the metasurface zone plate generating focused vector beams and (b) SEM picture of the fabricated sample; (c) simulated results of co-polarized and cross-polarized components under orthogonal circularly polarized incidence, with insets showing the corresponding phase distributions; (d) intensity distributions of four differently oriented linear polarization components in the output wave under linearly polarized incidence.

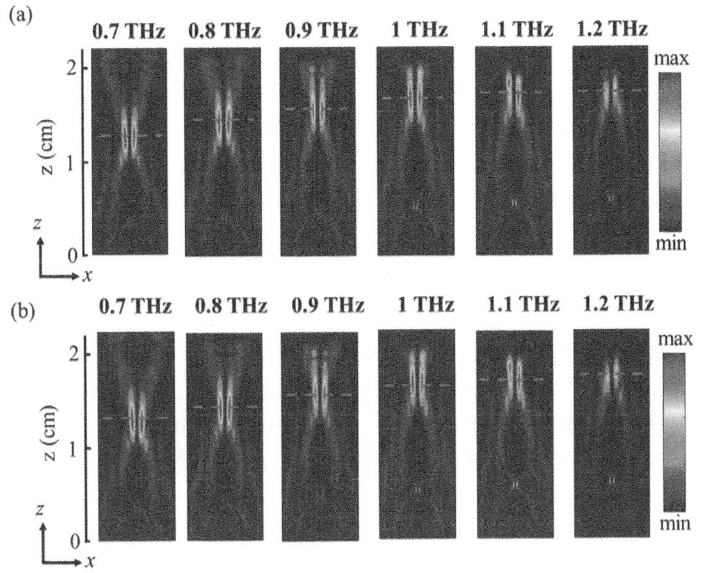

FIGURE 6.7 (a) Intensity distribution of the RCP output wave on the *xoz* plane under LCP incidence at different frequencies; (b) intensity distribution of the LCP output wave on the *xoz* plane under RCP incidence at different frequencies.

evident that the cross-polarized components carry OAM, while the co-polarized components are negligible, which aligns with the design. Furthermore, to characterize the output beam under linearly polarized incidence, the intensity distributions of linear polarization components with four orientations on the focal plane were analysed, as shown in Fig. 6.6(d). Both measured and simulated results reveal distinct electric field distributions at various orientations due to the generated radially polarized vector field under linearly polarized incidence. Note that, ideally, the field intensity should be uniform across all orientations. The non-uniform spot distribution observed here is primarily attributed to design errors in the unit cells. As shown in Fig. 6.7, to further characterize the properties of this metasurface zone plate, the output results of the cross-polarized components under circularly polarized incidence at different frequencies were analysed. The intensity distributions of the RCP and LCP components on the *xoz* plane, shown in Fig. 6.7(a) and (b) respectively, indicate that as the incident wavelength decreases, the distance between the focal point of the output beam and the metasurface zone plate gradually increases. The metasurface zone plate constructed based on the above scheme can generate focused vector beams within an incident frequency range of 0.7–1.2 THz, demonstrating its broadband capability for manipulating vector light fields.

6.3 CHAPTER SUMMARY

This chapter provides a comprehensive introduction to the principles underlying vector light field formation and the design of meta-devices. Vector light fields possess non-uniform polarization distributions across their wavefronts. The radially vector light field is a classic cylindrical symmetric linearly polarized vector light field, which can be transformed into azimuthally polarized fields and other generalized linearly polarized vector fields through rotation. Using generalized Poincaré spheres, more complex elliptically and circularly polarized vector light fields can also be designed. The design and analysis of a vector light field generator were demonstrated using the focused vector beam as an example. A strategy combining a Fresnel zone plate with a spin-decoupling metasurface was proposed to generate cross-polarized vortex beams carrying different OAM under orthogonal circularly polarized incidence. For linearly polarized incidence, since linear polarization can be decomposed into RCP and LCP components of equal intensity, the output wave can be viewed as the superposition of two orthogonal circularly polarized vortex beams with identical intensity. By appropriately selecting specific unit cells and constructing a metasurface zone plate with a defined phase distribution, it can generate RCP beams carrying OAM under LCP incidence and LCP beams carrying OAM under RCP incidence. Ultimately, this metasurface zone plate can produce focused vector light fields under linearly polarized incidence.

REFERENCES

1. Pattanayak, D.N. and Agrawal, G.P. (1980) 'Representation of vector electromagnetic beams', Physical Review A, 22(3), 1159–1164.
2. Snitzer, E. (1961) 'Cylindrical dielectric waveguide modes', Journal of the Optical Society of America, 51(5), 491–498.
3. Mushiake, Y., Matsumura, K. and Nakajima, N. (1972) 'Generation of radially polarized optical beam mode by laser oscillation', Proceedings of the IEEE, 60(9), 1107–1109.
4. Pohl, D. (1972) 'Operation of a ruby laser in the purely transverse electric mode TE01', Applied Physics Letters, 20(7), 266–267.
5. Youngworth, K.S. and Brown, T.G. (2000) 'Focusing of high numerical aperture cylindrical-vector beams', Optics Express, 7(2), 77–87.
6. Richards, B. and Wolf, E. (1959) 'Electromagnetic diffraction in optical systems. II. Structure of the image field in an aplanatic system', Proceedings of the Royal Society A: Mathematical, Physical and Engineering Sciences, 253(1274), 358–379.
7. Beckley, A.M., Brown, T.G. and Alonso, M.A. (2012) 'Full Poincaré beams II: partial polarization', Optics Express, 20(9), 9357–9362.
8. Beckley, A.M., Brown, T.G. and Alonso, M.A. (2010) 'Full Poincaré beams', Optics Express, 18(10), 10777–10785.
9. Lou, K., Qian, S.X., Ren, Z.C., et al. (2013) 'Femtosecond laser processing by using patterned vector optical fields', Scientific Reports, 3, 2281.
10. Cai, M.Q., Tu, C.H., Zhang, H.H., et al. (2013) 'Subwavelength multiple focal spots produced by tight focusing the patterned vector optical fields', Optics Express, 21(25), 31469–31482.
11. Zhan, Q.W. (2009) 'Cylindrical vector beams: from mathematical concepts to applications', Advances in Optics and Photonics, 1(1), 1–57.
12. Rosales-Guzmán, C., Ndagano, B. and Forbes, A. (2018) 'A review of complex vector light fields and their applications', Journal of Optics, 20(12), 123001.
13. Liang, M.W., Lu, D.Z. and Ma, Y.G. (2024) 'Vectorial optical fields manipulation via metasurfaces', Opto-Electronic Engineering, 51(8), 240068.
14. Wang, X.L., Ding, J.P., Ni, W.J., et al. (2007) 'Generation of arbitrary vector beams with a spatial light modulator and a common path interferometric arrangement', Optics Letters, 32(24), 3549–3551.
15. Milione, G., Sztul, H.I., Nolan, D.A., et al. (2011) 'Higher-order Poincaré sphere, stokes parameters, and the angular momentum of light', Physical Review Letters, 107(5), 053601.

16. Holleczek, A., Aiello, A., Gabriel, C., et al. (2011) 'Classical and quantum properties of cylindrically polarized states of light', Optics Express, 19, 9714–9736.
17. Chen, S.Z., Zhou, X.X., Liu, Y.C., et al. (2014) 'Generation of arbitrary cylindrical vector beams on the higher order Poincaré sphere', Optics Letters, 39(18), 5274–5276.
18. Liu, Y.C., Ling, X.H., Yi, X.N., et al. (2014) 'Realization of polarization evolution on higher-order Poincaré sphere with metasurface', Applied Physics Letters, 104(19), 191110.
19. Yi, X., Liu, Y., Ling, X., et al. (2015) 'Hybrid-order Poincaré sphere', Physical Review A, 91, 023801.
20. Ren, Z.C., Kong, L.J., Li, S.M., et al. (2015) 'Generalized Poincaré sphere', Optics Express, 23(20), 26586–26595.
21. Yue, Z., Liu, J., Li, J., et al. (2022) 'Vector beam generation based on spin-decoupling metasurface zone plate', Applied Physics Letters, 120, 191704.
22. Yoon, G., Jang, J., Mun, J., et al. (2019) 'Metasurface zone plate for light manipulation in vectorial regime', Communications Physics, 2, 156.

CHAPTER 7

Polarization Manipulation in 3D Space

Abstract

The preceding chapters primarily focused on polarization control of light in the transverse plane, where the mentioned polarization modulation was confined to the two-dimensional plane (i.e. the *x*-*y* plane), and the polarization state remained unchanged at each output plane along the propagation path. However, driven by the demands of optical manipulation and related applications, new spatial degrees of freedom have become increasingly important, and various meticulously designed optical fields with specific longitudinal variations have been continuously proposed. This chapter will further introduce the principles and device designs for photonic polarization manipulation along the propagation path (the *z*-axis), i.e. longitudinal polarization control of optical fields. This extends polarization manipulation from the two degrees of freedom (*x*-*y*) to three degrees of freedom (*x*-*y*-*z*), thereby enabling polarization control in three-dimensional space. The essence of three-dimensional spatial polarization control lies in managing the phase retardation between the left-handed circularly polarized (LCP) component and the right-handed circularly polarized (RCP) component in the incident wave, allowing the LCP and RCP components to converge along the *z*-axis with distinct phase delays. Along the *z*-axis propagation path, the phase difference between the LCP and RCP components varies with propagation distance, ultimately modifying the polarization state at every output plane along the propagation path.

7.1 OVERVIEW OF 3D SPATIAL POLARIZATION MANIPULATION

In recent years, beyond in-plane polarization structure manipulation, the control of polarization structures along the propagation direction has gradually gained attention. This is particularly true for vector optical fields formed by coherent superposition of different

DOI: 10.1201/9781003784821-7

eigenmodes, whose intensity, phase, and polarization structures all change during propagation. Understanding the evolution patterns of such optical field structures and developing methods to regulate this transmission evolution can facilitate the discovery of more potential applications. Whether vector optical fields are generated through coherent superposition of orthogonal polarization modes or by using polarization conversion elements, their polarization modulation mechanism can be understood by decomposing the incident optical field into left-handed circularly polarized (LCP) and right-handed circularly polarized (RCP) components. Assuming the incident scalar field is $\boldsymbol{E}_{\mathrm{i}}$, with LCP and RCP (spin) states represented as $E_{\mathrm{L,R}}\boldsymbol{e}_{\mathrm{L,R}}$, where $E_{\mathrm{L,R}}$ denote the complex amplitude distributions of the two spin states of the incident field, and $\boldsymbol{e}_{\mathrm{L,R}} = (1, \mp i)^T / \sqrt{2}$ are the unit vectors for the LCP and RCP states, respectively. After passing through a polarization modulation system $\boldsymbol{M}$, the output optical field is $\boldsymbol{E}_{\mathrm{o}} = \boldsymbol{M}\boldsymbol{E}_{\mathrm{i}} = E_{\mathrm{R}}\mathrm{e}^{i2\varphi}\boldsymbol{e}_{\mathrm{L}} + E_{\mathrm{L}}\mathrm{e}^{-i2\varphi}\boldsymbol{e}_{\mathrm{R}}$. In this polarization conversion process, the two spin states respond differently to the system, acquiring phases of $2\varphi(x, y)$ and $-2\varphi(x, y)$, respectively. This is the Pancharatnam-Berry (PB) phase.

Utilizing the PB phase allows not only wavefront shaping of the two spin components of the optical field but also control over their propagation. When the PB phase is chosen as a plane-wave phase, i.e. $2\varphi(x, y) = k_x x$, where k_x represents the wavevector component along the x-direction, the LCP and RCP components of the incident scalar field acquire tilted phases of $k_x x$ and $-k_x x$, respectively. Under this phase modulation, these components separate transversely during propagation – a phenomenon known as the photonic spin Hall effect [1]. In 2002, Bomzon et al. [2] proposed spatially varying subwavelength gratings and observed their polarization selectivity, whereby circularly polarized beams of opposite handedness diffract in different directions. In 2013, Yin et al. [3] realized an enhanced photonic spin Hall effect using optical metasurfaces composed of V-shaped gold nanoantennas with a transversely varying period. Linearly polarized light incident on such a metasurface acquires opposite transverse phase gradients for its two spin components, leading to transverse separation. In 2015, Ling et al. [4] used femtosecond laser writing to create periodically curved nanogratings in glass, achieving a significantly enhanced photonic spin Hall effect. If the incident optical field is chosen to be a vortex beam, Hermite-Gaussian beam, Airy beam, or even a field with an arbitrary intensity distribution, the output field can be split into two beams with identical complex amplitude distributions but opposite circular polarizations (LCP and RCP) [5].

When the PB phase is selected as a quadratic phase, there is $2\varphi(x, y) = \alpha r^2$, where α is a constant controlling the divergence or convergence of the optical field. If the left- and right-handed spin components of the beam are multiplied by phase factors $\exp(i\alpha r^2)$ and $\exp(-i\alpha r^2)$, respectively (meaning the left-handed component diverges while the right-handed component converges), then the two components will undergo spin-dependent focusing. In 2003, Hasman et al. [6] used radially varying nanogratings to realize a polarization-dependent dual-focus lens with focal lengths of $\pm f$. In 2014, Gu et al. [7] observed this dual-focus phenomenon in radially polarization-varying optical fields. In 2016, Liu et al. [1] modulated the polarization conversion characteristics of a beam using a Sagnac interferometer. By controlling the PB phase, they achieved not only transverse and longitudinal separation of the two spin states but also arbitrary three-dimensional spatial separation of spin states. For example, when the PB phase was set as $2\varphi(x, y) = k_x x + k_y y + \alpha r^2$, the two spin states focused at different transverse and longitudinal coordinates after passing through the polarization conversion system. Furthermore, the PB phase can carry a helical phase, thereby generating spin-dependent orbital angular momentum (OAM) optical fields with separated spin components. Ke et al. [8] utilized this spin-dependent focusing/defocusing effect to propose a photonic spin filter. The filter consists of two confocal metalenses. Under the action of the first lens, the two spin states exhibit convergent and divergent behaviour, respectively. A small hole in the Fourier plane performs low-pass filtering to remove the divergent spin component. The second metalens then reconstructs the transmitted spin state. Wang et al. [9] implemented spin-dependent focusing separation using metasurfaces and further applied it to achieve polarization-selective imaging.

In the aforementioned optical fields, due to spin separation, the polarization distribution changes with propagation distance. Notably, the intensity distribution structure of the optical field also undergoes significant changes during this process. If special optical fields with stable mode characteristics and longitudinally continuous polarization variation could be generated [10–17], this would undoubtedly further expand the application areas of vector optical fields. For example, such fields could modify the absorption peaks at different longitudinal positions in polarization-dependent media or excite specific nonlinear refractive index distributions. In 2013, Cardano et al. [18] generated Poincaré beams by coaxially superposing orthogonally polarized Gaussian and Laguerre-Gaussian (LG) beams. Due to the different mode characteristics of the two mode parameters, their Gouy phase shift difference varies with propagation distance. Within the Rayleigh range, as the light propagates, the polarization structure of the superimposed Poincaré beam rotates spatially, exhibiting multi-type topological structure rotation. In 2015, Moreno et al. [19] used Bessel beams to generate longitudinally modulated polarization structures that rotate linearly with propagation, creating diffraction-free fields with controlled longitudinal polarization. By using a conical birefringent prism to provide radially varying phase retardation, a radially varying polarization state was achieved. Then, leveraging the transverse-to-longitudinal mapping inherent in Bessel beam generation via an axicon, polarization control at different propagation distances was realized, i.e. a scalar diffraction-free field with longitudinally modulated polarization. Further, by combining a polarization conversion system with an axicon, radially linear phases with different slopes were introduced for the two spin components, generating vector Bessel beams whose polarization state and polarization order vary with propagation distance [20].

In 2016, Li et al. [21] proposed a diffraction-free optical field with longitudinally modulated polarization states, where the polarization state can vary along any arbitrary trajectory on the Poincaré sphere during propagation. This was achieved by spectrally modulating the longitudinal amplitude relationship of two orthogonal polarization components. According to the Poincaré sphere representation of polarization states, each point on the sphere corresponds to a specific polarization state, which can be decomposed into any two orthogonal polarization components symmetric about the sphere's centre. In other words, coaxially superposing two orthogonal polarization components with any amplitude ratio and phase difference yields the polarization state represented by any point on the Poincaré sphere. In 2020, by leveraging the Gouy phase shift characteristics of radially higher-order LG modes and Bessel modes, nonlinear longitudinal transformation of polarization states was further achieved [22]. Research shows that radially higher-order LG optical fields not only exhibit highly similar mode field distributions to Bessel beams of the same order but also possess quasi-diffraction-free properties, providing a foundation for transmission stability of the optical field mode. Furthermore, LG optical fields have Gouy phase shifts significantly different from those of Bessel optical fields. When radially higher-order LG optical fields are employed, the Gouy phase shift accumulates nonlinearly with increasing propagation distance. As the azimuthal order increases, this nonlinear accumulation intensifies. Thus, parameters such as polarization order and radial order can be adjusted to control the nonlinear variation rate.

The lens, as the most commonly used two-dimensional Fourier transform system, has been extensively studied for longitudinal control of optical fields near its focal plane. The use of high numerical aperture (NA) objective lenses for tight focusing has become a widely adopted experimental model. This approach not only allows for the introduction of substantial longitudinal phase variation rates but also induces strong longitudinal field components through polarization conversion under tight focusing conditions. Consequently, extensive research has been conducted on the longitudinal and three-dimensional structural control of intensity and polarization states in tightly focused fields [23–31]. In 2005, Zhao et al. [23] pioneered research on longitudinal control of focal fields by modulating the phase of incident radially polarized light using an annular diffractive optical element (DOE). This DOE consisted of three concentric annular zones, with the innermost and outermost rings being transparent, while the middle ring was opaque. Due to the Gouy phase shift between the focused fields generated by the two transparent rings, the amplitude/intensity of the coherent field varied significantly along the optical axis near the focal plane. Meanwhile, owing to tight focusing, the focal field exhibited rapid phase variations in the transverse plane as well. As a result, its three-dimensional intensity distribution displayed a quasi-periodic chain-like structure along the axial direction, forming what is known as an "optical chain". By adjusting the phase difference between the two transparent rings, movement and directional control of the optical chain could be achieved.

Subsequently, Kozawa and Sato [24] utilized cylindrical vector beams of higher-order radial modes to generate a longitudinal dark region on the order of a wavelength in length, referred to as an "optical cage". In 2009, Wang et al. [25] achieved controllable optical cage generation through tight focusing of dual-mode cylindrical vector beams. By modulating the local polarization direction of the incident dual-mode cylindrical vector beam, the shape of the optical cage could be altered. In 2015, Nie et al. [26] introduced a dual-ring vortex DOE to control azimuthally polarized Bessel-Gaussian beams. Under high-NA focusing, by adjusting the radius of each ring of the DOE, both purely transversely polarized optical chains and sub-diffraction-limited purely longitudinal "magnetic chains" with single channels could be simultaneously generated in the focal region. Furthermore, using cascaded DOEs, periodic bottle-shaped magnetic channels and longitudinal magnetic chains with dual channels could be produced. Such tunable hollow-structured focal fields have demonstrated significant potential for applications in particle trapping and directional transport.

Compared to a single lens, the 4π focusing system using dual lenses provides richer optical field modes for coherent superposition. For example, in 2012, Chen and Zhao [27] utilized 4π focusing of radially polarized vortex beams to achieve focal fields with special longitudinal intensity distributions under tight focusing conditions. This was realized by controlling the radial transmittance of the incident optical field and the polarization modes of the two counter-propagating focused beams. When first-order radially polarized LG beams with opposite polarization directions were focused, a spherical hollow focal field with dimensions of 0.32λ (transverse) and 0.34λ (longitudinal) was generated. When second-order radially polarized LG beams with opposite polarization directions were incident, the resulting hollow beam's dark channel could extend up to 30λ in length. Furthermore, Yu et al. [28] systematically studied the formation mechanism of multi-focus arrays with uniform axial intensity under 4π focusing conditions. They established an inversion mechanism for the required optical field structure at the entrance pupil to construct such focal fields. In contrast to the cases above, focal fields with uniform axial intensity hold significant application value in areas such as super-resolution imaging and microfabrication. Therefore, suppressing axial structural variations and constructing ultra-long "optical needle" fields with uniform intensity [29,30] has long been a key focus in longitudinal control of focal fields. In 2008, Wang et al. [31] pioneered the generation of a diffraction-free "optical needle" with a longitudinal length of approximately 4λ and a transverse full width at half maximum (FWHM) of about 0.43λ. This was achieved under tight focusing conditions using a phase-modulated DOE with an incident radially polarized Bessel-Gaussian beam. The DOE was divided into five concentric annular zones, each with an amplitude transmittance of 1, and adjacent rings exhibited a π-phase retardation. Modulating the radially polarized optical field with this DOE not only effectively suppressed the divergence behaviour of the tightly focused field near the focal plane but also enabled nearly complete conversion of the focal field's polarization state into longitudinal components.

With the advancement of meta-photonics, longitudinal polarization control of optical fields has ushered in new opportunities. A single metasurface can now integrate functionalities such as focusing, vector field generation, and longitudinal polarization manipulation [11,32,33], which is beyond the capabilities of traditional single optical elements. Most metasurfaces share a common limitation: their Jones transfer matrix designs focus on amplitude and phase variations for different polarization components within the x-y plane [34–36], often neglecting the factor of spatial propagation distance. Based on this premise, the phase variation of the transmitted wave with increasing propagation distance remains fixed. Consequently, traditional electromagnetic metasurfaces are typically confined to manipulating polarization states in two-dimensional (x-y) space, where the polarization state remains unchanged across each output plane along the propagation path. To achieve polarization manipulation in three-dimensional space, it is necessary to modify the polarization states of electromagnetic waves at different output planes along the propagation path (z-axis). The core concept is as follows: first, decompose the incident uniform scalar optical field into orthogonal circularly polarized components; then, independently modulate the spatiotemporal characteristics of each circularly polarized component to control their phase difference as it varies along the spatial propagation path; finally, recombine them to obtain different polarization states at different output planes.

This type of longitudinally variable polarization metasurface device offers users more flexible polarization customization schemes and has the potential to advance various applications. Some electromagnetic-responsive media (e.g. electro-optic crystals) are sensitive to both intensity and polarization of light. Such meta-devices can be used to spatially adjust optical waves, thereby modifying the stimulated intensity and modes in the medium, ultimately yielding different output information. In the field of communication, the longitudinally variable polarization property could enhance confidentiality, as intercepted polarization information differs at various positions along the propagation path, adding complexity to decryption. Moreover, the ability to achieve different polarization states at the terminal by adjusting the device distance itself could serve as a novel information transmission method. Since polarization transformation is a function of propagation distance, this feature may have potential applications in electromagnetic wave radar systems. Such radar systems could emit longitudinally variable polarized electromagnetic waves towards targets and identify target motion by detecting changes in the polarization of reflected waves. Additionally, light propagation is linked to the refractive index of the medium. By placing an additional medium along the propagation path, polarization states at a specific output plane could be altered, enabling new sensing schemes for detecting unknown refractive indices.

The concept of decomposing the incident optical field into circularly polarized components, independently modulating them, and then recombining them can be realized through spin-decoupling and dual-meta-atom interference techniques. In the following sections, this chapter will demonstrate the principles and device designs for three-dimensional polarization manipulation based on spin-decoupling and dual-meta-atom interference, respectively.

7.2 PRINCIPLE AND DESIGN OF 3D SPATIAL POLARIZATION MANIPULATOR

7.2.1 The 3D Spatial Polarization Manipulation Based on Spin-Decoupling

This section will demonstrate the manipulation of longitudinal vector optical fields based on spin-decoupling [37], achieving control over polarization states across the three dimensions (x-y-z). Two representative functionalities are designed: first, the generation of a vector field that varies continuously along the propagation direction; second, the independent control of polarization distributions across multiple focal planes. The former requires a design with a long depth of focus, while the latter necessitates multi-plane focusing. The device schematics and their functionalities are illustrated in Fig. 7.1(a, b). The designed fundamental principle of the device is to independently tailor the phase for orthogonal circularly polarized waves (LCP and RCP) by modifying the structural dimensions and spatial orientation of the meta-atoms, while introducing a phase difference that varies along the propagation direction.

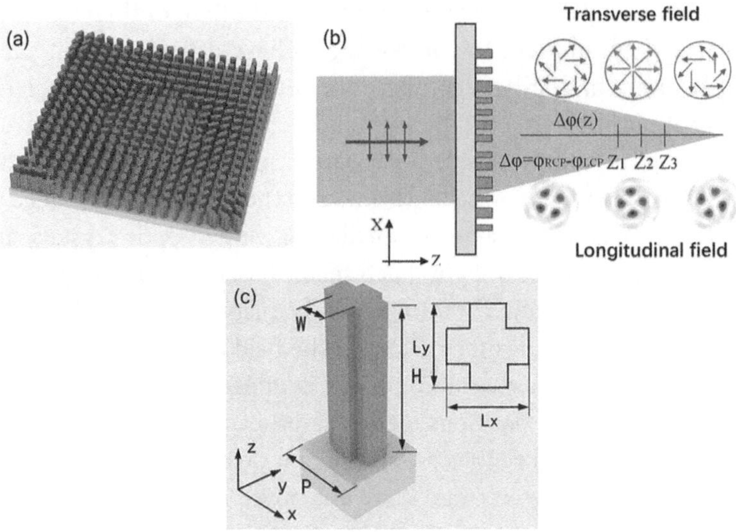

FIGURE 7.1 Longitudinal vector light field manipulation based on spin-decoupling: schematic diagram for (a) device, (b) function, and (c) the designed unit cell structure.

The polarization state of light in any *x-y* cross section (with light propagating along the *z*-axis) can be described using a 2 × 1 Jones matrix. For a vortex-type vector optical field within the *x-y* plane, assuming an initial phase of zero, the normalized electric field can be expressed as:

$$\mathbf{E}(r,\theta)=\begin{bmatrix}\cos(l\theta)\\ \sin(l\theta)\end{bmatrix}=\frac{1}{2}e^{il\theta}\begin{bmatrix}1\\ -i\end{bmatrix}+\frac{1}{2}e^{-il\theta}\begin{bmatrix}1\\ i\end{bmatrix}, \tag{7.1}$$

where the radial coordinate r is a position vector pointing to a specific polarization state, and its angle to the *x*-axis is θ, referred to as the azimuthal angle. Through Euler's formula, Eq. (7.1) decomposes a vector field within the *x-y* plane into a coherent superposition of LCP and RCP components. It can be observed that these components possess phase factors only related to the azimuthal angle θ and the topological charge l within the *x-y* plane, lacking any phase factor associated with the *z*-axis. Therefore, the electric field described by Eq. (7.1) remains identical across different output planes along the propagation path. To create a device where the polarization state varies along the propagation path, the values of the two elements in the 2 × 1 matrix must be modified at each output plane. This requires the introduction of a phase factor associated with the *z*-axis into Eq. (7.1), i.e. introducing a longitudinal phase factor. The introduction of this longitudinal phase factor inherently causes the LCP and RCP components to acquire an additional phase difference related to the *z*-axis, denoted as $\Delta\varphi(z)$. Assuming the phase delay of the LCP component is greater than that of the RCP component, this phase difference $\Delta\varphi(z)$ is appended to the LCP component. Equation (7.1) thus transforms into:

$$\begin{aligned}\mathbf{E}(\theta) &= \frac{1}{2}e^{i(l\theta+\Delta\varphi(z))}\begin{bmatrix}1\\-i\end{bmatrix}+\frac{1}{2}e^{-il\theta}\begin{bmatrix}1\\i\end{bmatrix}\\ &= \frac{1}{2}e^{i(l\theta+\Delta\varphi(z)/2)+i\Delta\varphi(z)/2}\begin{bmatrix}1\\-i\end{bmatrix}+\frac{1}{2}e^{-i(l\theta+\Delta\varphi(z)/2)+i\Delta\varphi(z)/2}\begin{bmatrix}1\\i\end{bmatrix}.\\ &= e^{i\Delta\varphi(z)/2}\begin{bmatrix}\cos(l\theta+\Delta\varphi(z)/2)\\ \sin(l\theta+\Delta\varphi(z)/2)\end{bmatrix}\end{aligned} \tag{7.2}$$

From Eq. (7.2), it can be seen that since $\Delta\varphi(z)$ is a quantity that varies with z, different polarization distributions can be obtained at different output planes along the propagation path. Consequently, the manipulation of vector optical fields is extended from the two-dimensional x-y plane to the three-dimensional x-y-z space.

As derived in the previous chapter, the phase condition for achieving spin-decoupling using a single birefringent meta-atom is

$$\begin{cases}\theta=(\varphi_{\mathrm{L}}-\varphi_{\mathrm{R}})/4\\ \varphi_{xx}=(\varphi_{\mathrm{L}}+\varphi_{\mathrm{R}})/2\\ \varphi_{yy}=(\varphi_{\mathrm{L}}+\varphi_{\mathrm{R}})/2-\pi\end{cases}, \tag{7.3}$$

where θ represents the rotation angle of the meta-atom. The φ_{R} and φ_{L} denote the phase delays for RCP and LCP waves, respectively. The φ_{xx} and φ_{yy} represent the co-polarized transmission phase delays for x- and y-polarized waves when the fast and slow axes of the birefringent meta-atom are aligned with the x- and y-coordinate axes, and $\varphi_{xx} - \varphi_{yy} = \pi$. Equation (7.3) indicates that once the independent φ_{R} and φ_{L} are designed, the θ, φ_{xx}, and φ_{yy} can be determined accordingly.

Two metasurfaces are designed here, each demonstrating a distinct functionality: the generation of continuously varying vector beams along the propagation direction and the independent control of polarization states on multiple focal planes (Device 1 and Device 2, respectively). Both devices utilize high-resistivity silicon with a relative permittivity of 11.9 and are simulated using the time-domain solver in CST Microwave Studio. Figure 7.1(c) illustrates the designed unit cell, which features a substrate thickness of 300 μm and a periodicity of P = 140 μm. The geometric structure of the unit cell is a cross-shaped dielectric pillar with a height of H = 200 μm, a rectangular width of W = 40 μm, and lengths in two orthogonal directions denoted as Lx and Ly. The specific values of the dimensional parameters Lx and Ly are determined through parameter sweeping. Figure 7.2 presents the simulated electric field transmission amplitude and phase variations for different Lx and Ly values. It can be observed that, within the size variation range of 40–140 μm, the transmission amplitude exceeds 0.7 in most regions (Fig. 7.2(a, b)). Moreover, the phases φ_{xx} and φ_{yy} can cover the 0–2π range while maintaining a π phase difference between them (Fig. 7.2(c, d)). Consequently, several sets of Lx and Ly parameters that meet these requirements can be selected as the fundamental unit cells for device design.

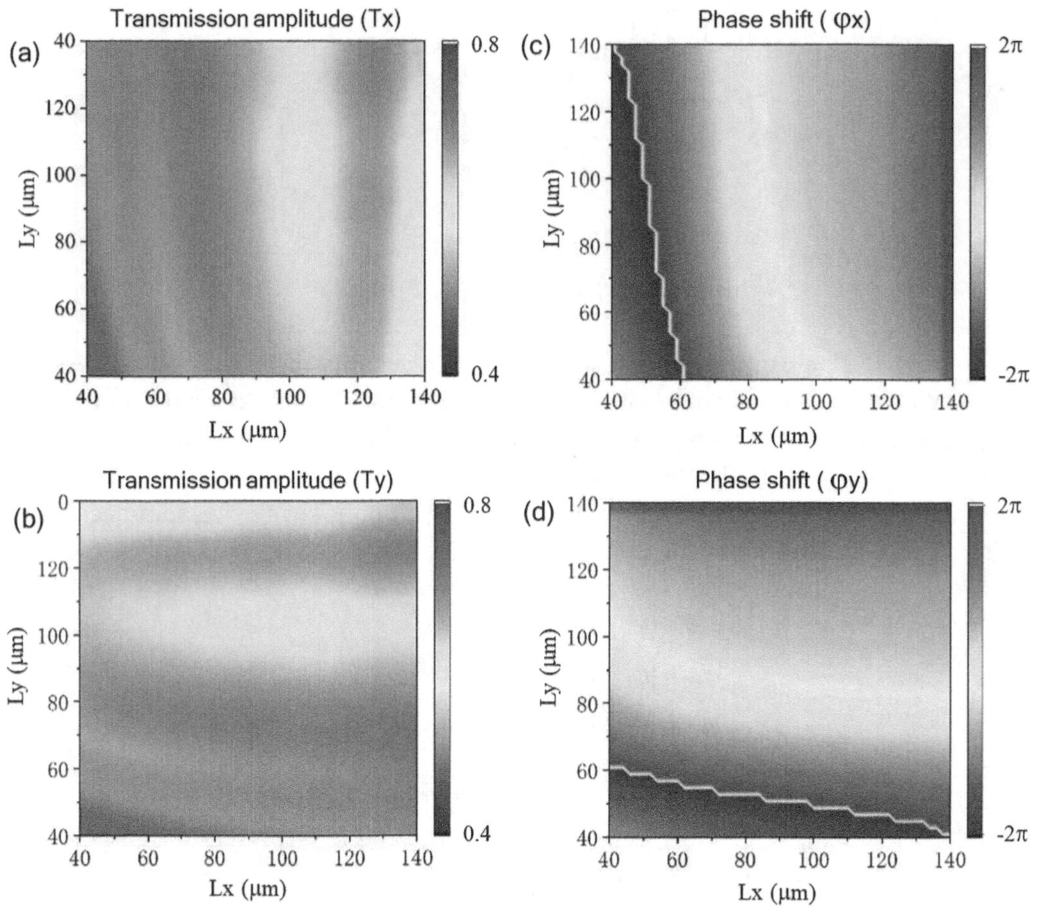

FIGURE 7.2 Sweeping results for the dimensional parameters *Lx* and *Ly*; (a) and (b), respectively, show the electric field transmission amplitude variations for the *x*- and *y*-polarized components under different size parameters; (c) and (d) display the corresponding phase variations.

For Device 1, which is designed to generate a vector field that continuously varies along the propagation direction, it is required to achieve an axially continuous phase difference $\Delta\varphi(z)$ while simultaneously controlling the three electric field components. Therefore, a focusing phase profile with a long focal depth is selected, and different focal lengths and positions are assigned to the RCP and LCP components. The phase distributions for the two channels are given as follows [38]:

$$\begin{cases} \varphi(r,\theta)_{\text{RCP}} = \dfrac{k \cdot r^2}{2\left[f_1 + \left(\Delta f_1 \cdot r^2 / R^2\right)\right]} + l\theta \\ \varphi(r,\theta)_{\text{LCP}} = \dfrac{k \cdot r^2}{2\left[f_2 + \left(\Delta f_2 \cdot r^2 / R^2\right)\right]} - l\theta \end{cases}, \tag{7.4}$$

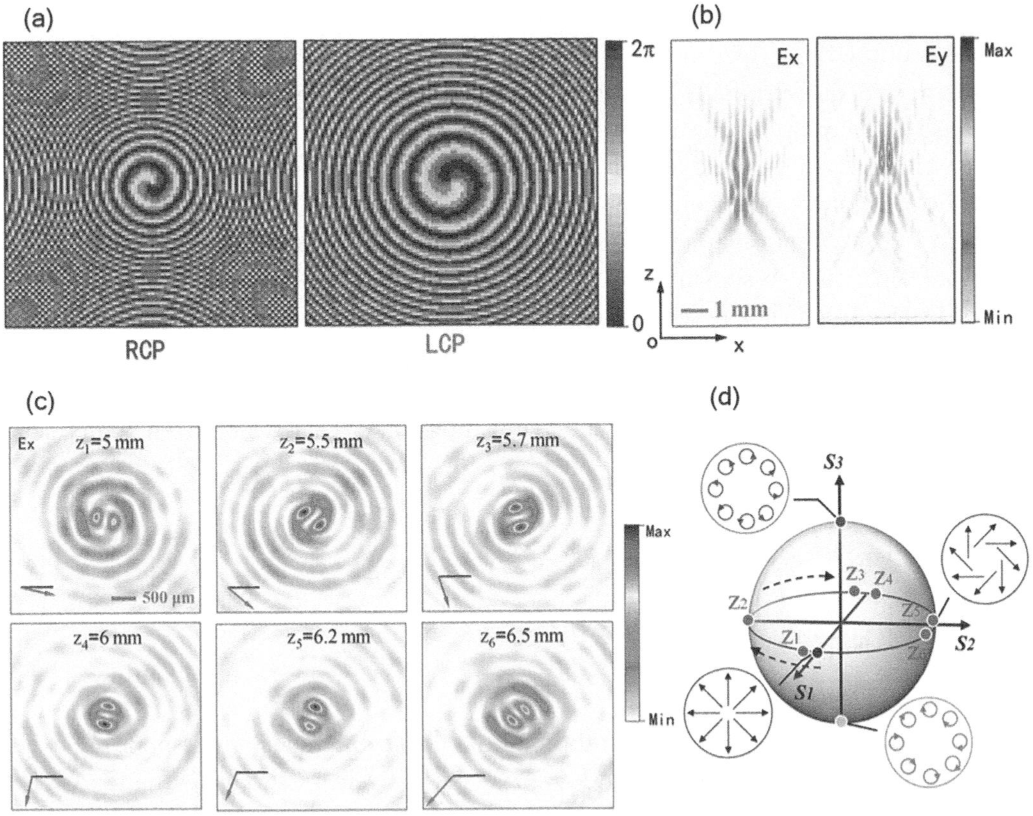

FIGURE 7.3 Device 1 performs the function of generating a continuously varying vector field along the propagation direction: (a) shows the phase configurations for the RCP and LCP waves; (b) displays the simulated electric field distributions of the *x*- and *y*-polarized components in the *x-z* plane; (c) presents the simulated transverse electric field components (*Ex*, *Ey*) at six different positions; and (d) illustrates the vector polarization states at these six positions on the Poincaré sphere.

where k is the free-space wavenumber, R is the radius of the metasurface, and the device consists of 90 × 90 unit cells. The f_i and Δf_i are the designed focal position and focal depth, respectively, with $f_1 = 6$ mm, $\Delta f_1 = 1.5$ mm, $f_2 = 4$ mm, $\Delta f_2 = 2$ mm. The l is the topological charge of the phase, designed as $l_{RCP} = 1$ and $l_{LCP} = -1$. The obtained phase distributions for the RCP and LCP waves are shown in Fig. 7.3(a). Using a full-wave simulation method, the three-dimensional transmitted optical field under *x*-polarized plane wave incidence is obtained. In the *x-z* plane, the electric field distributions of the *x*- and *y*-polarized components are shown in Fig. 7.3(b), with an operating frequency of 1 THz. It can be observed that the focal length of the *y*-polarized component is significantly elongated compared to that of the *x*-polarized component. As shown in Fig. 7.3(c), six different positions are selected to display the transverse electric field components (*Ex*, *Ey*), further illustrating the variation of the vector beam along the propagation direction. The arrows in the lower left corner of Fig. 7.3(c) indicate the orientation of the intensity, and the black solid lines serve as a reference. Clearly, as the observation distance from the device gradually increases, the intensity orientation of Ex rotates, indicating a gradual change in the polarization distribution of the three-dimensional vector optical field. This occurs because the designed device imparts different focal lengths and focal depths to the RCP and LCP components, meaning that the vortices accumulate a z-dependent phase difference along distinct paths during propagation. A more intuitive representation of the vector field evolution is provided by the higher-order Poincaré sphere. The vector polarization states at the six positions in Fig. 7.3(c) are mapped onto the Poincaré sphere, represented as points on the equator from z_1 to z_6, as shown in Fig. 7.3(d), where the polarization topological charge is 1.

For Device 2, which is designed to independently control the polarization distribution of the light field on multiple focal planes, it is necessary to design distinct phase differences $\Delta\varphi(z)$ at several discrete z-coordinates. Taking vertical dual-point focusing as an example, the designed phase distributions for the two orthogonal circularly polarized components in the metasurface are as follows:

$$\begin{cases} \varphi(r,\theta)_{\text{RCP}} = \arg\left(e^{k\cdot\sqrt{r^2+f_3^2}-l\theta} + e^{k\cdot\sqrt{r^2+f_4^2}-l\theta} \right) \\ \varphi(r,\theta)_{\text{LCP}} = \arg\left(e^{k\cdot\sqrt{r^2+f_3^2}+l\theta+\Delta\varphi_1} + e^{k\cdot\sqrt{r^2+f_4^2}+l\theta+\Delta\varphi_2} \right) \end{cases}, \tag{7.5}$$

where $f_3 = 5$ mm and $f_4 = 8.5$ mm, and the initial phase differences corresponding to the two focal planes are $\Delta\phi_1 = \pi/2$ and $\Delta\phi_2 = -\pi/2$, respectively. In this case, the topological charges of the phase for the RCP and LCP waves are $l_{\text{RCP}} = -1$ and $l_{\text{LCP}} = 1$. Figure 7.4(a)

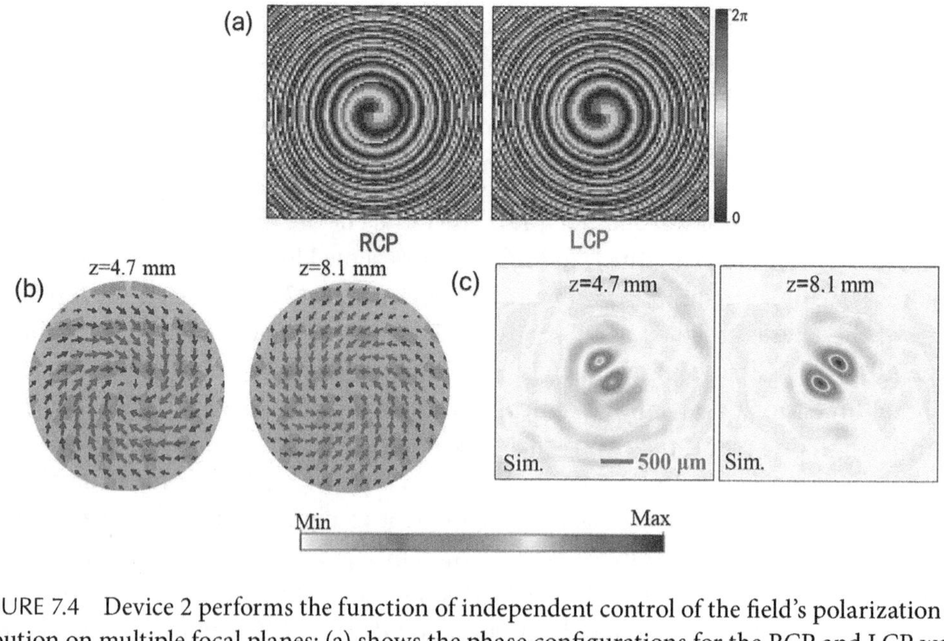

FIGURE 7.4 Device 2 performs the function of independent control of the field's polarization distribution on multiple focal planes: (a) shows the phase configurations for the RCP and LCP waves; (b) displays the simulated electric vector field distributions of the transmitted beam at the two output planes $z_1 = 4.7$ mm and $z_2 = 8.1$ mm; and (c) presents the corresponding electric field intensity results for the x-component.

illustrates the phase profiles of the RCP and LCP components calculated according to Eq. (7.5), showing opposite directions of phase variation in the two channels. Under the illumination of an x-polarized plane wave, the transmitted field of Device 2 is simulated. The vector field distributions of the transmitted beam at two planes $z_1 = 4.7$ mm and $z_2 = 8.1$ mm are presented in Fig. 7.4(b), while the electric field intensity of the x-component is shown in Fig. 7.4(c). As can be observed, the two radial vector fields exhibit polarization distributions corresponding to vortices of opposite handedness.

7.2.2 The 3D Spatial Polarization Manipulation Based on Dual-Meta-Atom Interference

Another more intuitive approach involves utilizing the concept of dual-meta-atom interference introduced in previous chapters. First, two metasurfaces are designed to independently manipulate LCP and RCP waves. These two metasurfaces are then simply combined according to their respective required phase distributions. This section will demonstrate longitudinal polarization control based on dual-meta-atom interference [39], with the desired functionality designed to achieve linear polarization rotation along the propagation path.

To realize linearly polarized states that rotate along the propagation path, it is necessary to revisit the composition and decomposition process between linear and circularly polarized waves. Assume an x-linearly polarized wave incident on an element is first decomposed into an LCP wave and an RCP wave. These waves then recombine during propagation to form a linearly polarized wave. If the effect of propagation (along the z-axis) is neglected – meaning the LCP and RCP waves experience identical phase delays at every point along the z-axis, denoted by a phase angle φ_z, and the normalized output wave Jones matrix can be expressed as:

$$\mathrm{LCP}+\mathrm{RCP}=\frac{1}{2}e^{i\varphi_z}\begin{pmatrix}1\\-i\end{pmatrix}+\frac{1}{2}e^{i\varphi_z}\begin{pmatrix}1\\i\end{pmatrix}=e^{i\varphi_z}\begin{pmatrix}1\\0\end{pmatrix}. \tag{7.6}$$

Equation (7.6) represents the simplest and most well-known form of combining circular polarizations into linear polarization. It shows that LCP and RCP waves combine to form an x-linearly polarized wave, and the polarization state remains identical at every point along the propagation path – a very ordinary scenario. Next, considering the effect of propagation, where LCP and RCP waves experience different phase accumulations along the z-axis, assume the phase angles of the two circularly polarized waves at a point z_0 differ by $\Delta\varphi$, with phase delays of $\exp(i\varphi z_0)$ and $\exp(i\varphi z_0+\Delta\varphi)$, respectively. Equation (7.6) then transforms into:

$$\begin{aligned}\mathrm{LCP}\big|_{z0}+\mathrm{RCP}\big|_{z0}&=\frac{1}{2}e^{i(\varphi_{z0}+\Delta\varphi)}\begin{pmatrix}1\\-i\end{pmatrix}+\frac{1}{2}e^{i\varphi_{z0}}\begin{pmatrix}1\\i\end{pmatrix}\\&=\frac{1}{2}e^{i\varphi_{z0}}e^{i\Delta\varphi/2}\left[e^{i\Delta\varphi/2}\begin{pmatrix}1\\-i\end{pmatrix}+e^{-i\Delta\varphi/2}\begin{pmatrix}1\\i\end{pmatrix}\right].\\&=e^{i\varphi_{z0}}e^{i\Delta\varphi/2}\begin{pmatrix}\cos(\Delta\varphi/2)\\\sin(\Delta\varphi/2)\end{pmatrix}\end{aligned} \tag{7.7}$$

The above equation represents a linear polarization at an angle of $\Delta\varphi/2$ relative to the x-axis. This can be viewed as an x-polarization rotated by $\Delta\varphi/2$ within the x-y plane. This implies that applying a phase difference between the LCP and RCP waves causes the combined linear polarization to deflect.

Clearly, if $\Delta\varphi$ is a constant value independent of z, the element values in the Jones matrix in Eq. (7.7) cannot be modified longitudinally, meaning the polarization state at each output plane remains identical. To make the linear polarization direction described by Eq. (7.7) continuously vary with light propagation, $\Delta\varphi$ must necessarily be a function of z, i.e. $\Delta\varphi = \Delta\varphi(z)$. This will be achieved by designing the optical paths for the LCP and RCP waves. As shown in Fig. 7.5(a), each unit cell can transmit LCP and RCP wave components with different phase delays to points along the z-axis. Every point within a segment of the z-axis collects LCP and RCP waves generated by different unit cells, and these circularly polarized waves recombine to form a linearly polarized wave. For example, in Fig. 7.5(a), by controlling the phase delays of LCP and RCP waves, the unit cells on the metasurface can transmit RCP waves (light lines with an angle of θ_R with the z-axis) and LCP waves (light lines with an angle of θ_L with the z-axis) to points z_2 and z_3 on the z-axis, respectively. The LCP wave from one unit cell and the RCP wave from another unit cell converge at z_3 to form a linearly polarized wave. The propagation paths for all LCP waves have the same slope, forming an angle θ_L to the z-axis. Similarly, all RCP wave paths have

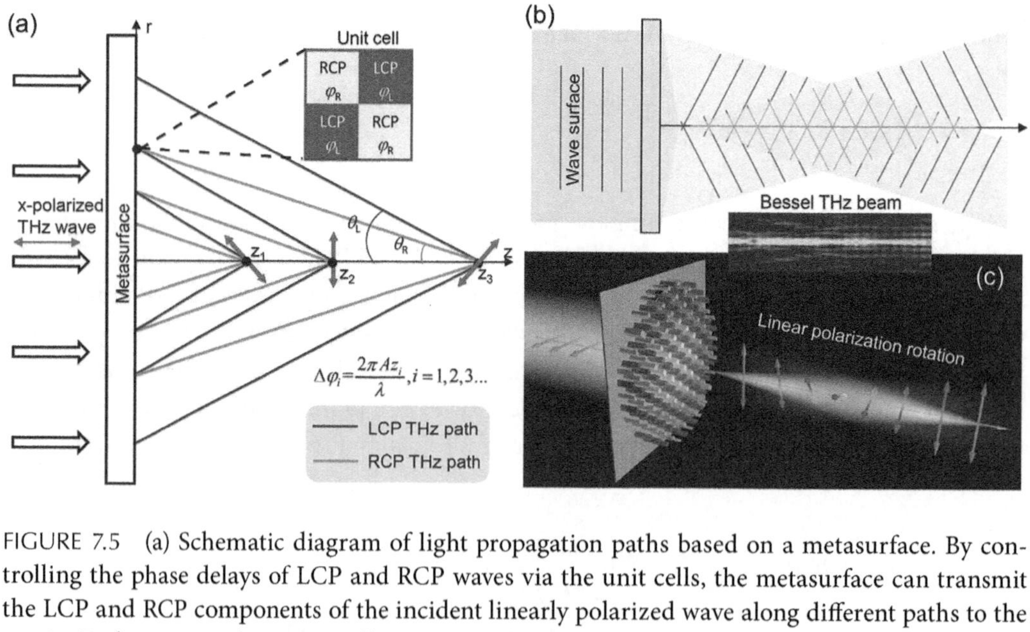

FIGURE 7.5 (a) Schematic diagram of light propagation paths based on a metasurface. By controlling the phase delays of LCP and RCP waves via the unit cells, the metasurface can transmit the LCP and RCP components of the incident linearly polarized wave along different paths to the z-axis. Each point on the z-axis collects LCP and RCP waves generated by different unit cells, and these two circularly polarized waves recombine to form a linearly polarized wave. The orientation angle of the output linear polarization varies with propagation distance. (b) The output wave forms a Bessel beam. (c) Functional schematic of the longitudinally polarization-variable device. Here, the output linearly polarized wave rotates along the propagation path.

the same slope, forming an angle θ_R to the z-axis, and $\theta_L \neq \theta_R$. Therefore, based on geometric relationships, the optical path lengths for the LCP and RCP waves reaching a point on the z-axis are $z/\cos\theta_L$ and $z/\cos\theta_R$, respectively. Consequently, the phase difference between the two waves is:

$$\Delta\varphi(z) = \varphi_{LCP} - \varphi_{RCP} = \frac{2\pi Az}{\lambda}, \tag{7.8}$$

where λ is the operating wavelength, and A is a constant given by $A = 1/\cos\theta_L - 1/\cos\theta_R$. The $\Delta\varphi(z)$ varies monotonically with z, causing the linear polarization described by Eq. (7.7) to gradually rotate as light propagates. The designed wavefront propagations for the LCP and RCP waves are shown in Fig. 7.5(b), where they form Bessel beams. The final functionality of this meta-device can be described as shown in Fig. 7.5(c): when a linearly polarized beam illuminates the device, the transmitted wave forms a Bessel beam, and polarization states with different rotation angles are obtained along the propagation direction, achieving polarization control along the propagation path.

By separating the propagation paths of the RCP and LCP waves, the geometric relationships in the diagram become clear, as shown in the upper part of Fig. 7.6. The optical path

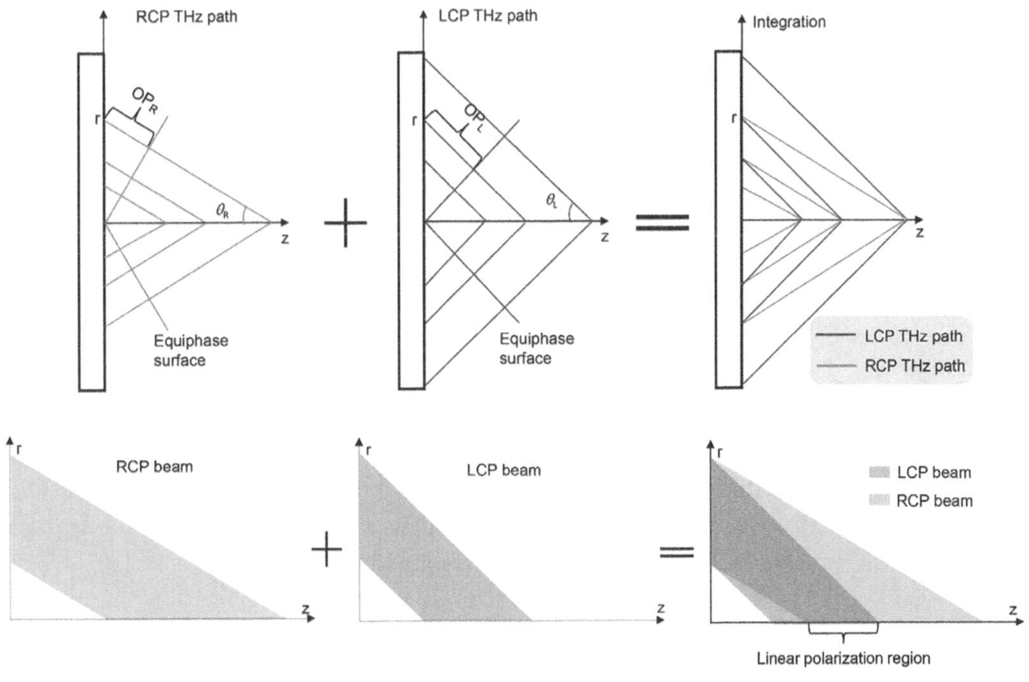

FIGURE 7.6 Schematic diagram showing the separation and combination of RCP and LCP wave propagation paths (upper part) and the corresponding wave beam representation (lower part). After being transmitted to the z-axis, the LCP and RCP waves superimpose within a segment along the z-axis and recombine into a linearly polarized wave.

lengths from a point on the metasurface to the equiphase surface for the LCP and RCP waves can be easily calculated as:

$$\begin{cases} OP_L = r\sin\theta_L \\ OP_R = r\sin\theta_R \end{cases}, \tag{7.9}$$

where r is the distance from a point on the metasurface to its centre. Since the metasurface lies in the x-y plane, there is $r = \sqrt{x^2 + y^2}$. Therefore, to achieve the optical path lengths described in Eq. (7.9), the required phase profile for the metasurface is:

$$\begin{cases} \phi_{\mathrm{LCP}} = -\dfrac{2\pi}{\lambda}(r\sin\theta_L) \\ \phi_{\mathrm{RCP}} = -\dfrac{2\pi}{\lambda}(r\sin\theta_R) \end{cases}. \tag{7.10}$$

For easier understanding, the propagation of LCP and RCP waves can be depicted using wave beams instead of rays, corresponding to the lower part of Fig. 7.6. It can be seen that after the LCP and RCP waves are transmitted to the z-axis, they superimpose within a segment along the z-axis, recombining to form a linearly polarized wave. Within this linearly polarized region, the LCP and RCP components at each point do not originate from the same unit cell but from different ones. Clearly, the LCP and RCP components at each point carry different phase delays, and their phase difference satisfies the description in Eq. (7.8).

The metasurface requires each unit cell to simultaneously manipulate the LCP and RCP components of the incident linearly polarized wave. To achieve this, a linear-to-circular polarization conversion unit cell must first be obtained. The theory of linear-to-circular polarization conversion based on birefringent rectangular bars was clarified in Chapter 3 and will not be reiterated here. Neglecting reflection and absorption by the silicon wafer, i.e. assuming a transmission coefficient of 1, thus Eq. (3.6) in Chapter 3 is referenced as:

$$\boldsymbol{E}_0^t = \frac{1}{2}e^{i\varphi}\begin{pmatrix} 2\cos\theta & -2i\sin\theta \\ -2i\sin\theta & 2\cos\theta \end{pmatrix}\begin{pmatrix} 1 \\ 0 \end{pmatrix} = \frac{1}{2}e^{i\varphi}\begin{pmatrix} 2\cos\theta \\ -2i\sin\theta \end{pmatrix}. \tag{7.11}$$

It is easily observed that when $\theta = \pi/4$ and $3\pi/4$, the above equation represents the LCP wave and the RCP wave, respectively. Simultaneously, simplified forms of the transfer matrices for generating LCP and RCP waves are obtained:

$$\boldsymbol{T}_L = Ce^{i\varphi}\begin{pmatrix} 1 & -i \\ -i & 1 \end{pmatrix}, \boldsymbol{T}_R = Ce^{i\varphi}\begin{pmatrix} -1 & -i \\ -i & -1 \end{pmatrix}, \tag{7.12}$$

where $C = \sqrt{2}/2$. The above derivation indicates that, in theory, x-polarized light passing through a rectangular bar rotated by $\pi/4$ can be converted into LCP light. The structure is shown in Fig. 7.7(a), with light incident from the bottom of the structure. To manipulate the

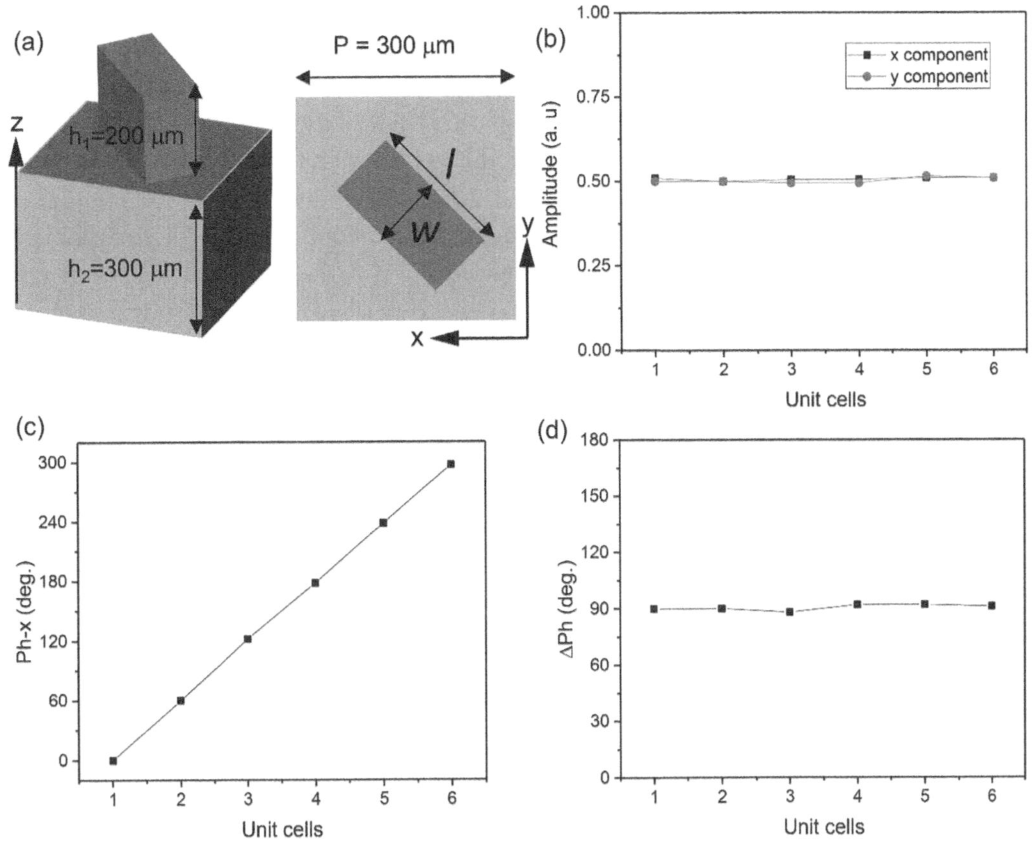

FIGURE 7.7 (a) Side and top views of the linear-to-circular polarization conversion sub-unit cell; six structures are selected in total, with parameters listed in Table 7.1. When a 1.05 THz *x*-polarized wave is incident from the bottom of these structures, the *x*- and *y*-components of the output wave are shown in (b), the relative phase of the corresponding *x* output component is shown in (c), and the phase difference between the *y*-component and the *x*-component is shown in (d).

wavefront, six structures are designed in total, with their parameters listed in Table 7.1. Six silicon rectangular bars tilted at $\pi/4$ in the *x*-*y* plane can achieve co-polarized and cross-polarized linear transmission with equal amplitude, as shown in Fig. 7.7(b). The phase covers the range of 0–2π, as shown in Fig. 7.7(c), while maintaining a phase difference where the *x*-component lags the *y*-component by $\pi/2$, as shown in Fig. 7.7(d). Thus, a LCP light outputs. Simply rotating these structures by $\pi/2$ further yields RCP light.

As mentioned above, a silicon rectangular bar tilted at $\pi/4$ in the *x*-*y* plane can achieve co-polarized and cross-polarized linear transmission with equal amplitude while maintaining

TABLE 7.1 Structural Parameters of Six Sub-Unit Cells
Structural Parameters of Six Sub-Unit Cells

No.	1st	2nd	3rd	4th	5th	6th
w (μm)	36	48	52	54	56	60
l (μm)	62	62	64	72	92	114

a phase difference of $\pi/2$ between them, ultimately resulting in circularly polarized output. This functionality is analogous to that of a quarter-wave plate, as illustrated in Fig. 7.8(a). Utilizing the PB phase characteristic, rotating the silicon rectangular bar by $\pi/2$ yields circularly polarized output with opposite handedness. For example, in Fig. 7.8(a), a rectangular bar that outputs RCP waves, when rotated by $\pi/2$, outputs LCP waves. The complete unit cell designed in this work consists of four linear-to-circular polarization conversion sub-unit cells arranged in an interleaved spatial pattern: two identical structures controlling the RCP component and the other two identical structures controlling the LCP component. The transmission matrix of such a unit cell is described as the superposition of two transmission matrices:

$$\boldsymbol{T}(z)=\boldsymbol{T}_L(z)+\boldsymbol{T}_R(z)=\frac{1}{2}Ce^{i\varphi_L}\begin{pmatrix} 1 & -i \\ -i & 1 \end{pmatrix}+\frac{1}{2}Ce^{i\varphi_R}\begin{pmatrix} -1 & -i \\ -i & -1 \end{pmatrix}. \tag{7.13}$$

In Eq. (7.13), the first matrix inside the parentheses represents the transfer matrix for outputting LCP waves, and the second matrix represents the transfer matrix for outputting

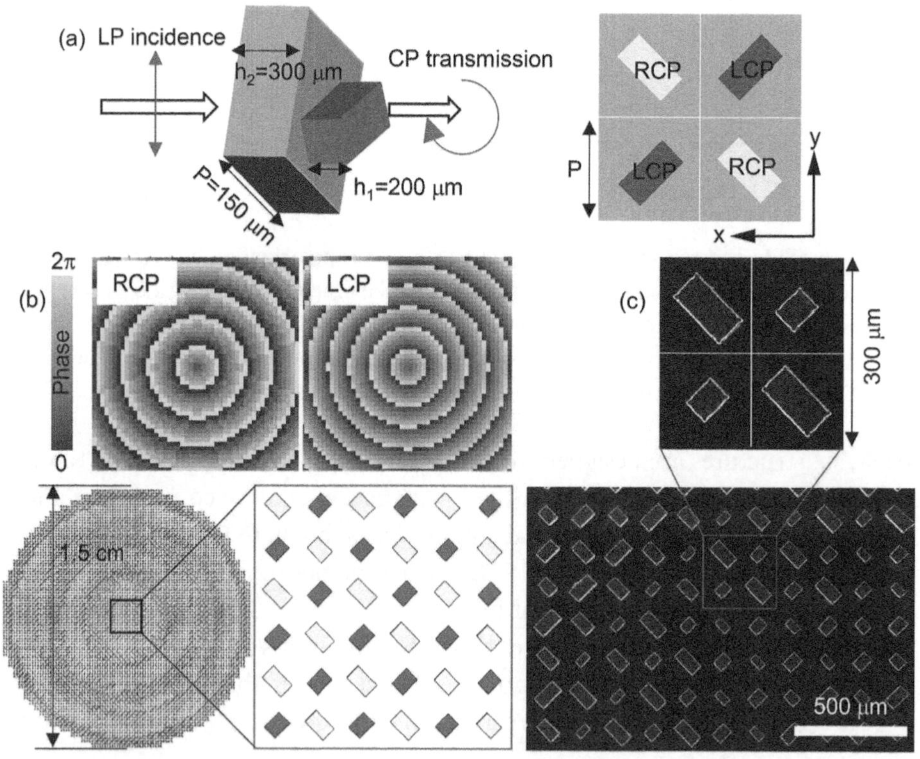

FIGURE 7.8 (a) Schematic for sub-unit cells of the device. A complete unit cell is formed by spatially interleaving four sub-unit cells: two identical purple unit cells control the propagation of the LCP component, and the other two identical yellow unit cells control the propagation of the RCP component. (b) The required phase profiles for the RCP and LCP waves constitute the device, along with a device schematic. The corresponding scanning electron microscopy image is shown in (c).

RCP waves. To achieve wavefront manipulation of the circularly polarized wave components, six fundamental structures from Table 7.1 are utilized, with their relative phase delays covering 0 to 2π at intervals of $\pi/3$. The required phase profiles for controlling the RCP and LCP components, along with a schematic of the device, are shown in Fig. 7.8(b). These phase profiles are determined by Eq. (7.10). This device operates at a frequency of 1.05 THz, with $\sin\theta_L = 0.38$ and $\sin\theta_R = 0.27$. The entire device has a diameter of 1.5 cm. The device is fabricated from 500 μm thick high-resistivity silicon (relative permittivity $\varepsilon = 11.9$, resistivity $\rho > 10{,}000\ \Omega{\cdot}\text{cm}$) using ion-beam etching with mask exposure technology. The physical device is shown in Fig. 7.8(c).

To determine the rotational variation of the transmitted linear polarization along the z-direction, first observe the intensity changes of the x- and y-polarization components in the w plane, as shown in Fig. 7.9(a). The longitudinal monitoring range is from $z = 8$ mm to $z = 19$ mm, at a frequency of 1.05 THz. The beam diameter is 300–400 μm, which is close

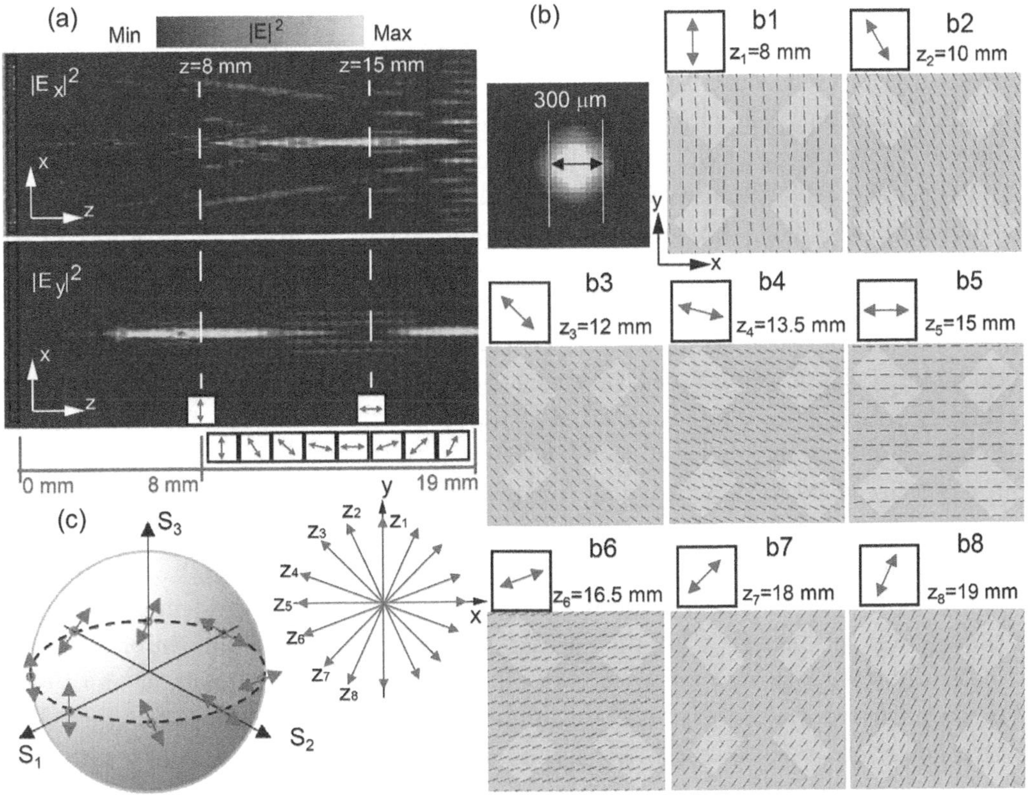

FIGURE 7.9 (a) Simulation results of the electric field intensity distributions for the x- and y-components of the output wave in the x-z plane. (b) Simulation results of the electric field intensity distribution of the output wave in the x-y plane, with a spot diameter of approximately 300–400 μm. (b1–b8) show the simulated polarization states of the spot at different positions along the propagation path, with the wave propagation direction perpendicularly pointing out of the page. (c) The corresponding polarization states from (b1–b8) plotted on the Poincaré sphere (left) and their two-dimensional view in the x-y plane (right).

to the size of a full unit cell, as illustrated in Fig. 7.9(b). As the transmitted linear polarization rotates, the intensity distributions of the x- and y-polarization components inevitably change. In Fig. 7.9(a), the polarization states at two points are most easily determined: $z = 8$ mm and $z = 15$ mm. At $z = 8$ mm, only the electric field intensity of the y-component $|E_y|^2$ is present, while the x-component electric field intensity is zero ($|E_x|^2 = 0$), clearly indicating a y-polarized output wave. At $z = 15$ mm, the y-component electric field intensity is zero, and only the x-component electric field intensity exists, meaning the output wave is x-polarized. From $z = 8$ mm to $z = 15$ mm, the output polarization state evolves from y-polarization to x-polarization. Examining the electric field line distributions at the beam centre on different planes provides further insight into the variation pattern of the output polarization. The simulation results are shown in Fig. 7.9(b1–b8). When projecting the polarization states from different planes onto an x-y plane, it is evident that the polarization state rotates counterclockwise along the propagation path, as shown on the right side of Fig. 7.9(c). Specifically, the output polarization states between 8 mm and 15 mm (z_1–z_5) transfer from y to x, with electric field oscillations passing through the second and fourth quadrants. The output polarization states between 15 mm and 19 mm (z_5–z_8) transfer from x to y, with electric field oscillations passing through the first and third quadrants. Further plotting these eight polarization states (z_1–z_8) on the Poincaré sphere, as shown on the left side of Fig. 7.9(c), reveals that they precisely cover the equator of the Poincaré sphere.

Observing the total intensity $|E_x|^2 + |E_y|^2$ in the x-z plane reveals a complete Bessel beam, as shown in Fig. 7.10(a). The electric field amplitude and phase distributions at several key points were tested to validate the simulation results, as shown in Fig. 7.10(b–e). At $z = 8$ mm, the simulation shows that the spot centre is a y-polarized wave, and the test results confirm the presence of only the y-component electric field intensity with no x-component electric field intensity (Fig. 7.10(b)). When the wave propagates to $z = 12$ mm, the simulation yields a linearly polarized wave tilted at $\pi/4$ and distributed in the second and fourth quadrants. The test results show nearly equal electric field intensities for the x- and y-components, with the y-polarized field lagging the x-polarized field by a phase of π (Fig. 7.10(c)). This precisely corresponds to a linearly polarized wave tilted at $\pi/4$ in the second and fourth quadrants, matching the simulation. At $z = 15$ mm, the output polarization rotates to x-polarization, so only the x-component shows significant electric field intensity, consistent between simulation and experiment (Fig. 7.10(d)). Furthermore, at $z = 18$ mm, the simulation shows the x-polarized wave rotating counterclockwise into the first and third quadrants, forming a linearly polarized wave tilted at $\pi/4$. This is also confirmed by the test results, which show nearly equal intensities for the x- and y-components and an almost zero phase difference between them (Fig. 7.10(e)).

7.3 CHAPTER SUMMARY

Three-dimensional polarization manipulation extends polarization control from the two degrees of freedom (x-y) to three degrees of freedom (x-y-z). This chapter first introduces the development and significance of 3D spatial polarization manipulation. It then demonstrates the principles and meta-device designs for 3D spatial polarization control based on spin-decoupling and dual-meta-atom interference. In terms of polarization manipulation in 3D

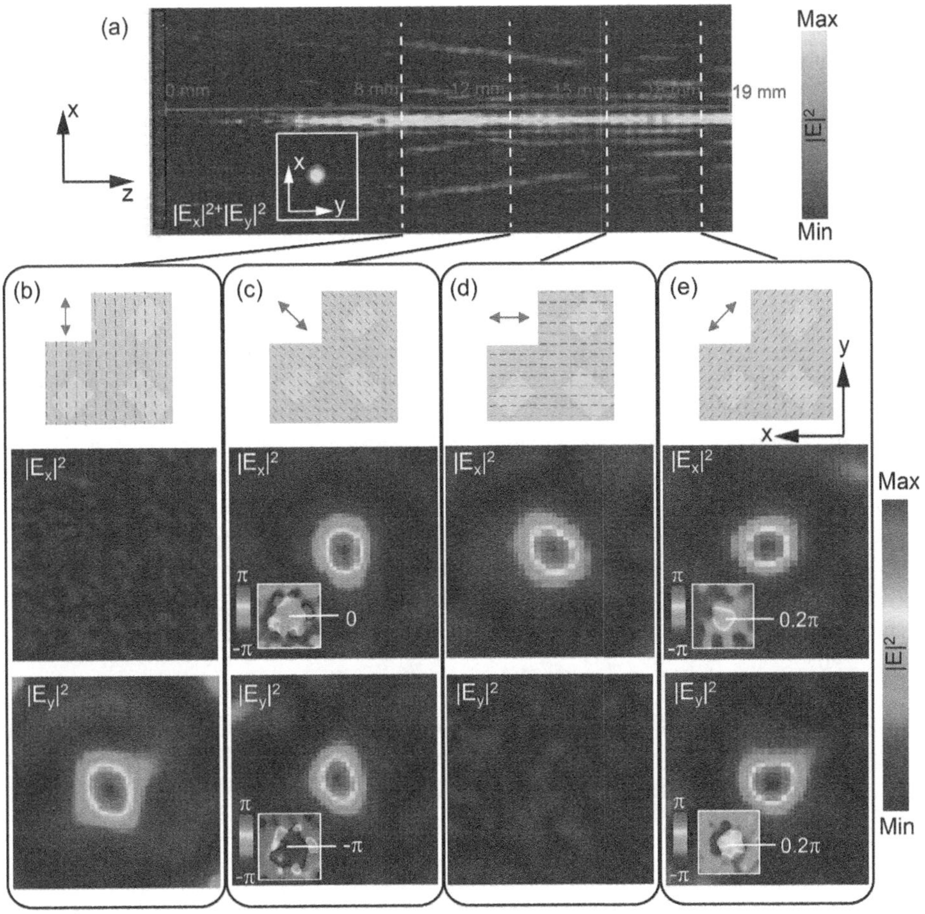

FIGURE 7.10 (a) Simulation results of the total electric field intensity distribution of the output wave in the *x*-*z* plane. (b–e) Test results of the electric field intensity and phase distributions for the *x*- and *y*-components at four points along the propagation path, determining the output polarization states. The propagation direction is perpendicularly pointing out of the page.

space, meta-photonic devices exhibit significant advantages over conventional optical components. They enable the complete integration of functions such as tightly focused beams, vector optical fields, and longitudinal polarization control on a single metasurface, whereas traditional techniques often require multiple optical components to achieve the same functionality. Using spin-decoupling, two functionalities are demonstrated: first, generating a vector field that continuously varies along the propagation direction; and second, independently controlling polarization distributions on multiple focal planes. Using dual-meta-atom interference, the transformation of an incident linearly polarized wave into a transmitted linearly polarized wave that continuously rotates along the propagation path is demonstrated. Regardless of the principle or functionality, the core process essentially involves: decomposing the incident linearly polarized wave into orthogonal circularly polarized components (LCP and RCP), applying a longitudinally variable phase difference to the LCP and RCP components, and then recombining them. For spin-decoupling, independent phase delays can be applied to the LCP and RCP components on a single structure, and a phase difference that varies along the propagation path can be imposed on these components according to the desired phase distribution. For dual-meta-atom interference, each unit cell is formed by spatially interleaving sub-unit cells that independently control the transmission of LCP and RCP waves. This allows each complete unit cell to decompose the incident linearly polarized wave into LCP and RCP waves and apply different phase delays to these two waves, thereby achieving a phase difference that changes along the propagation path.

REFERENCES

1. Liu, S., Li, P., Zhang, Y., et al. (2016) 'Longitudinal spin separation of light and its performance in three-dimensionally controllable spin-dependent focal shift', Scientific Reports, 6, 20774.
2. Bomzon, Z., Biener, G., Kleiner, V., et al. (2002) 'Space-variant Pancharatnam-Berry phase optical elements with computer-generated subwavelength gratings', Optics Letters, 27(13), 1141–1143.
3. Yin, X., Ye, Z., Rho, J., et al. (2013) 'Photonic spin Hall effect at metasurfaces', Science, 339(6126), 1405–1407.
4. Ling, X., Zhou, X., Yi, X., et al. (2015) 'Giant photonic spin Hall effect in momentum space in a structured metamaterial with spatially varying birefringence', Light: Science & Applications, 4(5), e290.
5. Ke, Y., Liu, Y., He, Y., et al. (2015) 'Realization of spin-dependent splitting with arbitrary intensity patterns based on all dielectric metasurfaces', Applied Physics Letters, 107(4), 041107.
6. Hasman, E., Kleiner, V., Biener, G., et al. (2003) 'Polarization dependent focusing lens by use of quantized Pancharatnam Berry phase diffractive optics', Applied Physics Letters, 82(3), 328–330.
7. Gu, B., Pan, Y., Wu, J.L., et al. (2014) 'Manipulation of radial-variant polarization for creating tunable bifocusing spots', Journal of the Optical Society of America A: Optics, Image Science and Vision, 31(2), 253–257.
8. Ke, Y., Liu, Y., Zhou, J., et al. (2015) 'Photonic spin filter with dielectric metasurfaces', Optics Express, 23(26), 33079–33086.
9. Wang, S., Wang, X., Kan, Q., et al. (2015) 'Spin-selected focusing and imaging based on metasurface lens', Optics Express, 23(20), 26434–26441.
10. Bouchard, F., Larocque, H., Yao, A.M., et al. (2016) 'Polarization shaping for control of nonlinear propagation', Physical Review Letters, 117(23), 233903.
11. Dorrah, A.H., Rubin, N.A., Zaidi, A., et al. (2021) 'Metasurface optics for on-demand polarization transformations along the optical path', Nature Photonics, 15(4), 287–296.
12. Zhong, R.Y., Zhu, Z.H., Wu, H.J., et al. (2021) 'Gouy-phase-mediated propagation variations and revivals of transverse structure in vectorially structured light', Physical Review A, 103(5), 053520.
13. Li, H.R., Zhao, P.Y. and Yin, J.P. (2021) 'Generation of diffraction-free vectorial elliptic hollow beams with space-varying inhomogeneous polarizations', Chinese Physics B, 30(8), 084204.
14. Otte, E., Rosales-Guzman, C., Ndagano, B., et al. (2018) 'Entanglement beating in free space through spin-orbit coupling', Light: Science & Applications, 7(5), 18009.
15. Vyas, S., Kozawa, Y. and Miyamoto, Y. (2015) 'Creation of polarization gradients from superposition of counter propagating vector LG beams', Optics Express, 23(26), 33970–33979.
16. Lu, J.Q., Wang, X.L., Zhang, G.L., et al. (2020) 'Bessel-like beams with controllable rotating local linear polarization during propagation', Optics Letters, 45(7), 1738–1741.
17. Corato-Zanarella, M., Dorrah, A.H., Zamboni-Racged, M., et al. (2018) 'Arbitrary control of polarization and intensity profiles of diffraction-attenuation-resistant beams along the propagation direction', Physical Review Applied, 9(2), 024013.

18. Cardano, F., Karimi, E., Marrucci, L., et al. (2013) 'Generation and dynamics of optical beams with polarization singularities', Optics Express, 21(7), 8815–8820.
19. Moreno, I., Davis, J.A., Sanchez-Lopez, M.M., et al. (2015) 'Nondiffracting Bessel beams with polarization state that varies with propagation distance', Optics Letters, 40(23), 5451–5454.
20. Davis, J.A., Moreno, I., Badkam, K., et al. (2016) 'Nondiffracting vector beams where the charge and the polarization state vary with propagation distance', Optics Letters, 41(10), 2270–2273.
21. Li, P., Zhang, Y., Liu, S.L., et al. (2016) 'Quasi-Bessel beams with longitudinally varying polarization state generated by employing spectrum engineering', Optics Letters, 41(20), 4811–4814.
22. Li, P., Fan, X., Wu, D., et al. (2020) 'Self-accelerated optical activity in free space induced by the Gouy phase', Photonics Research, 8(4), 475–481.
23. Zhao, Y.Q., Zhan, Q.W., Zhang, Y.L., et al. (2005) 'Creation of a three-dimensional optical chain for controllable particle delivery', Optics Letters, 30(8), 848–850.
24. Kozawa, Y. and Sato, S. (2006) 'Focusing property of a double-ring-shaped radially polarized beam', Optics Letters, 31(6), 820–822.
25. Wang, X.L., Ding, J., Qin, J.Q., et al. (2009) 'Configurable three-dimensional optical cage generated from cylindrical vector beams', Optics Communications, 282(17), 3421–3425.
26. Nie, Z., Ding, W., Shi, G., et al. (2015) 'Achievement and steering of light-induced sub-wavelength longitudinal magnetization chain', Optics Express, 23(16), 21296–21305.
27. Chen, Z. and Zhao, D. (2012) '4Pi focusing of spatially modulated radially polarized vortex beams', Optics Letters, 37(8), 1286–1288.
28. Yu, Y., Huang, H., Lin, S., et al. (2019) 'Generation of optical needles and bright spot arrays through reversing the radiation pattern of collinear antenna array', Journal of Optics, 21(7), 075602.
29. Dehez, H., April, A. and Piche, M. (2012) 'Needles of longitudinally polarized light: guidelines for minimum spot size and tunable axial extent', Optics Express, 20(14), 14891–14905.
30. Wang, S., Li, X., Zhou, J., et al. (2014) 'Ultralong pure longitudinal magnetization needle induced by annular vortex binary optics', Optics Letters, 39(17), 5022–5025.
31. Wang, H., Shi, L., Lukyanchuk, B., et al. (2008) 'Creation of a needle of longitudinally polarized light in vacuum using binary optics', Nature Photonics, 2(8), 501–505.
32. Zhang, F., Pu, M., Guo, Y., et al. (2022) 'Synthetic vector optical fields with spatial and temporal tenability', Science China-Physics, Mechanics & Astronomy, 65(5), 254211.
33. Zheng, C., Li, J., Liu, J., et al. (2022) 'Creating longitudinally varying vector vortex beams with an all-dielectric metasurface', Laser & Photonics Reviews, 16(10), 2200236.
34. Zhang, X., Tang, W., Jiang, W., et al. (2018) 'Light-controllable digital coding metasurfaces', Advanced Science, 5, 1801028.
35. Chen, W.T., Zhu, A.Y. and Capasso, F. (2020) 'Flat optics with dispersion-engineered metasurfaces', Nature Reviews Materials, 5, 604–620.
36. Guo, Y., Zhang, S., Pu, M., et al. (2021) 'Spin-decoupled metasurface for simultaneous detection of spin and orbital angular momenta via momentum transformation', Light: Science & Applications, 10, 63.
37. Li, J., Li, J.T., Yue, Z., et al. (2022) 'Structured vector field manipulation of terahertz wave along the propagation direction based on dielectric metasurfaces', Laser & Photonics Reviews, 16(12), 2200325.
38. Zang, X.F., Xu, W.W., Gu, M., et al. (2020) 'Polarization-insensitive metalens with extended focal depth and longitudinal high-tolerance imaging', Advanced Optical Materials, 8(2), 1901342.
39. Li, J., Liu, J., Yue, Z., et al. (2025) 'Polarization variable terahertz metasurface along the propagation path', Fundamental Research, 5, 124–131.

CHAPTER 8

BICs-Based Polarization Manipulation

Abstract

The polarization manipulation principles of the aforementioned meta-photonic devices can be traced back to the weak interaction between light and matter, which involves high-damping electromagnetic resonances with low quality (Q) factors. However, optical resonance systems also exhibit a special resonant state known as "Bound States in the Continuum (BICs)", which describes a physical system with an extremely tunable Q factor. BICs refer to a phenomenon where electromagnetic waves with frequencies below the diffraction limit of optical devices can still exist as bound states within the radiation continuum, thereby enabling strong interactions between electromagnetic waves and the devices. BICs possess infinite Q factors and zero-radiation linewidths, making them indistinguishable in spectra and thus not directly accessible. Nevertheless, by physically altering certain structural parameters of the devices, quasi-BICs with radiation leakage can be achieved. Quasi-BICs exhibit high Q factors and extremely narrow radiation linewidths. It has been found that both BICs and quasi-BICs possess intriguing polarization properties: (i) BICs exhibit polarization indeterminacy, yet the polarization around BIC points in momentum space displays vortex characteristics; (ii) once BICs transform into quasi-BICs, they manifest definite polarization states. By appropriately adjusting structural parameters, the polarization states in momentum space can cover the entire Poincaré sphere, thereby enabling the manipulation of arbitrary polarization states. This chapter will elucidate the fundamental concepts of BICs, introduce the polarization vortex characteristics and principles of BICs in momentum space within metasurfaces, and demonstrate a principle for photonic polarization manipulation and a device design method based on quasi-BICs.

DOI: 10.1201/9781003784821-8

8.1 THE CONCEPT AND POLARIZATION CHARACTERISTICS OF OPTICAL BICS

8.1.1 Concept and Classification of Optical BICs

According to classical mechanics, an electron will escape from its atomic system if it possesses sufficient kinetic energy, which lies within the system's scattering continuum. However, in reality, some electrons, despite having enough energy, remain bound within the system and are not liberated from the gravitational influence of their surroundings. This seemingly paradoxical yet real phenomenon can be explained by quantum mechanics, as elucidated by two pioneers in physics, John von Neumann and Eugene Paul Wigner, in 1929 [1]. The states in which these electrons reside are termed "bound states in the continuum", commonly abbreviated as BICs. Similar phenomena based on this concept are quite widespread and have been observed in electromagnetic waves [2], water waves [3–5], acoustic waves [6–8], and elastic waves in solids [9,10]. Entering the 20th century, optical BICs have been extensively studied in a wide range of media systems, including optical direct waveguides, optical fibres, photonic crystals, and optical metamaterials [11–16]. Conventionally, electromagnetic eigenstates within the radiation continuum are considered radiative, while bound states that can be completely confined by the medium are expected to lie below the radiation continuum. However, the emergence of BICs challenges this conventional understanding. Within the radiation continuum, BICs represent a class of electromagnetic eigenstates that are non-radiative. Despite their frequency and momentum potentially matching those of vacuum modes, they remain entirely confined within the resonant system.

As shown in Fig. 8.1(a), consider a periodic planar dielectric in vacuum. The wave vector $\boldsymbol{k}$ of a radiating electromagnetic wave can be decomposed into a vertical component $k_{\perp}$ and a horizontal component $k_{//}$. From geometric relationships, it is known that the frequency $\omega = c|\boldsymbol{k}|$ of any radiating electromagnetic wave is always higher than $c|k_{//}|$. Furthermore, for a given horizontal wave vector component $k_{//}$, the vertical component $k_{\perp}$ can take any value, making the frequency of radiating electromagnetic waves continuous. In the frequency-wave vector (ω-$k_{//}$) spectra, the frequencies of radiating electromagnetic waves occupy the region where $\omega > c|k_{//}|$ (also known as the light cone), i.e., the radiation continuum. Below the radiation continuum lies the region occupied by conventional bound states (CBSs). The eigensolutions of Maxwell's equations determine that the frequencies of bound-state electromagnetic waves in the medium are discrete. Conventionally, it is believed that within the radiation continuum, resonant modes all exhibit radiation leakage. However, BICs deviate from this principle: BICs are special resonant modes whose frequencies lie within the radiation continuum yet remain completely bound, with no energy leakage whatsoever [17].

Although BICs are located within the radiation continuum, they are completely decoupled from the far-field radiation. Based on the decoupling mechanism, they can be primarily categorized into two types [18]. As shown in Fig. 8.1(b), the first type of BICs arises from structural symmetry mismatch: a microstructure with C_2 symmetry can confine resonant modes in the radiation continuum that possess even symmetry. Such resonant modes are completely orthogonal to odd-symmetric radiation modes, thereby decoupling from far-field radiation and forming BICs. These are referred to as symmetry-protected BICs. Once

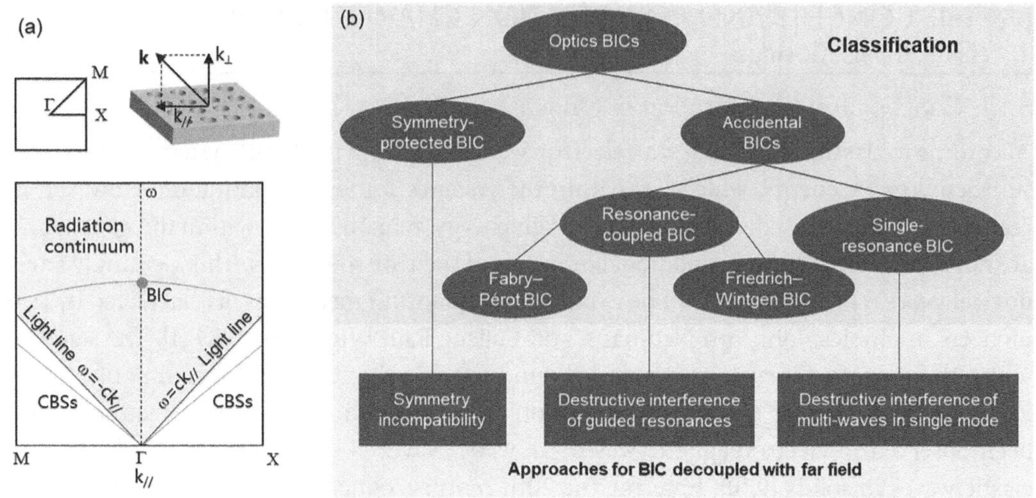

FIGURE 8.1 Concept and classification of optical BICs. (a) Schematic diagram of the photonic bands of radiation continuum, BICs, and CBSs in periodic optical systems. Taking symmetry-protected BICs as an example, they are located at high symmetry points (Γ-points) in the wave vector space. (b) Classification of optical BICs and their decoupling approach from far-field.

the C_2 symmetry of the structure is broken, the BICs will couple with radiation modes of different symmetries, leading to radiation leakage and the formation of quasi-BICs. The second type of BICs originates from destructive interference: the decoupling from far-field radiation is indirectly achieved through destructive interference among different electromagnetic modes in the structure or among different waves of the same electromagnetic mode. The former can be further divided into Fabry-Pérot BICs and Friedrich-Wintgen BICs, while the latter generally refers to single-resonance BICs. Adjusting structural parameters can modulate the degree of destructive interference. When radiation cannot be completely cancelled out, quasi-BICs with radiation leakage are formed. These BICs and quasi-BICs are realized by tuning structural parameters, and the exact parameter conditions for the emergence of BICs are difficult to predict accurately. Thus, they are also referred to as parameter-tuned BICs or accidental BICs.

In addition to the two types of BICs mentioned above, there are also some BICs that have not been precisely defined or even have academic controversies, such as BICs associated with Brillouin zone (BZ) folding, which can be generally referred to as folded BICs. Their characteristics for folded BICs are as follows [19]: by altering structural parameters to abruptly increase the lattice constant of the unit cell while maintaining its C_2 or C_4 symmetry, the corresponding BZ shrinks. As a result, CBSs originally located below the light line are instantaneously folded into the radiation continuum, transforming into radiative quasi-BICs.

8.1.2 Polarization Characteristics and Principles of Optical BICs

The application of photonic crystals or metasurfaces in polarization manipulation has garnered significant interest due to their simple fabrication, designable band structures, and complex polarization characteristics topologically linked to BICs in momentum space, all of which facilitate polarization modulation. In two-dimensional photonic crystals or metasurfaces, a series of Bloch resonant modes with different frequencies and wave vectors form photonic bands. The radiation polarization states of these modes on any band can be projected onto the structural plane and mapped onto the BZ, thereby defining a polarization field in momentum space. Most of these Bloch modes are radiative, except when destructive interference or symmetry mismatch renders them non-radiative, i.e. BICs. BICs manifest as vortex singularities in this field [20], where the polarization state is indeterminate, while the polarization states surrounding BICs in the BZ are linear and exhibit vortex variations (Fig. 8.2(a)). This linear polarization vortex characteristic can be decomposed into the superposition of a left-handed circularly polarized (LCP) wave and a right-handed circularly polarized (RCP) wave in momentum space, each carrying phase topological charges of equal magnitude but opposite direction.

Next, based on a circular-hole photonic crystal slab (with C_4 symmetry) supporting Γ-point BICs, the polarization vortex characteristics of BICs in momentum space are derived. The two sides of the photonic crystal slab along the normal direction (defined as the z-axis) are labelled as I and II (Fig. 8.2(a)). Only the frequency range below the diffraction limit is considered. For a guided resonance propagating in the direction $k_{//}$ within a photonic crystal slab possessing C_4 rotational symmetry, the scattering process can be represented by a 4 × 4 matrix as follows:

$$\begin{pmatrix} E_s^{\mathrm{I}-} \\ E_p^{\mathrm{I}-} \\ E_s^{\mathrm{II}-} \\ E_p^{\mathrm{II}-} \end{pmatrix} = \begin{pmatrix} r_{ss} & r_{sp} & t_{ss} & t_{sp} \\ r_{ps} & r_{pp} & t_{ps} & t_{pp} \\ t_{ss} & t_{sp} & r_{ss} & r_{sp} \\ t_{ps} & t_{pp} & r_{ps} & r_{pp} \end{pmatrix} \begin{pmatrix} E_s^{\mathrm{I}+} \\ E_p^{\mathrm{I}+} \\ E_s^{\mathrm{II}+} \\ E_p^{\mathrm{II}+} \end{pmatrix}, \tag{8.1}$$

where the superscript "I(II)" denotes the two ports of the photonic crystal, while the superscript "+(–)" represents the incoming (outgoing) direction. The subscripts "*sp*" in the

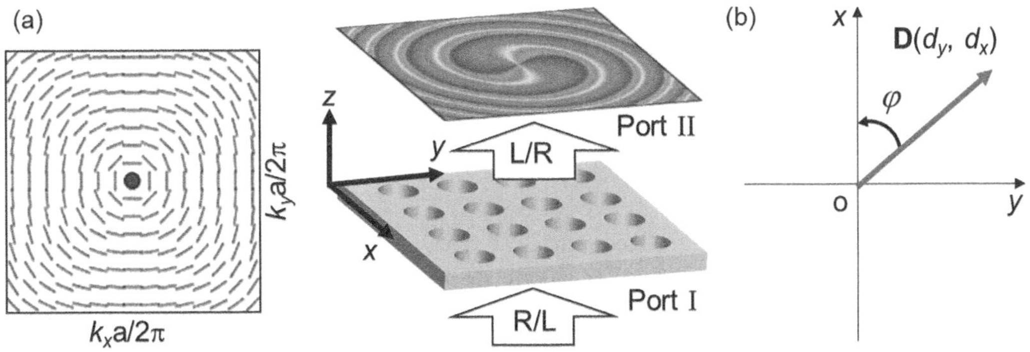

FIGURE 8.2 The polarization characteristics of BICs in momentum space. (a) Schematic diagram of far-field polarization vortex characteristics of BICs photonic crystal plate, where L/R represent LCP/RCP waves. (b) Schematic diagram of the mathematical image of the coupling coefficient D vector in y–x basis.

transmission (t) and reflection (r) coefficients indicate that the coefficient corresponds to s-polarized outgoing and p-polarized incoming waves. For simplicity, the following notation is used:

$$|O\rangle=\begin{pmatrix}E_s^{\text{I}-}\\E_p^{\text{I}-}\\E_s^{\text{II}-}\\E_p^{\text{II}-}\end{pmatrix},|I\rangle=\begin{pmatrix}E_s^{\text{I}+}\\E_p^{\text{I}+}\\E_s^{\text{II}+}\\E_p^{\text{II}+}\end{pmatrix},\mathbf{S}=\begin{pmatrix}r_{ss}&r_{sp}&t_{ss}&t_{sp}\\r_{ps}&r_{pp}&t_{ps}&t_{pp}\\t_{ss}&t_{sp}&r_{ss}&r_{sp}\\t_{ps}&t_{pp}&r_{ps}&r_{pp}\end{pmatrix}, \tag{8.2}$$

where $|O\rangle$ and $|I\rangle$ represent the outgoing and incoming wave amplitudes, respectively. $\mathbf{S}$ is the structural scattering matrix for a specific wave vector $k_{//}$, which simultaneously provides the transmission and reflection coefficients. To obtain explicit mathematical expressions, absorption loss is neglected here, and temporal coupled-mode theory (TCMT) is introduced, which will yield the transmission coefficient matrix.

A resonant mode of a resonator is a spatiotemporal dynamic field, from which an independent term $e^{-i\omega t}$ can be mathematically separated to represent the temporal characteristic (the absolute phase has no physical significance and could also be $e^{i\omega t}$). Let the amplitude of this dynamic field be A, whose magnitude is determined by the energy stored in the mode. First, assuming the system is closed, lossless, and energy-conserving, taking the time derivative yields the dynamic relation for its resonant amplitude, expressed as:

$$\frac{dA}{dt}=-i\omega A, \tag{8.3}$$

where ω represents the frequency. When two ports are introduced into such a resonant system, incident waves from the ports couple to the resonator, and the internal resonance of the resonator radiates towards the ports. The decay rate due to radiation is γ_0. The dynamic relation of the system can now be expressed by the following master equations:

$$\begin{cases}\dfrac{dA}{dt}=(-i\omega_0-\gamma_0)A+\mathbf{K}^T|I\rangle\\|O\rangle=\mathbf{S}|I\rangle=\mathbf{C}|I\rangle+\mathbf{D}A\end{cases}, \tag{8.4}$$

where ω_0 and γ_0 are the resonant frequency and decay rate of the mode. The coupling coefficients between the resonance and the incoming and outgoing plane waves are vectors $\mathbf{K}=(k_s^{\text{I}},k_p^{\text{I}},k_s^{\text{II}},k_p^{\text{II}})^T=(k_s,k_p,\alpha_z k_s,\alpha_z k_p)^T$ and $\mathbf{D}=(d_s^{\text{I}},d_p^{\text{I}},d_s^{\text{II}},d_p^{\text{II}})^T=(d_s,d_p,\alpha_z d_s,\alpha_z d_p)^T$, respectively, where α_z is a parity number linking radiation from the two ports due to the mirror symmetry of the photonic crystal slab. Owing to the up-down mirror symmetry of the system, the coupling coefficients for the two ports are constrained, differing only by a factor $\alpha_z=\pm1$. When $\alpha_z=1$, the mode is even symmetric with respect to the mirror plane; when $\alpha_z=-1$, the mode is odd symmetric. The $\mathbf{C}$ is a 4×4 background scattering coefficient matrix describing the scattering process in the absence of resonance, given by:

$$\mathbf{C}=\begin{pmatrix} r_s & 0 & t_s & 0 \\ 0 & r_p & 0 & t_p \\ t_s & 0 & r_s & 0 \\ 0 & t_p & 0 & r_p \end{pmatrix}. \tag{8.5}$$

For a lossless reciprocal system, the background scattering matrix **C** is unitary and symmetric. The physical meanings of the two dynamic equations in Eq. (8.4) are intuitive. The first dynamic equation indicates that the change of the resonant mode amplitude with time consists of three parts: resonance (the real part of the eigenfrequency ω_0), loss due to radiation into free-space plane waves (the imaginary part of the eigenfrequency γ_0), and gain from the input of free-space plane waves into the guided resonance. The second dynamic equation states that the outgoing plane wave electric field is the superposition of two parts: the directly scattered field of the incident plane wave through the non-resonant structure, and the field generated by the resonance in the structure coupling to free space.

Using Eqs. (8.3) and (8.4), the relationship between the resonant mode amplitude and the incoming field can be obtained:

$$A=\frac{\mathbf{K}^T|I\rangle}{i(\omega_0-\omega)+\gamma_0}. \tag{8.6}$$

Moreover, under the case without absorption loss, the system obeys energy conservation and time-reversal symmetry, which implies that **C**, **D**, and **K** are not independent. Consider a passive case where the input $|I\rangle=0$. In this scenario, the mode amplitude will gradually decay:

$$\begin{cases} \dfrac{dA}{dt}=(-i\omega_0-\gamma_0)A \\ |O\rangle=\mathbf{D}A \end{cases}. \tag{8.7}$$

From the first equality in Eq. (8.7), get the energy loss rate:

$$\begin{aligned} \frac{d|A|^2}{dt} &= \frac{d(A^*A)}{dt}=A^*\frac{dA}{dt}+A\frac{dA^*}{dt} \\ &=(-i\omega_0-\gamma_0)A^*A+(i\omega_0-\gamma_0)AA^*, \\ &=-2\gamma_0|A|^2 \end{aligned} \tag{8.8}$$

where A^* is the complex conjugate of A. The energy of the output wave given by the second equality in Eq. (8.7) is:

$$|O\rangle^+|O\rangle=(\mathbf{D}A)^+\mathbf{D}A=|A|^2\,\mathbf{D}^+\mathbf{D}, \tag{8.9}$$

where $\mathbf{D}^{+}$ is the Hermitian adjoint (conjugate transpose) of $\mathbf{D}$. Note that the output wave is a column vector, so its energy is calculated using the inner product of $|O\rangle$ and its conjugate transpose $|O\rangle^{+}$. Due to energy conservation, the sum of the energy lost by the resonance and the output energy should be zero, yielding the first condition:

$$-2\gamma_0|A|^2+|A|^2\mathbf{D}^{+}\mathbf{D}\rightarrow\mathbf{D}^{+}\mathbf{D}=2\gamma_0. \tag{8.10}$$

Further, the above decay process can be time-reversed: the outward radiation decay process as $|O\rangle$ at complex frequency $\omega=\omega_0-i\gamma_0$ is reversed into a gain process where there is only external input $|O\rangle^{*}$ but no radiation. According to the second equation in Eq. (8.7), it can be seen that using $|O\rangle^{*}$ as input will synchronize causing the time-reversal amplitude A^{*} (the temporal term $e^{-i\omega t}$ is changed as $e^{i\omega t}$), and thus the relationship between the amplitude of the resonance mode and the incoming field becomes:

$$\begin{aligned}&\frac{dA^{*}}{dt}=i\omega A^{*}=i(\omega_0-\gamma_0)A^{*}+\mathbf{K}^{T}|O\rangle^{*}\\&\Rightarrow A^{*}=\frac{\mathbf{K}^{T}|O\rangle^{*}}{i(\omega-\omega_0)+\gamma_0}\rightarrow A^{*}=\frac{\mathbf{K}^{T}|O\rangle^{*}}{2\gamma_0}=\frac{\mathbf{K}^{T}\mathbf{D}^{*}A^{*}}{2\gamma_0}\;.\\&\Rightarrow\mathbf{K}^{T}\mathbf{D}^{*}=2\gamma_0\rightarrow(\mathbf{K}^{+}\mathbf{D})^{*}=2\gamma_0\end{aligned} \tag{8.11}$$

Equation (8.11) indicates that $\mathbf{K}^{+}\mathbf{D}$ must also be a real number with a value of $2\gamma_0$. By comparing Eq. (8.10), it can be found that $\mathbf{K}$ and $\mathbf{D}$ are essentially the same, thus obtaining the second condition:

$$\mathbf{K}=\mathbf{D}. \tag{8.12}$$

Simultaneously, considering that the above time-reversal process does not allow radiation, the second equality in Eq. (8.4) should be written as:

$$0=\mathbf{C}|O\rangle^{*}+\mathbf{D}A^{*}=\mathbf{C}\mathbf{D}^{*}A^{*}+\mathbf{D}A^{*}. \tag{8.13}$$

This yields the third condition:

$$\mathbf{C}\mathbf{D}^{*}=-\mathbf{D}. \tag{8.14}$$

Combined with the symmetry of $\mathbf{C}$ matrix ($\mathbf{C}^{T}=\mathbf{C}$), it further follows that:

$$\mathbf{C}\mathbf{D}^{*}=-\mathbf{D}\rightarrow(\mathbf{D}^{*})^{T}\mathbf{C}^{T}=-\mathbf{D}^{T}\rightarrow-\mathbf{D}\mathbf{D}^{+}\mathbf{C}=\mathbf{D}\mathbf{D}^{T}. \tag{8.15}$$

Combining Eqs. (8.3), (8.4), (8.10), (8.12), (8.14), and (8.15), the scattering matrix can be obtained:

$$\mathbf{S}=\mathbf{C}+\frac{\mathbf{D}\mathbf{D}^{T}}{i(\omega_0-\omega)+\gamma_0}=\left[\mathbf{I}-\frac{\mathbf{D}\mathbf{D}^{+}}{i(\omega_0-\omega)+\gamma_0}\right]\mathbf{C}. \tag{8.16}$$

From this, the transmission coefficient is calculated as:

$$\mathbf{S}=\begin{bmatrix} r_s-\frac{(r_s+\alpha_z t_s)|d_s|^2}{\gamma-i(\omega-\omega_0)} & -\frac{d_s d_p^*(r_p+\alpha_z t_p)}{\gamma-i(\omega-\omega_0)} & t_s-\frac{(t_p+\alpha_z r_p)|d_p|^2}{\gamma-i(\omega-\omega_0)} & -\frac{d_s d_p^*(t_p+\alpha_z r_p)}{\gamma-i(\omega-\omega_0)} \\ -\frac{d_p d_s^*(r_s+\alpha_z t_s)}{\gamma-i(\omega-\omega_0)} & r_p-\frac{(r_s+\alpha_z t_s)|d_s|^2}{\gamma-i(\omega-\omega_0)} & -\frac{d_p d_s^*(t_s+\alpha_z r_s)}{\gamma-i(\omega-\omega_0)} & t_p-\frac{(t_p+\alpha_z r_p)|d_p|^2}{\gamma-i(\omega-\omega_0)} \\ t_s-\frac{(t_s+\alpha_z r_s)|d_s|^2}{\gamma-i(\omega-\omega_0)} & -\frac{d_s d_p^*(t_p+\alpha_z r_p)}{\gamma-i(\omega-\omega_0)} & r_s-\frac{(r_s+\alpha_z t_s)|d_s|^2}{\gamma-i(\omega-\omega_0)} & -\frac{d_s d_p^*(r_p+\alpha_z t_p)}{\gamma-i(\omega-\omega_0)} \\ -\frac{d_p d_s^*(t_s+\alpha_z r_s)}{\gamma-i(\omega-\omega_0)} & t_p-\frac{(t_s+\alpha_z r_s)|d_s|^2}{\gamma-i(\omega-\omega_0)} & -\frac{d_p d_s^*(r_s+\alpha_z t_s)}{\gamma-i(\omega-\omega_0)} & r_p-\frac{(r_s+\alpha_z t_s)|d_s|^2}{\gamma-i(\omega-\omega_0)} \end{bmatrix}. \tag{8.17}$$

Extracting the transmission coefficient matrix for one port (taking incoming from port I and outgoing from port II as an example) gives:

$$\begin{pmatrix} t_{ss} & t_{sp} \\ t_{ps} & t_{pp} \end{pmatrix}=\begin{bmatrix} t_s-\frac{(t_s+\alpha_z r_s)|d_s|^2}{\gamma-i(\omega-\omega_0)} & -\frac{d_s d_p^*(t_p+\alpha_z r_p)}{\gamma-i(\omega-\omega_0)} \\ -\frac{d_p d_s^*(t_s+\alpha_z r_s)}{\gamma-i(\omega-\omega_0)} & t_p-\frac{(t_p+\alpha_z r_p)|d_p|^2}{\gamma-i(\omega-\omega_0)} \end{bmatrix}. \tag{8.18}$$

Near the Γ point, the background scattering coefficients approximately satisfy $r_s = r_p$ and $t_s = t_p$. Further combining the condition $\mathbf{C}\mathbf{D}^* = -\mathbf{D}$, it can be deduced that the **D** vector is nearly linearly polarized. In the case of linear polarization, the coupling coefficients d_s and d_p in **D** have a phase difference of 0 or π, and their combined vector $\mathbf{D}$ (d_s, d_p) equals a linear real vector multiplied by an arbitrary phase factor. This phase factor is unimportant, so **D** can be directly assumed to be a real vector; i.e. the coupling coefficients are set as real numbers. To simplify the derivation, it is further assumed that the resonance lies along the Γ–X direction $(k_x, 0)$. For a square lattice, the reciprocal lattice vector in momentum space along the k_x axis is perpendicular to the direct lattice vector along the y-axis in real space. Therefore, the Γ–X direction is parallel to the x-axis in real space. For wave vectors in this direction (considering near-normal incidence), s/p polarization is equivalent to y/x polarization. Thus, the transmission matrix can be rewritten as:

$$\mathbf{T}_{\Gamma\text{-X}}=\begin{pmatrix} t_{yy} & t_{yx} \\ t_{xy} & t_{yy} \end{pmatrix}=\begin{bmatrix} t_y-\frac{(t_y+\alpha_z r_y)d_y^2}{\gamma-i(\omega-\omega_0)} & -\frac{d_y d_x(t_x+\alpha_z r_x)}{\gamma-i(\omega-\omega_0)} \\ -\frac{d_x d_y(t_y+\alpha_z r_y)}{\gamma-i(\omega-\omega_0)} & t_x-\frac{(t_x+\alpha_z r_x)d_x^2}{\gamma-i(\omega-\omega_0)} \end{bmatrix}. \tag{8.19}$$

Now, consider the far-field radiation characteristics of the eigenstate. When there is no input, the eigen-radiation is given by the second equality in Eq. (8.7); i.e. the eigen-radiation field is $\mathbf{D}A = (d_y, d_x)^{\mathrm{T}}A$. Note that mathematically this is an eigen-radiation field described in the y–x basis (as shown in Fig. 8.2(b)). Some transformations are needed to describe it in the conventional x–y basis. For convenience in subsequent operations, first rotate the polarization state in the y–x basis so that it aligns with the x-axis. This requires a rotation matrix $\mathbf{R}(\varphi)$, expressed as:

$$\mathbf{R}(\varphi) = \begin{pmatrix} \cos\varphi & -\sin\varphi \\ \sin\varphi & \cos\varphi \end{pmatrix} = \begin{pmatrix} \dfrac{d_x}{\sqrt{d_x^2+d_y^2}} & -\dfrac{d_y}{\sqrt{d_x^2+d_y^2}} \\ \dfrac{d_y}{\sqrt{d_x^2+d_y^2}} & \dfrac{d_x}{\sqrt{d_x^2+d_y^2}} \end{pmatrix}. \tag{8.20}$$

Applying the rotation matrix $\mathbf{R}(\varphi)$ to $\mathbf{T}_{\Gamma\text{-X}}$ gives:

$$\mathbf{R}(\varphi)\mathbf{T}_{\Gamma\text{-X}}\mathbf{R}^{-1}(\varphi)$$

$$= \begin{bmatrix} \dfrac{d_y^2 t_x + d_x^2 t_y}{d_x^2+d_y^2} & -\dfrac{d_x d_y (t_x - t_y)}{d_x^2+d_y^2} \\ \dfrac{\alpha_z d_x d_y (r_x - r_y) + d_x d_y (t_x - t_y)}{\gamma - i(\omega-\omega_0)} - \dfrac{d_x d_y (t_x - t_y)}{d_x^2+d_y^2} & -\dfrac{\alpha_z (d_x^2 r_x + d_y^2 r_y) + (d_x^2 t_x + d_y^2 t_y)}{\gamma - i(\omega-\omega_0)} + \dfrac{d_x^2 t_x + d_y^2 t_y}{d_x^2+d_y^2} \end{bmatrix}. \tag{8.21}$$

Note that for near-normal incidence, it is approximately true that $r_x = r_y = r$ and $t_x = t_y = t$. Therefore, Eq. (8.21) simplifies to:

$$\mathbf{R}(\varphi)\mathbf{T}_{\Gamma\text{-X}}\mathbf{R}^{-1}(\varphi) = \begin{bmatrix} t & 0 \\ 0 & t - \dfrac{(\alpha_z r + t)(d_x^2+d_y^2)}{\gamma - i(\omega-\omega_0)} \end{bmatrix}. \tag{8.22}$$

By directly swapping the diagonal elements, the transmission coefficients can be switched from the y–x basis to the x–y basis:

$$\mathbf{T}'_{\Gamma\text{-X}} = \begin{pmatrix} t_{xx} & t_{xy} \\ t_{yx} & t_{yy} \end{pmatrix} = \begin{bmatrix} t - \dfrac{(\alpha_z r + t)(d_x^2+d_y^2)}{\gamma - i(\omega-\omega_0)} & 0 \\ 0 & t \end{bmatrix}. \tag{8.23}$$

Next, Jones calculus is introduced. For a monochromatic electromagnetic wave in free space, the complex amplitude of the electric field is:

$$\mathbf{E} = \begin{pmatrix} E_x(t) \\ E_y(t) \\ E_z(t) \end{pmatrix} = \begin{pmatrix} E_{0x} e^{i(\mathbf{k}\cdot\mathbf{r}-\omega t+\phi_x)} \\ E_{0y} e^{i(\mathbf{k}\cdot\mathbf{r}-\omega t+\phi_y)} \\ E_{0z} e^{i(\mathbf{k}\cdot\mathbf{r}-\omega t+\phi_z)} \end{pmatrix} = \begin{pmatrix} E_{0x} e^{i\phi_x} \\ E_{0y} e^{i\phi_y} \\ E_{0z} e^{i\phi_z} \end{pmatrix} e^{i(\mathbf{k}\cdot\mathbf{r}-\omega t)}, \tag{8.24}$$

where **k** is the wave vector of the wave, and $\phi_{x,y,z}$ represents the phase shifts of the x, y, z components. Assuming the plane wave propagates along the z-axis, the time- and space-independent part in the x-y plane can be extracted to describe the polarization state, known as the Jones vector:

$$\mathbf{E} = \begin{pmatrix} E_{0x}e^{i\phi_x} \\ E_{0y}e^{i\phi_y} \end{pmatrix}. \tag{8.25}$$

This form of the Jones vector is expressed on a linear polarization (x–y) basis. On this basis, common polarization states can be easily expressed, for example: x-direction linear polarization $\begin{pmatrix} 1 \\ 0 \end{pmatrix}$, y-direction linear polarization $\begin{pmatrix} 0 \\ 1 \end{pmatrix}$, RCP $\frac{1}{\sqrt{2}}\begin{pmatrix} 1 \\ i \end{pmatrix}$, and LCP $\frac{1}{\sqrt{2}}\begin{pmatrix} 1 \\ -i \end{pmatrix}$. Using Jones vectors, polarization elements can be represented as operators in the form of Jones matrices **J**. For instance, a right-handed circular polarizer can be expressed as $\frac{1}{2}\begin{pmatrix} 1 & -i \\ i & 1 \end{pmatrix}$. The transmitted polarization state will be the incident Jones vector left-multiplied by the corresponding Jones matrix. Furthermore, the basis can be changed via linear transformations. Now, consider another basis: the circular polarization (R–L) basis. On such a basis, RCP and LCP will be $\begin{pmatrix} 1 \\ 0 \end{pmatrix}$ and $\begin{pmatrix} 0 \\ 1 \end{pmatrix}$, respectively. Previous chapters have shown that using a transformation matrix $\Lambda = \frac{1}{\sqrt{2}}\begin{pmatrix} 1 & 1 \\ -\mathrm{i} & i \end{pmatrix}$ can convert the linear polarization basis to the circular polarization basis. The new Jones matrix on the circular polarization basis is $\mathbf{J}' = \Lambda^{-1}\mathbf{J}\Lambda$. Given the above, Eq. (8.23) can be rewritten in a more general form on the x–y basis:

$$\mathbf{T}'_{\Gamma\text{-X}} = \begin{pmatrix} T_x e^{i\varphi_x} & 0 \\ 0 & T_y e^{i\varphi_y} \end{pmatrix}, \tag{8.26}$$

where T_x and T_y represent the real amplitude parts of the transmission coefficients in the x and y directions, respectively, with their corresponding complex phase factors being φ_x and φ_y. Note that the preceding derivation has restricted this matrix to the case where the radiation polarization orientation of the guided resonance is parallel to the x-axis (equivalent to parallel to k_x). For an arbitrary resonance winding around (k_x, k_y), a rotation matrix $\mathbf{R}(\theta)$ can be introduced to write a general expression: $\mathbf{J} = \mathbf{R}(\theta)\mathbf{T}'_{\Gamma\text{-X}}\mathbf{R}^{-1}(\theta)$, where θ is the rotation angle of the resonance at (k_x, k_y) relative to k_x-axis. Further, using the transformation matrix, the transmission coefficient matrix for an arbitrary resonance in the linear polarization (k_x–k_y) basis is converted to the circular polarization (R–L) basis matrix:

$$\mathbf{J}' = \Lambda^{-1}\mathbf{J}\Lambda = \Lambda^{-1}\mathbf{R}(\theta)\mathbf{T}'_{\Gamma\text{-X}}\mathbf{R}^{-1}(\theta)\Lambda. \tag{8.27}$$

Substituting the specific forms of $\mathbf{R}(\theta)$, Λ and $\mathbf{T}'_{\Gamma-X}$ into Eq. (8.27) yields:

$$\mathbf{J}'=\begin{pmatrix} \frac{1}{2}(e^{i\varphi_x}T_x+e^{i\varphi_y}T_y) & \frac{1}{2}(e^{i\varphi_x}T_x-e^{i\varphi_y}T_y)(\cos\theta-i\sin\theta)^2 \\ \frac{1}{2}(e^{i\varphi_x}T_x-e^{i\varphi_y}T_y)(\cos\theta+i\sin\theta)^2 & \frac{1}{2}(e^{i\varphi_x}T_x+e^{i\varphi_y}T_y) \end{pmatrix}. \tag{8.28}$$

Equation (8.28) can be separated into two simple terms, written as:

$$\frac{1}{2}(e^{i\varphi_x}T_x+e^{i\varphi_y}T_y)\begin{pmatrix} 1 & 0 \\ 0 & 1 \end{pmatrix}+\frac{1}{2}(e^{i\varphi_x}T_x-e^{i\varphi_y}T_y)\begin{pmatrix} 0 & \exp(-2i\theta) \\ \exp(2i\theta) & 0 \end{pmatrix}. \tag{8.29}$$

Thus, the transmitted fields $|E_{\text{out}}\rangle$ under RCP and LCP incidence are:

$$|E_{out}\rangle=\mathbf{J}'|E_{in}\rangle=\begin{cases} \frac{1}{2}(e^{i\varphi_x}T_x+e^{i\varphi_y}T_y)\begin{pmatrix} 1 \\ 0 \end{pmatrix}+\frac{1}{2}(e^{i\varphi_x}T_x-e^{i\varphi_y}T_y)e^{-2i\theta}\begin{pmatrix} 0 \\ 1 \end{pmatrix} \\ \frac{1}{2}(e^{i\varphi_x}T_x+e^{i\varphi_y}T_y)\begin{pmatrix} 0 \\ 1 \end{pmatrix}+\frac{1}{2}(e^{i\varphi_x}T_x-e^{i\varphi_y}T_y)e^{2i\theta}\begin{pmatrix} 1 \\ 0 \end{pmatrix} \end{cases}, \tag{8.30}$$

where $|E_{in}\rangle$ represents the incident electric field in the circular polarization basis. In the circular polarization basis, the RCP and LCP incident fields should be expressed as $\begin{pmatrix} 1 \\ 0 \end{pmatrix}$ and $\begin{pmatrix} 0 \\ 1 \end{pmatrix}$, respectively. In Eq. (8.30), the first equation corresponds to RCP incidence, and the second equation corresponds to LCP incidence. It can be clearly seen that the transmitted light contains both co-polarized and cross-polarized components relative to the incident light. The transmitted cross-polarized component additionally acquires an extra geometric phase factor $e^{\mp 2i\theta}$. The sign of the acquired geometric phase depends on the polarization state of the incident beam. This geometric phase is the Pancharatnam–Berry (PB) phase. Here, the resonant state naturally winds around the Γ-point BICs (Γ-BICs) in momentum space. It should be noted that although metasurfaces in real space may have transmission coefficient terms similar in form to $e^{i\varphi_x}T_x-e^{i\varphi_y}T_y$, they are typically spatially invariant. In photonic crystal slabs, however, such terms may vary in momentum space. In certain specific cases, such terms may introduce winding phase factors. However, from Eq. (8.23), it can be seen that this term is:

$$e^{i\varphi_x}T_x-e^{i\varphi_y}T_y=-\frac{(d_x^2+d_y^2)(\alpha_z r+t)}{\gamma-i(\omega-\omega_0)}. \tag{8.31}$$

This is clearly a constant. Therefore, under the constraints proposed in this section, it cannot introduce additional winding phase factors. That is, the helical phase arises purely from the geometric structure of the polarization state. Furthermore, it is important to note that the resonance factor (Eq. (8.31)) also causes the centre of the far-field beam profile to become darker when the operating wavelength is below the band edge (i.e. the operating frequency lies within the photonic bands, and the photonic crystal is conductive at this wavelength). This is because the resonant mode exhibits radiation loss at this point, which is distinct from the effects caused by topologically protected phase singularities.

8.2 POLARIZATION CHARACTERISTICS AND DEVICE DESIGN OF QUASI-BICS

8.2.1 Polarization Characteristics of Quasi-BICs

To characterize the polarization properties of the system, the polarization states of Bloch modes can be mapped onto the Poincaré sphere, with coordinates specified by the Stokes parameters S_0, S_1, S_2, and S_3. The coverage of the Poincaré curve characterizes the polarization properties of the system. Photonic crystal slabs based on BICs only support resonant states that are nearly linearly polarized. The corresponding polarization map projected from the BZ onto the normalized Poincaré sphere appears as a band near the equator plus a singularity. The momentum-space polarization states of BICs-based photonic crystal slabs cover only a small region of the Poincaré sphere, lacking coverage at the poles (circular polarization states). This indicates their limited capability for full Stokes polarization modulation, which restricts the applications of photonic crystal slabs. A team led by Professor Jian Zi at Fudan University eliminated the vortex polarization singularities on photonic bands by breaking the in-plane C_2 symmetry of the photonic crystal slab (Fig. 8.3(a)) [21]. This process generated linearly polarized states at the original BIC positions. When the singularities are disrupted, the winding of the polarization principal axes is preserved, leading to the emergence

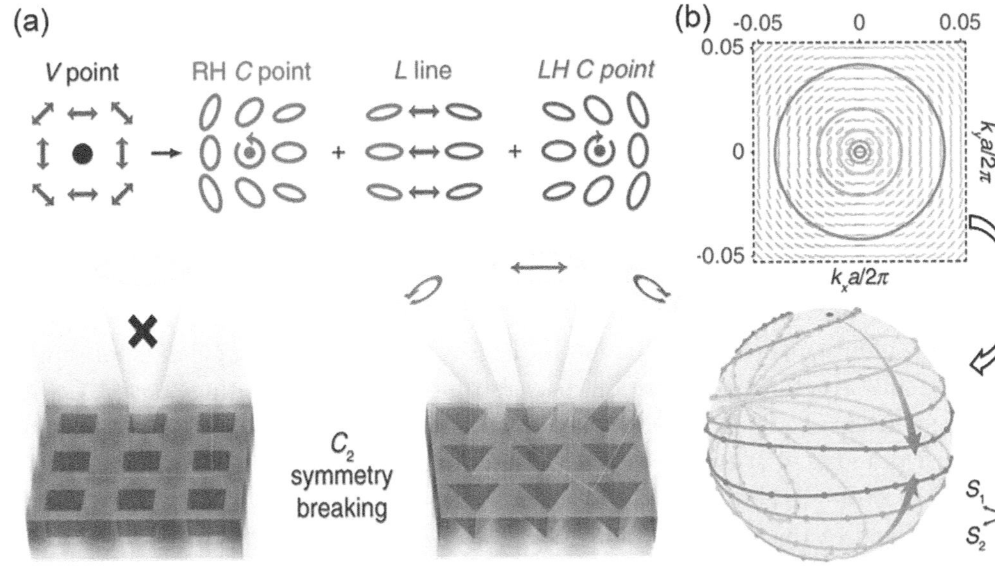

FIGURE 8.3 Radiation polarization characteristics of quasi-BICs in momentum space. (a) Schematic diagram of the change in momentum space radiation polarization state of quasi-BICs after breaking the C_2 symmetry of the structure. (b) The momentum space radiation polarization state of quasi-BICs can cover the entire Poincaré sphere.

of paired circular polarization states near the Γ point. These generated circular polarization states can even achieve full coverage of the Poincaré sphere (Fig. 8.3(b)). This phenomenon provides a new degree of freedom for polarization modulation.

Once BICs leak, they form quasi-BICs, and simultaneously, the singularities are disrupted, resulting in uniquely determined radiation polarization states. Currently, among the studies on polarization manipulation based on quasi-BICs, circularly polarized chiral transmission is the most widely reported. Teams led by Maxim V. Gorkunov [22] and Adam Overvig [23] successively published theoretical research in this field in 2020–2021. To enhance the feasibility of experimentally fabricating chiral quasi-BICs, researchers proposed a flat chiral quasi-BIC structure in 2022 [24]. The unit cell of this structure is an asymmetric dielectric cylinder, and experiments successfully verified the narrowband asymmetric transmission capability of chiral quasi-BICs. Quasi-BICs with frequencies below the first-order diffraction limit possess only a single zeroth-order radiation channel, with a substantial portion of electromagnetic energy localized within the metasurface, resembling the role of a resonant cavity. Building on chiral quasi-BICs and their high-intensity resonant characteristics, circularly polarized micro-laser emission devices can be achieved. Based on this, Professor Qinghai Song from Harbin Institute of Technology collaborated with Professor Yuri Kivshar from the Australian National University in 2022 and proposed a tilted elliptical pillar structure based on doped TiO_2 that supports quasi-BIC modes [15]. Under pulsed laser pumping, this quasi-BIC metasurface reflects circularly polarized laser light with high quality factors. Subsequently, outstanding research achievements on chiral circular polarization manipulation using quasi-BICs have continued to emerge [25,26]. Compared to circular polarization manipulation, Γ-point quasi-BICs radiation tends to form linearly polarized states more readily. Researchers have developed a variety of device structures to excite different electromagnetic multipole modes for achieving efficient quasi-BICs linearly polarized radiation. Examples include a circular hole structure offset from the centre of a square dielectric block to excite toroidal dipole modes [27], an asymmetric L-shaped aperture or two rectangular holes of different sizes in a dielectric to excite electric quadrupole modes [28], or directly constructing two dielectric cylinders with different diameters to excite toroidal dipole modes [29]. Next, this section will detail the design principles and analysis methods for quasi-BIC polarization devices.

8.2.2 Device Design and Scattering Principle

The polarization state of a non-radiation Γ-BIC is uncertain, i.e. it is a polarization singularity, but any structural perturbation may transform it into a radiating quasi-BIC with a uniquely determined polarization state. This implies that quasi-BICs with polarization characteristics located at the Γ point (corresponding to normally incident light) can be designed accordingly. Next, starting with symmetry-protected BICs (supported by a rectangular structure), this section demonstrates the design process and principles for polarization control from Γ-BICs to quasi-BICs [30]. Based on the quasi-BIC effect, it is possible to manipulate not only the linear polarization state of electromagnetic waves but also their elliptical and circular polarization states. This section primarily demonstrates the ability of quasi-BICs to manipulate circularly polarized waves.

As shown in Fig. 8.4(a–d), the radiation polarization state distributions of unit cells with different symmetries in the Brillouin zone (BZ) are presented. The polarization state at each wave vector point can be plotted by monitoring the electric field vector of the radiation leakage field. The symmetry distribution of the radiation polarization state in the BZ is closely related to the symmetry of the unit cell in real space. For example, a unit cell with C_2 symmetry also exhibits a C_2 symmetric distribution of radiation polarization states in the BZ. However, in the BZ of a C_2-symmetric unit cell, the radiation polarization states are almost linear, and the polarization of Γ-BICs remains uncertain, as shown in Fig. 8.4(b). This means that Γ-BICs lack polarization manipulation capabilities, and even in the entire momentum space, their polarization control ability is quite limited. To achieve a radiation polarization state at the Γ point, it is necessary to obtain

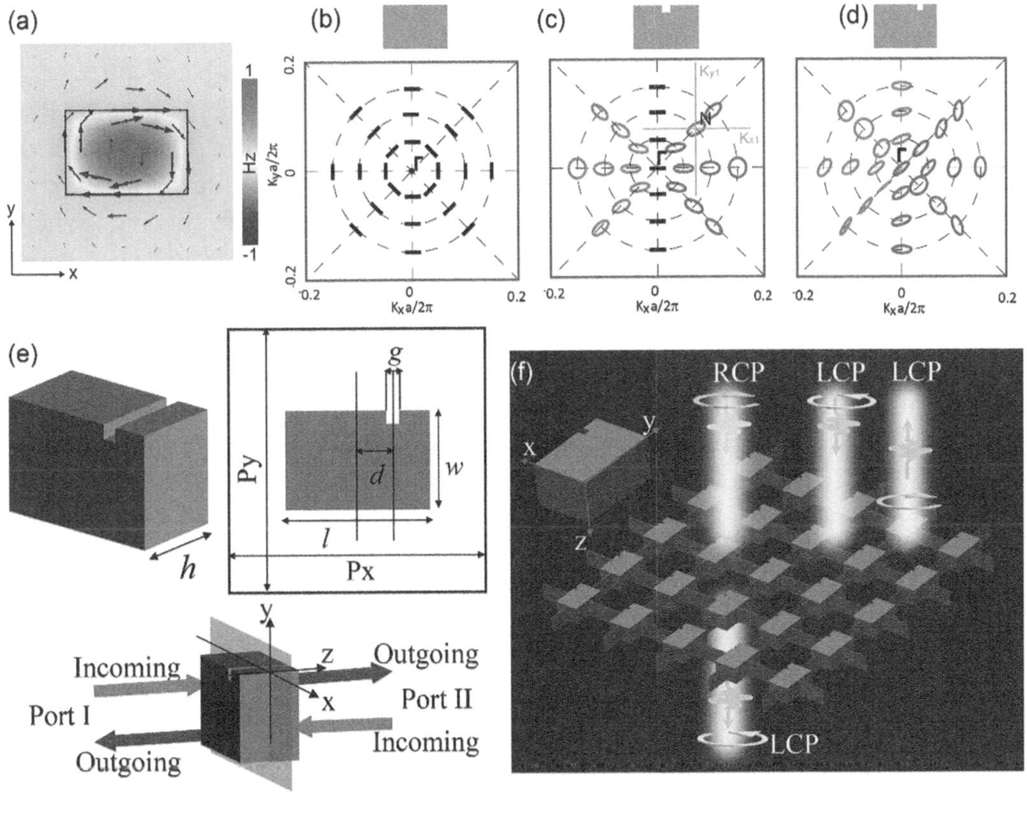

FIGURE 8.4 (a) Electric and magnetic field distributions of BICs in the C_2 symmetric structure. (b) Polarization distribution in momentum space of the BICs metasurface. (C) Polarization distribution in momentum space when the C_2 symmetry is broken by introducing an opening while preserving mirror symmetry. (d) Polarization distribution in momentum space after both the in-plane C_2 symmetry and mirror symmetry are broken. The ellipses on the left and right sides of the k_x=0 axis in (c, d) represent the elliptical polarization states with electric field vectors rotating counterclockwise and clockwise, respectively. (e) Designed three-dimensional quasi-BIC structure with neither C_2 symmetry nor mirror symmetry. (f) Schematic diagram of the circular polarization manipulation function supported by this structure.

a quasi-BIC located at the Γ point. The most intuitive method is to introduce an opening along the central axis of the structure. This operation breaks the C_2 symmetry but retains the mirror symmetry in the *x-y* plane, as illustrated in Fig. 8.4(c). At this point, since mirror symmetry with respect to the *y*-axis is preserved in real space, the polarization state distribution in the BZ is also mirror-symmetric with respect to k_y. However, the radiation at the Γ point is linearly polarized, and circular polarization states emerge at other positions where $k_x \neq 0$. This indicates that linear polarization waves can be manipulated at the Γ point, while circular polarization waves can be controlled at other locations in momentum space.

For the quasi-BIC shown in Fig. 8.4(c), the circular polarization state at the Γ point remains fully coupled, and circular dichroism (CD) is not achieved. However, it is evident that by varying the wave vector along the direction where $k_x \neq 0$, circular and elliptical polarization modes can be obtained (e.g. at point N), providing the possibility to achieve CD. In the BZ of Fig. 8.4(c), if the Γ point is shifted to point N, the symmetry distribution of the polarization states no longer possesses C_2 symmetry or any in-plane mirror symmetry. As mentioned earlier, the symmetry of the polarization distribution in the BZ is closely related to the structural symmetry of the unit cell in real space. Therefore, in real space, achieving CD under normal light incidence (corresponding to the Γ point) can rely on a structure that lacks both C_2 symmetry and any in-plane mirror symmetry, as shown in Fig. 8.4(d). In such a structure, the far-field radiation at the Γ point is no longer linearly polarized but instead exhibits decoupled circular polarization, thereby enabling circular polarization manipulation based on a Γ-point quasi-BIC.

Commercial CST Microwave Studio was used to perform structural design and numerical simulations of scattering spectra by solving Maxwell's equations with the frequency-domain solver. The eigenfields were solved using the frequency-domain eigenmode solver in Comsol. The designed structural unit is based on a rectangular silicon pillar (with a relative permittivity of 11.9 for silicon and a conductivity less than 0.03 S/m) that supports symmetry-protected BICs. When no opening is introduced, the structure exhibits in-plane C_2 symmetry, resulting in complete coupling of electromagnetic wave spins and an inability to achieve spin-selective propagation. When a square notch is created at a suitable position to break both the in-plane C_2 symmetry and the in-plane mirror symmetry of the structure, the BICs transform into quasi-BICs, as shown in Fig. 8.4(e). The lattice constants of the unit cell are $Px = Py = 160$ μm, the rectangular pillar length is $l = 90$ μm, the width is $w = 60$ μm, and the height is $h = 50$ μm. The side length of the square notch is $g = 8$–20 μm, and its position is offset from the centre of the rectangular pillar by $d = 18$ μm. This structure enables manipulation of the circular polarization state of light waves at the quasi-BIC point. Specifically, the port preferentially transmits the cross-polarized wave of incident waves with a certain helicity, while reflecting the co-polarized wave of input waves with the opposite helicity, as shown in Fig. 8.4(f). Here, LCP and RCP waves are defined as waves whose electric field vectors rotate counterclockwise and clockwise, respectively, along the direction of light propagation.

Figure 8.4(d) shows that the radiation polarization state of the quasi-BIC at the Γ-point has non-zero ellipticity. Within the framework of TCMT, the dynamic relationship governing the time-evolution of the resonance amplitude in a resonator with ports (Fig. 8.4(e)) is as follows:

$$\frac{d\mathbf{a}_c}{dt} = [i\omega_0 - \gamma_0]\mathbf{a}_c + \mathbf{D}_c^T |I_{cir}\rangle, \tag{8.32}$$

where ω_0 and γ_0 are the resonance centre frequency and radiation damping rate of the quasi-BIC, respectively. The $\boldsymbol{a}_c$ is the scalar amplitude defined by the energy stored in the eigenmode, and $\mathbf{D}_c$ describes the coupling between the resonator and the ports. Compared to Eq. (8.4), "$-i\omega_0$" is intentionally written here as "$i\omega_0$" because, as previously noted, the resonant mode can also be mathematically expressed with a separate $e^{i\omega t}$ term to represent its temporal characteristics, which is equivalent to $e^{-i\omega t}$. The $|I_{cir}\rangle = \begin{bmatrix} I_R^{\mathrm{I}} & I_L^{\mathrm{I}} & I_R^{\mathrm{II}} & I_L^{\mathrm{II}} \end{bmatrix}^{\mathrm{T}}$ represents the amplitudes of the incident RCP (or LCP) waves from Port I and Port II. The scattering matrix of the metasurface in the quasi-BIC mode is derived next.

First, considering the outgoing wave $|O''_{cir}\rangle$ from the metasurface under mode coupling, it arises from two contributions. The first term $\mathbf{C}_{cir}|I_{cir}\rangle$ originates from background scattering caused by ordinary non-chiral responses, where incoming and outgoing waves are coupled via a direct path. The second term $|\mathrm{O}_{cir}^{D}\rangle = \mathbf{D}_c\boldsymbol{a}_c$ comes from the coupling between the quasi-BIC resonance mode and the ports. The final output wave can be written as:

$$|O''_{cir}\rangle = \mathbf{C}_{cir}|I_{cir}\rangle + \mathbf{D}_c\boldsymbol{a}_c = \mathbf{S}''_{cir}|I_{cir}\rangle, \tag{8.33}$$

where $|O''_{cir}\rangle = \begin{bmatrix} O_R^{\mathrm{I}} & O_L^{\mathrm{I}} & O_R^{\mathrm{II}} & O_L^{\mathrm{II}} \end{bmatrix}^{\mathrm{T}}$ is the total output wave amplitude accounting for both background scattering and resonance-port coupling, and the corresponding total scattering matrix is $\mathbf{S}''_{cir}$. The background scattering can be expressed as:

$$\mathbf{C}_{cir} = \begin{bmatrix} r_{RR}^{\mathrm{I}} & r_{RL}^{\mathrm{I}} & t_{RR}^{\mathrm{II}} & t_{RL}^{\mathrm{II}} \\ r_{LR}^{\mathrm{I}} & r_{LL}^{\mathrm{I}} & t_{LR}^{\mathrm{II}} & t_{LL}^{\mathrm{II}} \\ t_{RR}^{\mathrm{I}} & t_{RL}^{\mathrm{I}} & r_{RR}^{\mathrm{II}} & r_{RL}^{\mathrm{II}} \\ t_{LR}^{\mathrm{I}} & t_{LL}^{\mathrm{I}} & r_{LR}^{\mathrm{II}} & r_{LL}^{\mathrm{II}} \end{bmatrix} = \begin{bmatrix} r & r' & t & t' \\ r' & r & t' & t \\ t & t' & r & r' \\ t' & t & r' & r \end{bmatrix}, \tag{8.34}$$

where the reflection and transmission coefficients r, r', t, t' for different ports are complex numbers, with phase delays $\exp(i\varphi_1)$, $\exp(i\varphi_2)$, $\exp(i\varphi_3)$, and $\exp(i\varphi_4)$, respectively, and satisfy $|r|^2 + |r'|^2 + |t|^2 + |t'|^2 = 1$. In the equation, $\boldsymbol{r}_{jk}^{\mathrm{I/II}}$ and $\boldsymbol{t}_{jk}^{\mathrm{I/II}}$ ($j, k \in \{L, R\}$) represent the reflection and transmission coefficients, respectively, for a k-component wave incident from Port I/II, resulting in a j-component wave at Port I/II. Considering the time-reversal symmetry of the system, the background scattering matrix should be unitary, i.e. $\mathbf{C}_{cir}^{+}\mathbf{C}_{cir} = \mathbf{I}$. Furthermore, for the resonance-port coupling output term $|O_{cir}^{D}\rangle = \mathbf{D}_c\boldsymbol{a}_c$, the eigenstate with amplitude $\boldsymbol{a}_c$ couples to the circularly polarized wave incident on one side of the metasurface, with the coupling coefficient matrix $\mathbf{D}_c$. The wave output to the port after coupling is denoted as $|\mathbf{O}_{cir}^{D}\rangle = \begin{bmatrix} O_R^{D-\mathrm{I}} & O_L^{D-\mathrm{I}} & O_R^{D-\mathrm{II}} & O_L^{D-\mathrm{II}} \end{bmatrix}^{\mathrm{T}}$, where $O_{L/R}^{D-\mathrm{I/II}}$ represents the wave amplitude resulting from the coupling between the resonance and Port I/II and can be written as:

$$\begin{bmatrix} O_R^{D-\mathrm{I}} \\ O_L^{D-\mathrm{I}} \\ O_R^{D-\mathrm{II}} \\ O_L^{D-\mathrm{II}} \end{bmatrix} = \begin{bmatrix} d_R^{\mathrm{I}} \\ d_L^{\mathrm{I}} \\ d_R^{\mathrm{II}} \\ d_L^{\mathrm{II}} \end{bmatrix} [a], \tag{8.35a}$$

$$\mathbf{D} = \begin{bmatrix} d_R^{\mathrm{I}} \\ d_L^{\mathrm{I}} \\ d_R^{\mathrm{II}} \\ d_L^{\mathrm{II}} \end{bmatrix}, \tag{8.35b}$$

where $d_j^{\mathrm{I/II}}$ ($j \in \{L, R\}$) represents the coupling coefficient between the resonant mode and the incident wave at Port I/II, forming the j-component wave. Since the radiation rate is related to the resonance-port coupling, the $\mathbf{D}_c$ matrix is not arbitrary and satisfies:

$$\mathbf{D}_c^{+}\mathbf{D}_c = 2\gamma_0. \tag{8.36}$$

The specific form of the $\mathbf{D}_c$ matrix is derived next. It is important to note that circular polarization in this system is not fully decoupled, meaning each eigenmode coupled to the port outputs both LCP and RCP components. Due to the mirror symmetry of the structure with respect to the z-plane, the total output of LCP and RCP waves from the eigenmode coupled to the two ports is the same, i.e.:

$$d_L^{\mathrm{I}} + d_R^{\mathrm{I}} = d_L^{\mathrm{II}} + d_R^{\mathrm{II}}. \tag{8.37}$$

Additionally, the two chiral-flipped circular polarization states coupled to the two ports should have the same coupling coefficients:

$$\begin{cases} d_L^{\mathrm{I}} = d_R^{\mathrm{II}} = m \\ d_R^{\mathrm{I}} = d_L^{\mathrm{II}} = n \end{cases}. \tag{8.38}$$

Therefore, the $\mathbf{D}_c$ matrix can be written as:

$$\mathbf{D}_c = \begin{bmatrix} m \\ n \\ n \\ m \end{bmatrix}. \tag{8.39}$$

The resonance-port coupling involves a phase angle, so each term in Eq. (8.39) is a complex number. According to the relationship in Eq. (8.36), it follows that:

$$|m|^2+|n|^2=\gamma_0. \tag{8.40}$$

Solving this equation yields:

$$\begin{cases} m=\exp(i\theta_m)\sqrt{\beta\gamma_0} \\ n=\exp(i\theta_n)\sqrt{\alpha\gamma_0} \end{cases}, \tag{8.41}$$

where $0<\beta,\alpha<1$ and $\beta+\alpha=1$, and $\exp(i\theta)$ represents the coupling phase. The parameters β and α essentially relate to the proportion of energy in the LCP and RCP output waves coupled from the resonant mode to the port. For example, the total output wave from the resonance coupled to Port I contains both RCP and LCP components, with the RCP wave energy accounting for a proportion β (amplitude proportion $\sqrt{\beta}$), and the LCP wave energy accounting for a proportion α (amplitude proportion $\sqrt{\alpha}$). Clearly, there is $\beta+\alpha=1$. Moreover, there should be $\beta,\alpha\neq 0.5$. If $\beta,\alpha=0.5$, it would mean the output RCP and LCP waves have equal energy, resulting in linear polarization.

Combining Eqs. (8.32) and (8.33), the total scattering matrix $\mathbf{S}''_{cir}$ can be obtained:

$$\mathbf{S}''_{cir}=\mathbf{C}_{cir}+\frac{\mathbf{D}_c\mathbf{D}_c^T}{i(\omega-\omega_0)+\gamma_0}. \tag{8.42}$$

Substituting Eqs. (8.34) and (8.39) into Eq. (8.42) gives:

$$\mathbf{S}''_{cir}=\begin{bmatrix} r_{RR}^{\mathrm{I}} & r_{RL}^{\mathrm{I}} & t_{RR}^{\mathrm{II}} & t_{RL}^{\mathrm{II}} \\ r_{LR}^{\mathrm{I}} & r_{LL}^{\mathrm{I}} & t_{LR}^{\mathrm{II}} & t_{LL}^{\mathrm{II}} \\ t_{RR}^{\mathrm{I}} & t_{RL}^{\mathrm{I}} & r_{RR}^{\mathrm{II}} & r_{RL}^{\mathrm{II}} \\ t_{LR}^{\mathrm{I}} & t_{LL}^{\mathrm{I}} & r_{LR}^{\mathrm{II}} & r_{LL}^{\mathrm{II}} \end{bmatrix}$$

$$=\begin{bmatrix} r+\frac{m^2}{i(\omega-\omega_0)+\gamma_0} & r'+\frac{mn}{i(\omega-\omega_0)+\gamma_0} & t+\frac{mn}{i(\omega-\omega_0)+\gamma_0} & t'+\frac{m^2}{i(\omega-\omega_0)+\gamma_0} \\ r'+\frac{mn}{i(\omega-\omega_0)+\gamma_0} & r+\frac{n^2}{i(\omega-\omega_0)+\gamma_0} & t'+\frac{n^2}{i(\omega-\omega_0)+\gamma_0} & t+\frac{mn}{i(\omega-\omega_0)+\gamma_0} \\ t+\frac{mn}{i(\omega-\omega_0)+\gamma_0} & t'+\frac{n^2}{i(\omega-\omega_0)+\gamma_0} & r+\frac{n^2}{i(\omega-\omega_0)+\gamma_0} & r'+\frac{mn}{i(\omega-\omega_0)+\gamma_0} \\ t'+\frac{m^2}{i(\omega-\omega_0)+\gamma_0} & t+\frac{mn}{i(\omega-\omega_0)+\gamma_0} & r'+\frac{mn}{i(\omega-\omega_0)+\gamma_0} & r+\frac{m^2}{i(\omega-\omega_0)+\gamma_0} \end{bmatrix}. \tag{8.43}$$

Through a series of mathematical operations on Eq. (8.43), the reflection and transmission coefficients can be simplified into a Fano lineshape function form with centre frequency ω_0 and radiation damping rate γ_0, denoted as $(M + iK)/[i(\omega - \omega_0) + \gamma_0]$, where M and K are real numbers. The scattering matrix is as follows:

$$\begin{bmatrix} r_{RR}^{\mathrm{I}} & r_{RL}^{\mathrm{I}} & t_{RR}^{\mathrm{II}} & t_{RL}^{\mathrm{II}} \\ r_{LR}^{\mathrm{I}} & r_{LL}^{\mathrm{I}} & t_{LR}^{\mathrm{II}} & t_{LL}^{\mathrm{II}} \\ t_{RR}^{\mathrm{I}} & t_{RL}^{\mathrm{I}} & r_{RR}^{\mathrm{II}} & r_{RL}^{\mathrm{II}} \\ t_{LR}^{\mathrm{I}} & t_{LL}^{\mathrm{I}} & r_{LR}^{\mathrm{II}} & r_{LL}^{\mathrm{II}} \end{bmatrix} = \begin{bmatrix} \frac{M_1+iK_1}{i(\omega-\omega_0)+\gamma_0} & \frac{M_2+iK_2}{i(\omega-\omega_0)+\gamma_0} & \frac{M_4+iK_4}{i(\omega-\omega_0)+\gamma_0} & \frac{M_6+iK_6}{i(\omega-\omega_0)+\gamma_0} \\ \frac{M_2+iK_2}{i(\omega-\omega_0)+\gamma_0} & \frac{M_3+iK_3}{i(\omega-\omega_0)+\gamma_0} & \frac{M_5+iK_5}{i(\omega-\omega_0)+\gamma_0} & \frac{M_4+iK_4}{i(\omega-\omega_0)+\gamma_0} \\ \frac{M_4+iK_4}{i(\omega-\omega_0)+\gamma_0} & \frac{M_5+iK_5}{i(\omega-\omega_0)+\gamma_0} & \frac{M_3+iK_3}{i(\omega-\omega_0)+\gamma_0} & \frac{M_2+iK_2}{i(\omega-\omega_0)+\gamma_0} \\ \frac{M_6+iK_6}{i(\omega-\omega_0)+\gamma_0} & \frac{M_4+iK_4}{i(\omega-\omega_0)+\gamma_0} & \frac{M_2+iK_2}{i(\omega-\omega_0)+\gamma_0} & \frac{M_1+iK_1}{i(\omega-\omega_0)+\gamma_0} \end{bmatrix}. \tag{8.44}$$

In Eq. (8.44), the frequency and radiation loss for all scattering coefficients correspond to those of the quasi-BIC. By adjusting the Q factor of the quasi-BIC, the radiation loss can be reduced to an extremely low level. Substituting the elements of the $\mathbf{C}_{cir}$ and $\mathbf{D}_c$ matrices into Eq. (8.44), all scattering coefficients can be expressed in more specific forms:

$$\frac{M_j+iK_j}{i(\omega-\omega_0)+\gamma_0} = \frac{[\gamma_0 A_j+(\omega-\omega_0)B_j]+i[\gamma_0 C_j+(\omega-\omega_0)D_j]}{i(\omega-\omega_0)+\gamma_0}, j \in \{1,2,3,4,5,6\}, \tag{8.45}$$

where the values of A_j, B_j, C_j, and D_j can be obtained by fitting Eq. (8.45) with simulation results.

8.2.3 Polarization Spectral Analysis of Devices

As shown in Fig. 8.5, the structure exhibits three quasi-BICs within the range of 1.1–1.5 THz. At each quasi-BIC frequency point, terahertz wave spin decoupling occurs, enabling chiral transmission. The CD spectrum reveals that the frequencies supporting this chiral transmission have extremely narrow linewidths (full width at half maximum), which do not exceed 1 GHz in this design.

To identify the type of BICs, an in-plane mirror symmetry structure (with $d = 0$ μm) is first constructed. This structure only breaks the in-plane C_2 symmetry, providing a convenient framework for identifying symmetry-protected BICs. Figure 8.6(a, b) shows the photonic band structure and radiation Q factor for the transverse electric (TE) mode in a rectangular silicon structure without openings. A resonant mode with an infinite Q factor is found at the Γ point (1.09 THz) on the TE_1 band. This bound electromagnetic state corresponds to a magnetic dipole mode with even symmetry and no out-of-plane radiation, as illustrated in Fig. 8.6(b). Furthermore, Fig. 8.6(c, d) investigates the transmission spectra as a function of the opening size. As the size of the central opening increases (enhancing the breaking of C_2 symmetry), the radiation Q factor (defined as $\omega_0/2\gamma_0$) rapidly decreases

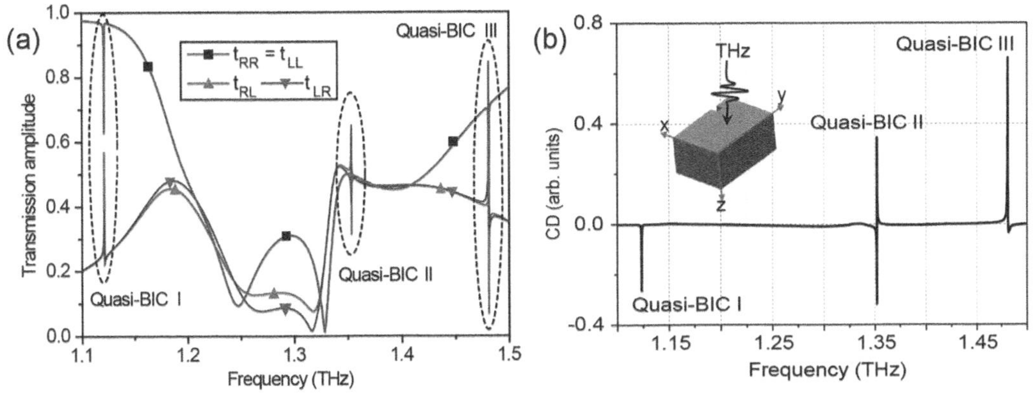

FIGURE 8.5 (a) The circularly polarized wave transmission spectra of a metasurface ($g = 10$ μm) supporting quasi-BICs, and its (b) circular dichroism (CD); three chiral quasi-BICs were displayed in the spectra.

following a $Q \propto \alpha^{-2}$ relationship (Fig. 8.6(d, e)), where α^{-2} describes the degree of asymmetry and is defined as the ratio of the opening area to the initial area ($\Delta S/S$). These characteristics indicate that the first quasi-BIC (Quasi-BIC I) belongs to the symmetry-protected type (this BIC corresponds to the TE mode). As shown in Fig. 8.6(f, g), a second BIC is identified from the photonic band structure and radiation Q factor results for the transverse magnetic (TM) mode in the rectangular silicon structure. A resonant mode with an infinite Q factor is found at the Γ point (1.34 THz) on the TM_3 band. This bound electromagnetic mode corresponds to an electric dipole mode with even symmetry and no out-of-plane radiation, as depicted in Fig. 8.6(g). Under the breaking of C_2 symmetry, the radiation Q factor also decreases rapidly following a $Q \propto \alpha^{-2}$ relationship, as observed in the transmission spectra shown in Fig. 8.6(h–j). Clearly, the second quasi-BIC (Quasi-BIC II) is also classified as the symmetry-protected type (this BIC corresponds to the TM mode).

In the aforementioned structure where in-plane C_2 symmetry is broken but in-plane mirror symmetry is preserved, a third resonant mode with an infinite Q factor corresponding to the third BIC is found at another Γ point (1.465 THz) in the TM photonic band, as shown in Fig. 8.7(a–c). However, even though C_2 symmetry has been broken, the third BIC located within the 1.45–1.5 THz range still does not couple to the far field. Clearly, the third BIC is not symmetry-protected. Furthermore, when the opening position is shifted along the *x*-axis, the structure loses its in-plane mirror symmetry, and this quasi-BIC (Quasi-BIC III) is formed, as illustrated in Fig. 8.7(d–f). As the opening continues to shift to a distance of 35 μm from the centre, the quasi-BIC transitions back into a BIC state (Fig. 8.7(e, f)). To further identify the type of this BIC, more details are provided in Fig. 8.7(g–i). First, as seen in Fig. 8.7(c), the electric field lines exhibit a toroidal distribution in the *y-z* plane, while the magnetic field is distributed in the *x-y* plane. A clearer schematic is shown in Fig. 8.7(g), where the electromagnetic field mode essentially consists of two magnetic dipoles with opposite magnetic moments. Both magnetic moments lie in the *x-y* plane, radiating electromagnetic

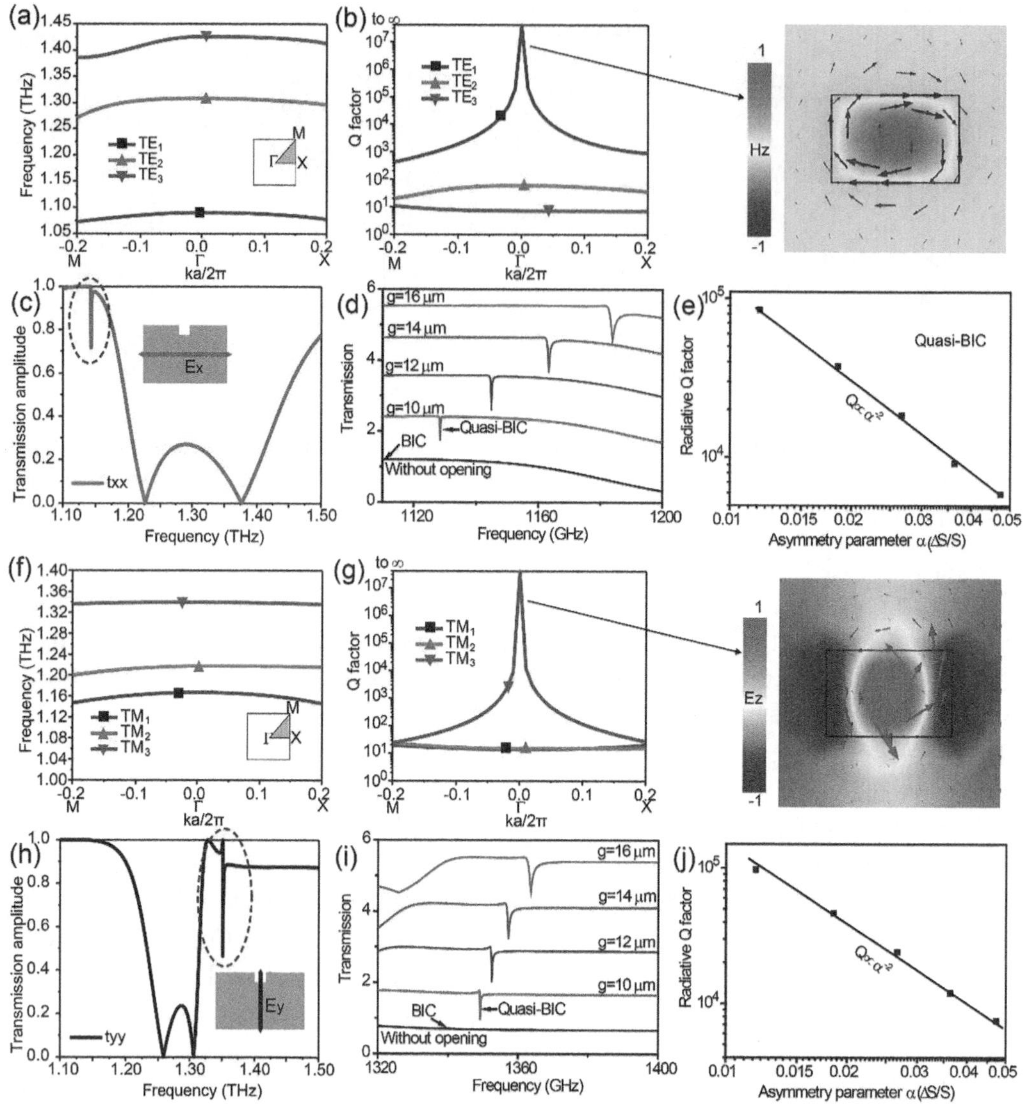

FIGURE 8.6 (a–b) The photonic band structure and radiation Q factor of TE mode in rectangular silicon without openings. The illustration on the right in (b) represents the normalized magnetic field distribution corresponding to the *z*-component of the first BIC, where the arrows represent the electric field vectors. (c) For the structure with $d = 0$ μm and $g = 12$ μm, the full spectrum in the range of 1.1–1.5 THz under *x*-polarized wave excitation. (d) Fine transmission spectra with different opening size structures corresponding to (e) changes in radiation Q factor. (f, g) The photonic band and radiation Q factor of TM mode in rectangular silicon without openings. The illustration on the right in (g) represents the normalized electric field distribution corresponding to the *z*-component of the second BIC, where the arrows represent the magnetic field vectors. (h) For the structure with $d = 0$ μm and $g = 12$ μm, the full spectrum in the range of 1.1–1.5 THz under *y*-polarized wave excitation. (i) Fine transmission spectra with different opening size structures corresponding to (j) changes in radiation Q factor.

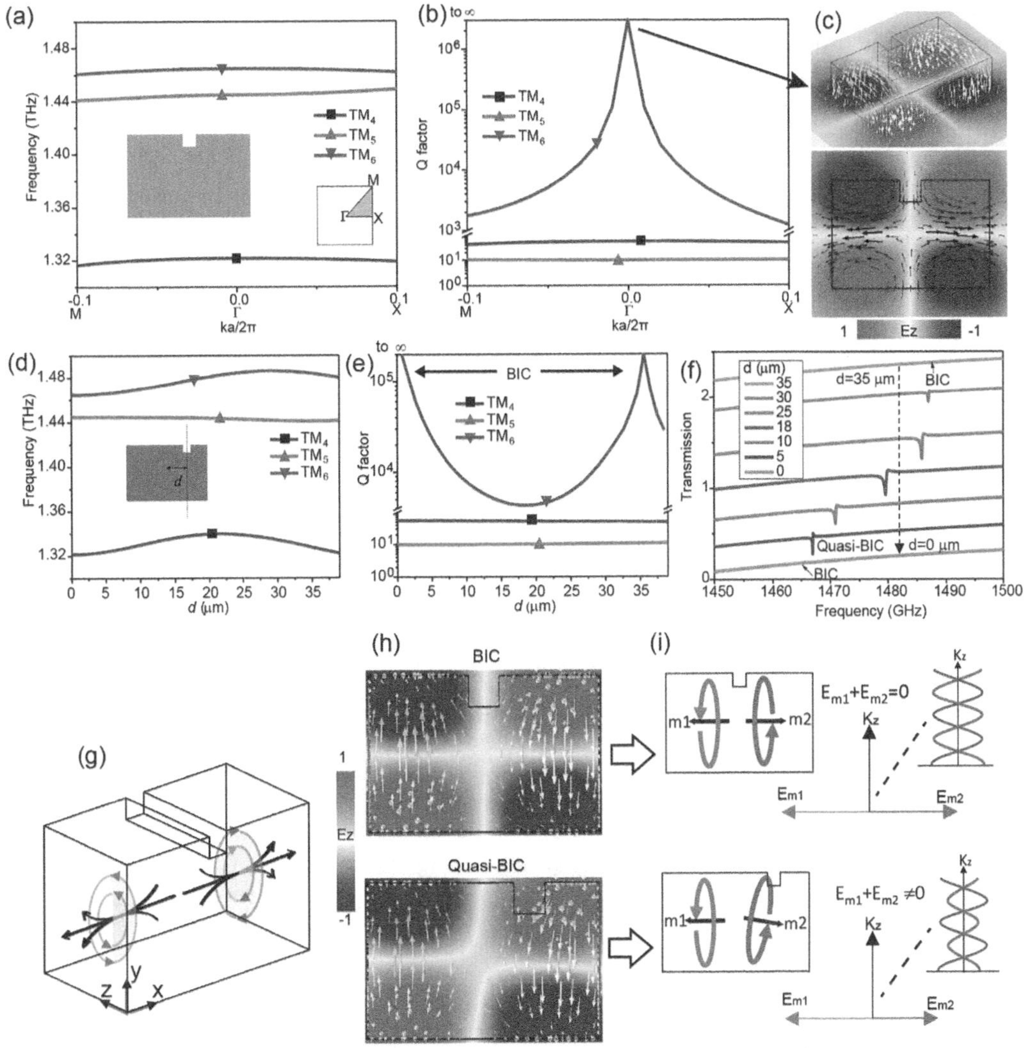

FIGURE 8.7 (a, b) The photonic band structure and radiation Q factor of TM mode in the structure where the in-plane C_2 symmetry is disrupted but the in-plane mirror symmetry is retained. (c) Normalized electric field distribution corresponding to the z-component of the third BIC, where white arrows represent electric field lines and black arrows represent magnetic field lines. (d, e) The variation of Γ point frequency and radiation Q factor on the TM band with the opening position corresponds to the (f) fine transmission spectrum under x-polarized wave excitation. (g) Schematic diagram of multipole distribution for the third quasi-BIC (Quasi BIC III). (h) The electric field changes with the position of the opening, where white arrows represent electric field lines. (i) As the BIC transitions to a quasi-BIC, two magnetic moments and out-of-plane radiation changes are illustrated, with ring-shaped arrows representing electric field lines and black arrows representing magnetic moments.

fields out of the plane. The radiation intensities of the two magnetic dipoles are equal, but their magnetic moments are completely opposite, resulting in a fixed phase difference of π (denoted as m1 and m2 in the figure). This leads to complete cancellation of radiation in each channel ($E_{m1} + E_{m2} = 0$), which is the reason for the formation of the non-radiative BIC (Fig. 8.7(i)). When the opening position is altered, Fig. 8.7(h) shows a slight deflection in the electric field distribution of one of the magnetic dipoles, causing an imbalance between the two magnetic dipoles. Consequently, the sum of radiation from the two magnetic dipoles is no longer zero ($E_{m1} + E_{m2} \neq 0$), forming the quasi-BIC state (Fig. 8.7(i)). This type of BIC is based on the coupling of all radiation waves in the resonant mode and belongs to the single-resonance type. Single-resonance BICs are not guaranteed by symmetry. BICs without symmetry incompatibility also lead to tunable light trapping, as discovered by Chia Wei Hsu et al. in 2013 [31]. Coupled-wave theory is typically used to explain such BICs, where a single resonance itself is considered as the destructive interference of multiple sets of radiation waves [32]. By adjusting given parameters, these radiation waves can be combined to cancel each other out, achieving suppressed total radiation (nearly zero radiation). This differs from symmetry-protected BICs, where zero-radiation arises from symmetry incompatibility that decouples bound states from the radiation continuum.

Now, returning once again to the discussion on the CD and scattering properties of the metasurface. The third quasi-BIC point is selected for further investigation of its fine reflection and transmission characteristics (it should be noted that the discussion on the CD and scattering properties of the remaining two quasi-BICs is similarly general and will not be elaborated here separately). The opening size is reduced to 8 μm to achieve a higher Q factor and narrower linewidth. Considering the scenario where a terahertz wave is incident from Port I, the detailed spectra are shown in Fig. 8.8. For convenience in description, its scattering analytical expression is reiterated here. According to Eq. (8.44), the scattering coefficient matrix for incidence from Port I is obtained as:

$$\begin{bmatrix} r_{RR} & r_{RL} \\ r_{LR} & r_{LL} \end{bmatrix} = \begin{bmatrix} \dfrac{M_1 + iK_1}{i(\omega-\omega_0)+\gamma_0} & \dfrac{M_2 + iK_2}{i(\omega-\omega_0)+\gamma_0} \\ \dfrac{M_2 + iK_2}{i(\omega-\omega_0)+\gamma_0} & \dfrac{M_3 + iK_3}{i(\omega-\omega_0)+\gamma_0} \end{bmatrix}. \tag{8.46}$$

$$\begin{bmatrix} t_{RR} & t_{RL} \\ t_{LR} & t_{LL} \end{bmatrix} = \begin{bmatrix} \dfrac{M_4 + iK_4}{i(\omega-\omega_0)+\gamma_0} & \dfrac{M_5 + iK_5}{i(\omega-\omega_0)+\gamma_0} \\ \dfrac{M_6 + iK_6}{i(\omega-\omega_0)+\gamma_0} & \dfrac{M_4 + iK_4}{i(\omega-\omega_0)+\gamma_0} \end{bmatrix}. \tag{8.47}$$

For transmission, the co-polarized waves have identical transmission coefficients ($t_{RR} = t_{LL}$), whereas the cross-polarized waves have different transmission coefficients ($t_{RL} \neq t_{LR}$), as illustrated in Fig. 8.8(a). For reflection, the cross-polarized waves share identical reflection coefficients ($r_{RL} = r_{LR}$), while the co-polarized waves exhibit different reflection coefficients ($r_{LL} \neq r_{RR}$), as shown in Fig. 8.8(b). In this design, when RCP terahertz waves at the

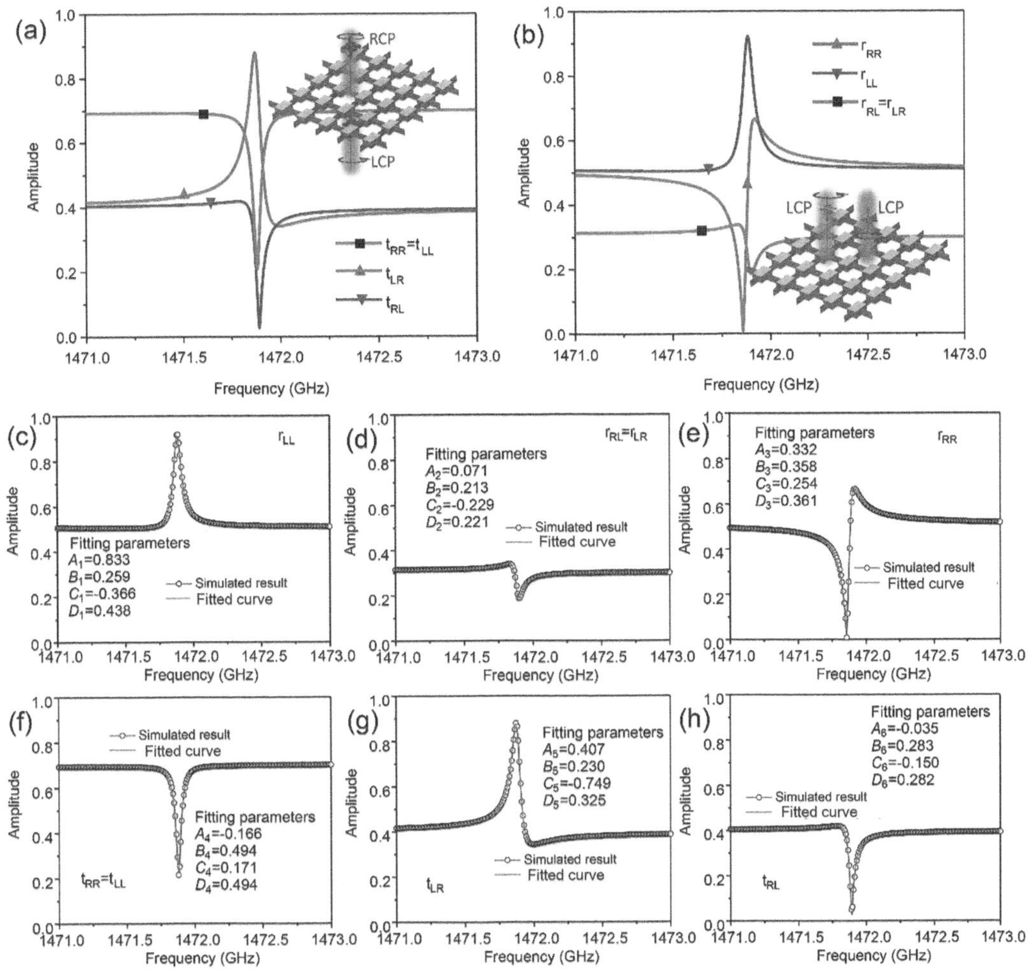

FIGURE 8.8 For Quasi BIC III, (a, b) shows the circularly polarized transmission and reflection spectra based on the structure with an opening size of 8 μm. (c–e) Comparison of circularly polarized reflectance spectra and fitting structures. (f–h) Comparison of circularly polarized transmission spectra and fitting results.

quasi-BIC frequency are incident, most of them will be converted into LCP waves and transmitted (comparing t_{RR} and t_{LR} in Fig. 8.8(a)), while reflection is relatively weak (comparing r_{RR} and r_{LR} in Fig. 8.8(b)). However, when LCP terahertz waves at the quasi-BIC frequency are incident on the metasurface, both the co-polarized t_{LL} and cross-polarized t_{RL} components exhibit low transmission (Fig. 8.8(a)), and most of them are reflected as co-polarized waves (comparing r_{LL} and r_{RL} in Fig. 8.8(b)). In short, for waves incident from Port I onto this quasi-BIC metasurface, RCP waves are primarily transmitted, while LCP waves are largely reflected. All simulation results in Fig. 8.8(a, b) are consistent with the theoretical trends described by Eqs. (8.46) and (8.47). Furthermore, the simulation results can be perfectly fitted using Eq. (8.45), and the values of the real parameters A_j, B_j, C_j, and D_j (j = 1, 2, 3, 4, 5, 6) are determined, as shown in Fig. 8.8(c–h).

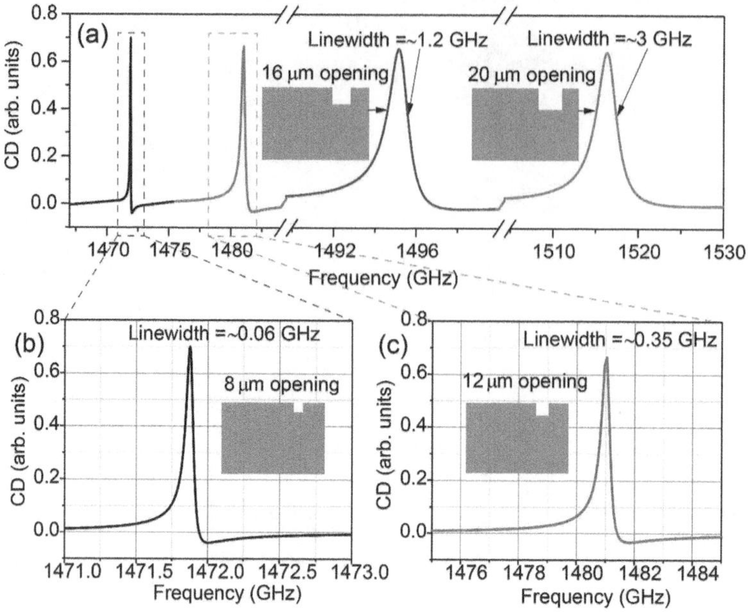

FIGURE 8.9 For Quasi BIC III, (a) shows a comparison of circular dichroism with different opening sizes, and (b, c) show their fine spectra correspondingly.

Figure 8.9 further presents a comparison of the transmission CD for metasurfaces with different openings. The CD of the all-silicon metasurface modulated by the quasi-BIC reaches as high as 0.7, as shown in Fig. 8.9(a). It should be noted that the quasi-BIC frequency shifts with variations in the opening, which is attributed to frequency detuning caused by changes in the degree of structural asymmetry. Modulated by the quasi-BIC radiation rate, the CD linewidth largely aligns with the spectral linewidth of each circularly polarized transmission component. Even for a 20 μm opening, the CD linewidth does not exceed 3 GHz, with a Q factor on the order of 10^2. As the opening size is reduced to 12 μm, the CD linewidth decreases to approximately 0.35 GHz, and the Q factor improves to the order of 10^3, as shown in Fig. 8.9(c). When the opening size is 8 μm, the radiation linewidth is only about 0.06 GHz, and the Q factor is enhanced to the order of 10^4, as shown in Fig. 8.9(b).

8.3 CHAPTER SUMMARY

This chapter summarizes the concepts of optical BICs and quasi-BICs and their polarization characteristics, elucidating the formation mechanism of polarization vortex in momentum space based on BICs. Optical BICs are a special type of electromagnetic resonance that reside within the radiation continuum yet do not radiate. By altering certain structural parameters, BICs can be induced to leak, thereby forming quasi-BICs. Optical BICs are commonly classified as symmetry-protected or parameter-tuned (in addition to a small number of unclassified or controversial BICs). Since BICs are non-radiative, they correspond to polarization singularities. However, regions in momentum space surrounding BICs can radiate linear polarization with vortex-like characteristics, which can be explained based on TCMT. Although BICs are polarization singularities, once they leak and become quasi-BICs, they exhibit well-defined polarization states. Leveraging the quasi-BIC effect, it is possible to manipulate not only linear polarization states of light waves but also elliptical and circular polarization states. Therefore, the design principles and methods for polarization devices based on quasi-BICs are further demonstrated. The demonstrated device is a metasurface with an opening rectangular block as its basic unit. Although the C_2 symmetry of the device is broken, mirror symmetry with respect to the *z*-plane is preserved. This device supports multiple BICs, including two symmetry-protected BICs and one single-resonance parameter-tuned BIC, all of which possess circular polarization manipulation capabilities. Based on TCMT, the mechanism by which the quasi-BIC resonances induce circular polarization scattering is demonstrated.

REFERENCES

1. Von Neumann, J. and Wigner, E. (1929) 'Über merkwürdige diskrete Eigenwerte', Physikalische Zeitschrift, 30, 465–467.
2. Khee-Kyun, V. (2008) 'Trapped electromagnetic modes in forked transmission lines', Wave Motion, 45, 795–803.
3. Cobelli, P.J., Pagneux, V., Maurel, A., et al. (2009) 'Experimental observation of trapped modes in a water wave channel', Europhysics Letters, 88, 20006.
4. Cobelli, P.J., Pagneux, V., Maurel, A., et al. (2011) 'Experimental study on water-wave trapped modes', Journal of Fluid Mechanics, 666, 445–476.
5. Linton, C.M. and McIver, P. (2007) 'Embedded trapped modes in water waves and acoustics', Wave Motion, 45, 16–29.
6. Callan, M., Linton, C.M. and Evans, D.V. (1991) 'Trapped modes in two-dimensional waveguides', Journal of Fluid Mechanics, 229, 51–64.
7. Hein, S., Koch, W. and Nannen, L. (2012) 'Trapped modes and Fano resonances in two-dimensional acoustical duct–cavity systems', Journal of Fluid Mechanics, 692, 257–287.
8. Lyapina, A.A., Maksimov, D.N., Pilipchuk, A.S., et al. (2015) 'Bound states in the continuum in open acoustic resonators', Journal of Fluid Mechanics, 780, 370–387.
9. Alshits, V.I., Darinskii, A.N. and Shuvalov, A.L. (1992) 'Elastic waves in infinite and semi-infinite anisotropic media', Physica Scripta, T44, 85–93.
10. Every, A.G. (2008) 'Guided elastic waves at a periodic array of thin coplanar cavities in a solid', Physical Review B, 78(17), 174104.
11. Bulgakov, E.N. and Sadreev, A.F. (2008) 'Bound states in the continuum in photonic waveguides inspired by defects', Physical Review B, 78(7), 075105.
12. Marinica, D.C., Borisov, A.G. and Shabanov, S.V. (2008) 'Bound states in the continuum in photonics', Physical Review Letters, 100(18), 183902.
13. Longhi, S. (2009) 'Optical analog of population trapping in the continuum: classical and quantum interference effects', Physical Review A, 79(2), 023811.
14. Dong, Z., Mahfoud, Z., Paniagua-Domínguez, R., et al. (2022) 'Nanoscale mapping of optically inaccessible bound-states-in-the-continuum', Light: Science & Applications, 11, 20.
15. Zhang, X., Liu, Y., Han, J., et al. (2022) 'Chiral emission from resonant metasurfaces', Science, 377, 1215–1218.
16. Dong, Z., Jin, L., Rezaei, S.D., et al. (2022) 'Schrödinger's red pixel by quasi-bound-states-in-the-continuum', Science Advances, 8, eabm4512.
17. Hsu, C.W., Zhen, B., Stone, A.D., et al. (2016) 'Bound states in the continuum', Nature Reviews Materials, 1, 1–13.

18. Yao, J., Li, J., Zhang, Y., et al. (2023) 'Bound states in continuum in periodic optical systems', Chinese Optics, 16(1), 1–23.
19. Wang, W., Srivastava, Y.K., Tan, T.C.W., et al. (2023) 'Brillouin zone folding driven bound states in the continuum', Nature Communications, 14, 2811.
20. Zhen, B., Hsu, C.W., Lu, L., et al. (2014) 'Topological nature of optical bound states in the continuum', Physical Review Letters, 113, 257401.
21. Liu, W., Wang, B., Zhang, Y., et al. (2019) 'Circularly polarized states spawning from bound states in the continuum', Physical Review Letters, 123, 116104.
22. Gorkunov, M.V., Antonov, A.A. and Kivshar, Y.S. (2020) 'Metasurfaces with maximum chirality empowered by bound states in the continuum', Physical Review Letters, 125(9), 093903.
23. Overvig, A., Yu, N. and Alù, A. (2021) 'Chiral quasi-bound states in the continuum', Physical Review Letters, 126(7), 073001.
24. Shi, T., Deng, Z.L., Geng, G., et al. (2022) 'Planar chiral metasurfaces with maximal and tunable chiroptical response driven by bound states in the continuum', Nature Communications, 13, 4111.
25. Chen, Y., Deng, H., Sha, X., et al. (2023) 'Observation of intrinsic chiral bound states in the continuum', Nature, 613, 474–478.
26. Zhang, D., Liu, T., Lei, L., et al. (2024) 'Tailoring intrinsic chirality in a two-dimensional planar waveguide grating via quasibound states in the continuum', Physical Review B, 109, 205403.
27. Zhou, C., Huang, L., Jin, R., et al. (2023) 'Bound states in the continuum in asymmetric dielectric metasurfaces', Laser & Photonics Reviews, 17, 2200564.
28. Huang, L., Li, S., Zhou, C., et al. (2024) 'Realizing ultrahigh-Q resonances through harnessing symmetry-protected bound states in the continuum', Advanced Functional Materials, 34, 2309982.
29. Zhong, H., Huang, L., Li, S., et al. (2024) 'Toroidal dipole bound states in the continuum in asymmetric dimer metasurfaces', Applied Physics Reviews, 11, 031404.
30. Li, J., Yue, Z., Li, J., et al. (2023) 'Ultra-narrowband terahertz circular dichroism driven by planar metasurface supporting chiral quasi bound states in continuum', Optics and Laser Technology, 161, 109173.
31. Hsu, C.W., Zhen, B., Chua, S.L., et al. (2013) 'Bloch surface eigenstates within the radiation continuum', Light: Science & Applications, 2, e84.
32. Yang, Y., Peng, C., Liang, Y., et al. (2014) 'Analytical perspective for bound states in the continuum in photonic crystal slabs', Physical Review Letters, 113, 037401.

CHAPTER 9

Polarization Manipulation in Non-Hermitian Systems

Abstract

This chapter introduces polarization manipulation based on non-Hermitian metasurfaces. Rooted in non-Hermitian physics, non-Hermitian metasurfaces break the energy conservation framework of traditional Hermitian systems by incorporating gain and/or loss as core degrees of freedom for control. Leveraging singularities and parity-time properties, they provide a novel approach for manipulating light polarization states. This field is built on non-Hermitian Hamiltonians and temporal coupled-mode theory. When system parameters evolve to an exceptional point, both eigenvalues and eigenstates degenerate simultaneously, accompanied by unique effects such as topological phase transitions and asymmetric transmission.

9.1 THEORETICAL BASIS OF NON-HERMITIAN POLARIZATION META-PHOTONICS

When a physical system satisfies energy conservation and time-reversal symmetry, its eigenvalues are real, and a set of standard orthonormal bases can be found. In such cases, the corresponding Hamiltonian is Hermitian. Hermiticity, as a cornerstone of various models in quantum mechanics and electromagnetics, is typically established under the ideal condition that the system is completely isolated from the external environment [1]. However, real-world systems often involve energy exchange and dissipation, leading to interactions with the environment that result in energy gain or loss. Under such circumstances, a non-Hermitian Hamiltonian is required for description. Non-Hermitian metasurfaces [2–4], grounded in non-Hermitian physics, are novel meta-devices that have been developed over the past decade. Conventional perspectives regard loss in electromagnetic metasurfaces as an undesirable outcome, thus aiming to minimize it as

DOI: 10.1201/9781003784821-9

much as possible. In contrast, non-Hermitian metasurfaces break away from this conventional design paradigm by treating loss as a new degree of freedom to tailor metasurface functionalities.

Compared with traditional Hermitian systems, non-Hermitian systems exhibit characteristics such as complex eigenvalues and non-orthogonal eigenstates and feature a special class of degenerate points known as exceptional points (EPs) [5–7]. When system parameters evolve to an EP, two or more eigenvalues and their corresponding eigenstates degenerate simultaneously, giving rise to physical effects such as topological phase singularities, asymmetric transmission, and ultra-sensitive responses [8,9]. Hermitian systems adhere to energy conservation and typically prohibit energy exchange, whereas non-Hermitian systems permit energy input and output. By precisely tailoring the gain and loss within the system, flexible manipulation of its physical properties can be achieved, thereby opening a new design dimension for the performance optimization of polarization metasurfaces [10]. Non-Hermitian metasurfaces, by combining the polarization degree of freedom with the characteristics of EPs, can overcome the limitations of polarization manipulation in traditional Hermitian systems. To some extent, this advancement drives the development of high-performance photonic devices towards integration, multifunctionality, and dynamic tunability.

Non-Hermitian systems are associated with spatial inversion and time-reversal symmetries [11–13], namely, considering the parity-time (PT) symmetry of the system, where P represents the spatial parity operator and T denotes the time-reversal operator. Taking a one-dimensional system as an example, when the parity operator P acts on a quantum mechanical wave function $\psi(x, t)$, it flips the sign of the spatial coordinate, yielding $P\psi(x, t) = \psi(-x, t)$. The time-reversal operator T, on the other hand, represents reversing the direction of time $\psi(x, t) \rightarrow \psi(x, -t)$, which for spinless particles corresponds to complex conjugation, i.e. $T\psi(r) = \psi^*(r)$. In 1998, Bender and Boettcher [14] proposed that under complex potential conditions, if a non-Hermitian Hamiltonian remains invariant under the combined operation of parity (P) and time reversal (T), i.e. it satisfies $\psi(r) = \psi^*(-r)$, it can also exhibit entirely real eigenvalues, analogous to Hermitian systems. The corresponding operator is referred to as a PT-symmetric operator.

Based on the characteristics of PT-symmetric operators and the spatially symmetric distribution of gain and loss, combined with temporal coupled-mode theory (TCMT), the temporal evolution of resonant mode fields can be described. TCMT [15–18] simplifies the complex behaviour of electromagnetic systems into the dynamical evolution of a finite set of eigenmodes by constructing a system of coupled ordinary differential equations, thereby characterizing energy exchange between the resonant system and its environment. By employing such coupled differential equations, the requirements of PT symmetry can be translated into specific constraints on system parameters.

In optical resonators, resonant modes manifest as dynamic electromagnetic fields with spatiotemporal distributions. We denote the electric amplitude of the dynamic field by $\boldsymbol{\alpha}$, which includes the time-dependent term of $\exp(-i\omega t)$. The magnitude of this amplitude depends on the energy stored in the mode. For an ideal closed system with no energy exchange with the external environment and satisfying total energy conservation, the time derivative of the amplitude $\boldsymbol{a}$ yields the dynamic relation of the resonant amplitude [19]:

$$\frac{d\boldsymbol{a}}{dt} = -i\boldsymbol{\Omega}\boldsymbol{a}. \tag{9.1}$$

In Eq. (9.1), $\boldsymbol{\Omega}$ represents a frequency matrix. If the system contains two or more resonant modes, $\boldsymbol{\Omega}$ is at least a second-order matrix. When input and output ports are introduced into the resonant system, each mode may experience internal energy loss, as well as radiation or energy attenuation through the ports. We denote this loss rate by $\boldsymbol{\Gamma}$. At this point, the resonant mode exchanges energy with the outside world, and the system becomes an open system. The dynamic relation of the system is then expressed as:

$$\frac{d\boldsymbol{a}}{dt} = \left[-i\boldsymbol{\Omega} - \boldsymbol{\Gamma}\right]\boldsymbol{a} + G^T \boldsymbol{I}, \tag{9.2}$$

where $\boldsymbol{I}$ represents the incident wave from the ports, and G^T is the coupling coefficient matrix between the port incident wave and the resonance. Equation (9.2) is a common form for resonant systems in TCMT, describing not only resonance and attenuation but also the effects of external excitation.

For a system of two coupled resonant modes, assume their resonant frequencies are ω_1 and ω_2, and their loss rates are γ_1 and γ_2, respectively. Under near-field interaction, these two resonators couple with each other. Let k be the coupling parameter between the modes. The Hamiltonian of the system can then be written as:

$$H_i = -i\boldsymbol{\Omega} - \boldsymbol{\Gamma} = \begin{bmatrix} -i\omega_1 - \gamma_1 & k \\ k & -i\omega_2 - \gamma_2 \end{bmatrix}. \tag{9.3}$$

By performing a simple mathematical manipulation – specifically, factoring out the imaginary unit i to render the mode frequency terms real, the Hamiltonian can be rewritten as:

$$H = \boldsymbol{\Omega} - i\boldsymbol{\Gamma} = \begin{bmatrix} \omega_1 - i\gamma_1 & k \\ k & \omega_2 - i\gamma_2 \end{bmatrix}. \tag{9.4}$$

Note that, considering the coupling parameters on the off-diagonals remain identical after factoring out the imaginary unit i, for the convenience of subsequent discussion, the coupling parameter between modes is still denoted as k. Solving for the eigenvalues of the matrix in Eq. (9.4) yields:

$$\omega_\pm = \frac{(\omega_1 + \omega_2) - i(\gamma_1 + \gamma_2) \pm \sqrt{4k^2 + \left[(\omega_1 - \omega_2) + i(\gamma_1 - \gamma_2)\right]^2}}{2}. \tag{9.5}$$

When $\omega_+ = \omega_-$, the two eigenvalues of the system coincide. If the corresponding eigenstates also are the same, both the eigenvalues and eigenstates of the system degenerate simultaneously. This point is referred to as an EP of the system. As illustrated in Fig. 9.1(a),

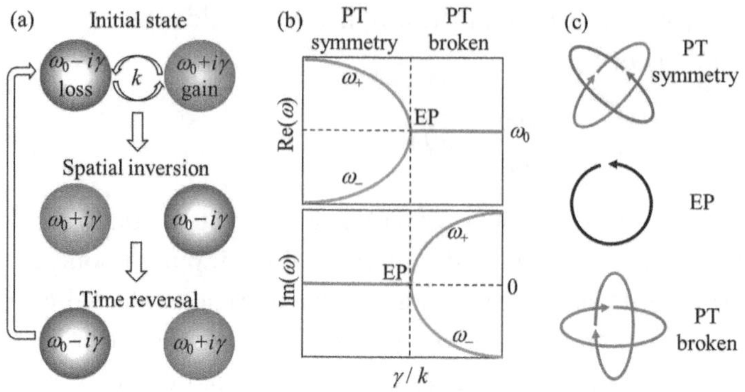

FIGURE 9.1 Schematic diagram of the PT-symmetric system and its phase transition process. (a) Illustration of a PT-symmetric system. (b) Evolution trajectory of eigenvalues as a function of system parameters. (c) Evolution trajectory of eigen-polarization modes at different phase transition stages.

assuming the two resonances in the system have identical frequencies, i.e. $\omega_1 = \omega_2 = \omega_0$, and the system possesses balanced gain and loss, i.e. $\gamma_1 = \gamma_2 = \gamma$, a simplified and idealized PT-symmetric model can be constructed. Under these conditions, the Hamiltonian in the PT-symmetric framework is expressed as:

$$H^{PT} = \begin{bmatrix} \omega_0 - i\gamma & k \\ k & \omega_0 + i\gamma \end{bmatrix}. \tag{9.6}$$

This Hamiltonian satisfies PT$H^{PT} = H^{PT}$, meaning that after applying spatial inversion and time-reversal operations to H^{PT}, the Hamiltonian remains unchanged. The corresponding eigenvalues are:

$$\omega_{\pm} = \omega_0 \pm \sqrt{k^2 - \gamma^2}. \tag{9.7}$$

As parameters vary in a system described by these operators, the EP serves as a critical boundary, dividing the system into a PT-symmetric phase and a PT-broken phase. As shown in Fig. 9.1(b), when $|k| > |\gamma|$, the eigenvalues form a pair of real numbers symmetric about ω_0. In this regime, PT symmetry is preserved, and the corresponding eigenstates remain orthogonal. When $|k| < |\gamma|$, the eigenvalues become a pair of complex conjugates, PT symmetry is spontaneously broken, and the eigenstates become non-orthogonal. At the EP, $|k| = |\gamma|$, the coupling strength equals the gain/loss strength. Here, both eigenvalues and eigenstates degenerate, and the electromagnetic response undergoes an abrupt transition. In the field of polarization control, the eigen-polarization states degenerate at EPs [20], as illustrated in Fig. 9.1(c). In early studies, such degenerate states were predominantly circularly polarized – corresponding to the poles of the Poincaré sphere. However, with continued research, it has become possible to realize EPs associated with arbitrary polarization states, including linear and elliptical polarizations, thereby enabling full coverage of the entire Poincaré sphere. This advancement has substantially broadened the application scope of non-Hermitian polarization metasurfaces.

9.2 POLARIZATION MANIPULATION IN NON-HERMITIAN METASURFACES

9.2.1 Geometry and Material Parameter Control

In the research field of non-Hermitian polarization metasurfaces, the coupling strength, loss distribution, and chiral characteristics of the system can be tuned by modifying the geometric configuration, arrangement, and physical sizes of the metasurface unit cells [21–23], or by introducing gain and loss media into the system. Whether through substrate gain modulation to tailor the spatial distribution of EPs or via optical excitation to dynamically adjust the material loss rate, gain-loss engineering has significantly enhanced the functional tailorability of non-Hermitian metasurfaces. In conventional optical system design, achieving strong chiral optical responses typically requires the construction of complex three-dimensional helical or tilted nanostructures. Oh et al. [24] exploited the simultaneous degeneracy of eigenvalues and eigenstates at EPs in non-Hermitian systems and designed a non-Hermitian metasurface by incorporating gain and loss. As shown in Fig. 9.2, the unit cell of this metasurface consists of two orthogonally placed plasmonic nanorod resonators

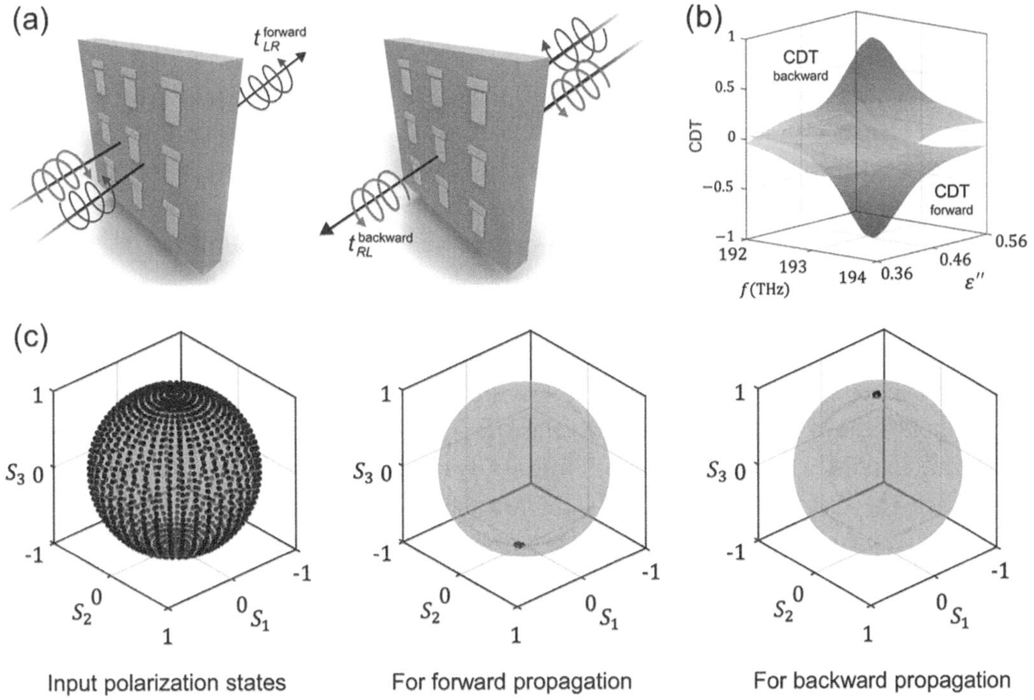

FIGURE 9.2 Design and polarization conversion characteristics of gain-type non-Hermitian chiral metasurfaces with complete asymmetric polarization conversion [24]. (a) Schematic diagram for function of metasurface, where a unit cell consists of a gain-type resonator and a passive resonator. The device achieves the conversion from right-handed circular polarization (RCP) to left-handed circular polarization (LCP) during forward propagation, and completely suppresses LCP incidence, vice versa. (b) Circular polarization conversion difference (CDT) in dual-parameter space; (c) Polarization state mapping on Poincaré sphere.

made of gold and silver. These unit cells are arranged in a two-dimensional periodic array and placed on a gain substrate, forming the final metasurface. Notably, the two plasmonic nanorods are not arranged centrosymmetrically but instead feature a preset lateral offset, thereby introducing planar chirality. Based on TCMT, the authors identified the conditions required to observe fully asymmetric polarization conversion with zero eigenvalues at a chiral EP. By leveraging the disparity in the complex permittivity of gold and silver in the target wavelength range, distinct intrinsic loss rates were obtained. By tuning the incident light frequency and the substrate gain in parameter space, an EP with zero eigenvalue was observed. At this zero-eigenvalue EP, the two orthogonal resonant modes degenerate due to strong coupling, and their associated eigen-polarization states collapse into a single circular polarization state. Owing to the planar chiral configuration of the metasurface, this fully asymmetric polarization conversion effect exhibits propagation-direction dependence: forward and backward incidences correspond to opposite circular polarization helicity preferences, thereby yielding circular dichroism responses approaching the theoretical limit in the transmission spectrum. In summary, the metasurface designed by Oh et al. integrates dissimilar loss materials, controllable gain media, and asymmetric geometric coupling to construct a polarization-selective topological singularity at the EP of a non-Hermitian system. By combining gain-loss balance, mode degeneracy, and chiral response, this approach enables high-efficiency, high-contrast polarization manipulation at subwavelength scales.

In 1900, Max Planck, a professor at the University of Berlin in Germany, proposed the theory of blackbody radiation [25], which subsequently ushered in a quantum revolution. Over the following decades, a series of distinguished physicists gradually established the theoretical framework of quantum physics based on Hermitian operator algebra [26–29]. Nearly a century later, with the introduction of a non-Hermitian Hamiltonian, quantum theory was further expanded to accommodate systems satisfying PT symmetry and exhibiting real-valued spectra [30]. Although there has been a breakthrough in the concept, the application and full implications of PT symmetry within quantum physics remain at the forefront of current research. Addressing the phenomenon of PT symmetry breaking in the polarization space of terahertz metasurfaces, Lawrence et al. [31] designed, for the first time in 2014, a metasurface exhibiting PT symmetry in the polarization domain, by leveraging the vectorial properties of light and the inherent anisotropy of artificial meta-atoms, as shown in Fig. 9.3(a, b). This terahertz non-Hermitian metasurface comprises periodically arranged meta-molecular units, and each meta-molecular unit contains two split-ring resonators (SRRs) with identical resonant frequencies. These two sets of resonators are fabricated using metals with drastically different loss characteristics (silver and lead). The selection of these materials enables the dissipation contrast required for PT symmetry under passive conditions, without the need for introducing active optical gain. Furthermore, the orthogonal orientation of these two sets of resonators allows for the independent detection of the radiation field emanating from each antenna. Within this metasurface, varying the spacing between the SRRs enables control over the non-Hermitian parameter, facilitating the observation of the complete PT symmetry phase transition in the polarization space. The inter-resonator distance dictates the near-field coupling

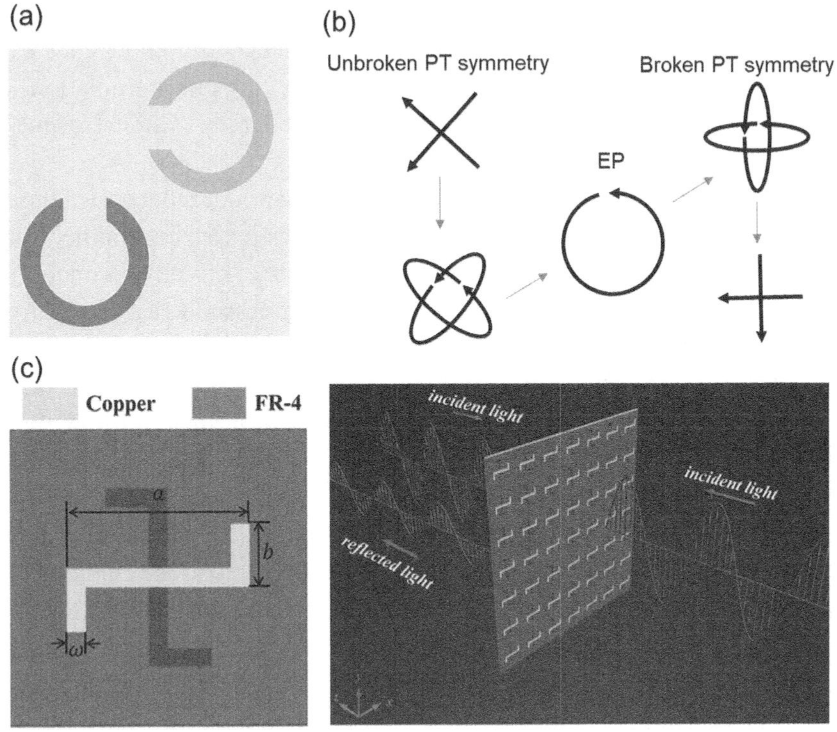

FIGURE 9.3 (a) A non-Hermitian polarization metasurface unit cell based on two SRRs with orthogonal orientation, and its (b) schematic diagram of polarization state evolution. (c) Non-Hermitian polarization metasurface based on a bilayer complementary Z-shaped copper structure achieves unidirectional light propagation without reflection. At the EP, the system exhibited ideal reverse-direction reflectionless for x-polarized incident light, while demonstrating forward-direction reflectionless for y-polarized incident light [33].

strength between the two orthogonal SRRs, thus governing the system's phase transition based on the relationship between the coupling strength and the inherent loss disparity of the resonators. When the coupling strength significantly exceeds the loss difference, the system resides in the unbroken PT-symmetric phase. In this regime, the eigenmodes undergo frequency splitting due to coupling, manifesting in the polarization state as a pair of elliptically polarized waves with identical rotation directions and whose major axes align with the system's symmetry axes. When the coupling strength and loss difference achieve a specific balanced condition, the system reaches an EP. At this point, the two eigen-polarization states coalesce into a single circular polarization state, and their corresponding eigenvalues become degenerate. In the transmission spectrum, this phenomenon is specifically observed as the complete merging of two resonant peaks. Further increasing the distance between the resonators, such that the coupling strength falls below the loss difference, causes the system to spontaneously enter the broken PT-symmetric phase. In this phase, the two eigenmodes share identical resonant frequencies but exhibit markedly different decay rates. Concurrently, the corresponding eigen-polarization states undergo an abrupt transformation, evolving into elliptical polarization states with their major axes oriented along the x- and y-directions, respectively. By utilizing the highly controllable platform of terahertz metasurfaces, Lawrence and colleagues experimentally observed, for the first time, the PT symmetry-breaking phase transition within the polarization space and successfully witnessed the degeneracy of non-Hermitian eigenstates at the EP.

The phenomenon of unidirectional light propagation without reflection [32] is a quintessential characteristic exhibited at the EPs of non-Hermitian systems and holds significant importance for realizing unidirectional photonic devices. In conventional approaches, achieving a unidirectional reflectionless effect typically necessitates a system with strict PT symmetry and a precise balance between gain and loss, which is a requirement often challenging to fulfil in practical implementations. In contrast, within non-Hermitian metasurfaces, EPs can be constructed more directly by optimizing structural parameters, even without satisfying PT symmetry, thereby enabling a more straightforward realization of unidirectional reflectionless. Building on this concept, Gao et al. [33] proposed a polarization-sensitive radio-frequency metasurface based on a bilayer complementary Z-shaped copper structure, as shown in Fig. 9.3(c). This metasurface employs FR-4 as the dielectric substrate, with the copper foil patterns on the top and bottom layers arranged in a rotationally complementary configuration. This architecture introduces anisotropy, endowing the system with chiral characteristics that enable differentiated responses to incident light of varying polarizations. Based on coupled-mode theory, they established a two-port system model, derived the relationship between the scattering matrix and the reflection coefficients, and defined a reflection asymmetry factor as the criterion for identifying an EP. By adjusting and optimizing the geometric parameters of the metasurface according to the derived results, they successfully constructed an EP within this loss-assisted system. At the EP, the system exhibited ideal reverse-direction reflectionless for x-polarized incident light, while demonstrating forward-direction reflectionless for y-polarized incident light. Further investigation revealed that by tuning the polarization angle of the incident light to a specific value, bidirectional and unidirectional reflectionless could be simultaneously observed at two distinct characteristic frequencies. Their research demonstrates that, without requiring a strictly balanced gain and loss, unidirectional reflectionless effects based on EPs can be achieved solely by modulating the structural parameters to alter the system's loss distribution. This provides a viable experimental pathway for the development of integrated photonic devices such as non-reciprocal filters and perfect absorbers.

The inherent robustness of the topological phase provides an anti-interference solution for polarization control. Near the wavelength of the EP, the conversion between circularly polarized lights is accompanied by a 2π phase change [34,35], a characteristic that is independent of the closed-loop path, laying the foundation for the design of topologically protected polarization devices. In 2021, Song et al. [36] first proposed a topological phase control strategy by encircling the EP. The team employed a plasmonic topological metasurface with a metal-dielectric-metal structure. The bottom metal layer blocks transmitted light, ensuring the metasurface operates only in reflection mode. The top metal layer consists of a two-dimensional chiral structure array, where each unit cell is formed by the near-field coupling between an L-shaped nanostructure and a straight nanorod. By modifying the geometric parameters of the L-shaped nanostructure and the straight nanorod, the value of the complex coefficient is tuned. Utilizing the complex coefficient values to encircle the EP within the geometric parameter space generates a 2π phase shift. Combined with the PB phase, this achieves completely independent control of the polarization channels. If the degenerate eigenstate is RCP, only the right-handed circularly polarized light is preserved, and the other polarization conversion channel disappears. An EP (zero point) is obtained at this location, corresponding to a perfect absorption scenario. The chiral response of the surface current distribution at the EP can induce asymmetry in the conversion between the two circular polarizations and plays a crucial role in achieving polarization-dependent reflectivity. According to topological protection, as long as the motion encircles the EP within the parameter space, the closed path of the motion does not affect the 2π phase change in the reflection phase diagram.

In the field of metasurface coding [37], a discrete reflection phase binary encoding method has been proposed in recent years, using "0" and "1" as coding primitives, offering an alternative approach for electromagnetic wave manipulation. However, significant challenges remain in the independent control of circularly polarized light using phase-encoding metasurfaces, particularly concerning the direction conversion of cross-polarization. For linearly polarized light, modulation can be achieved relatively easily by modifying the geometric parameters of the metasurface unit structure, whereas achieving modulation for circularly polarized light still presents considerable difficulties. The emergence of exceptional topological phase control provides more optional methods for phase manipulation, and this approach can also be used in conjunction with traditional phase control, significantly expanding the design and application scenarios of metasurfaces. Based on this method of exceptional topological phase control, Li et al. [38] proposed a metasurface structure satisfying PT symmetry. The EP of a PT-symmetric metasurface possesses a characteristic: encircling the EP along any closed path in the parameter space accumulates a 2π phase change, and moreover, this phase does not exhibit coupling interference with the conventional PB phase. The metasurface structure they designed consists of two lossy aluminium layers and a silicon dioxide spacer layer, as shown in Fig. 9.4. The bottom metal layer suppresses electromagnetic wave transmission, ensuring the metasurface operates in reflection mode, while the top-layer chiral structure array can be excited by polarization waves in the x and y directions, respectively. Based on this metasurface, they selected two sets of coding units near the EP, defining them as binary "0" and "1". When RCP light is incident, due to the presence of the EP in the system, a π phase delay occurs between the two sets of coding units, with nearly identical reflection amplitudes. Conversely, when LCP light is incident, the EP is absent in the system, resulting in identical phase values for the two sets of coding units. This difference enables the independent coding of circularly polarized light. This metasurface, employing composite phase coding, achieves independent control over RCP and LCP light, simplifying the structural design and phase calculation processes compared to traditional phase-encoding metasurfaces, and demonstrates excellent performance in application scenarios such as beam deflection and multi-beam generation.

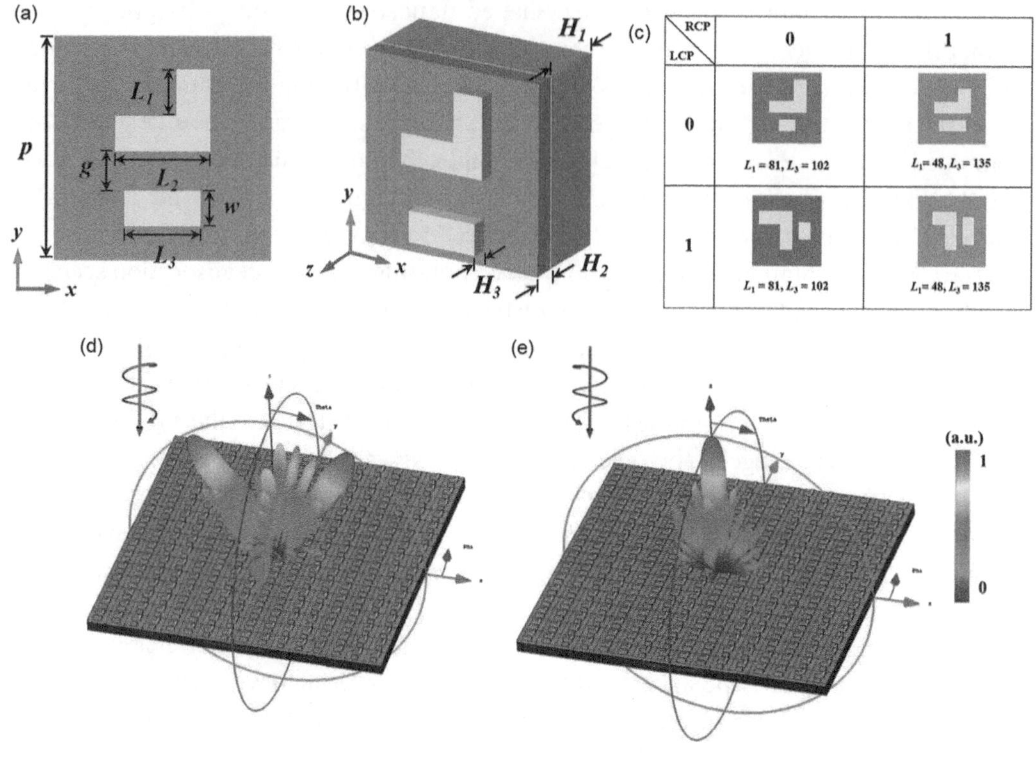

FIGURE 9.4 Coding metasurface combining the exceptional topological phase and PB phase [38]. (a, b) Top view and three-dimensional schematic diagram of the metasurface unit cell; (c) 2 × 2 polarization-phase coding unit library, achieving independent control of LCP and RCP components through the combination of different units; (d) far-field scattering patterns of the 1-bit coding metasurface under RCP and LCP light incidence.

Chiral materials or structures in optics can exhibit different responses to LCP and RCP light, giving rise to the phenomenon of circular dichroism [39,40]. In non-Hermitian polarization metasurfaces, the chiral response at the EP originates from the degeneracy characteristics of the system's eigenstates. Theoretically, the eigenstate at the EP should converge to a single circular polarization state, at which point the metasurface exhibits a strongly selective response to this specific circular polarization state [41,42]. However, due to the coexistence of conversion components and parallel components, achieving ideal circular dichroism remains challenging. Consequently, Li et al. [43] proposed a PT-symmetric active metasurface composed of orthogonally oriented nano-resonators with local gain, as shown in Fig. 9.5. By tuning the coupling strength between the resonators and the gain-to-loss ratio, they achieved PT symmetry phase transitions and a super chiral EP. They discovered that in non-Hermitian systems, besides the EP curve, there exists another independent parameter curve along which two parallel transmission components vanish simultaneously. At this super chiral EP, only a single circular polarization transmission component exists, enabling near-ideal circular polarization conversion and circular dichroism, accompanied by a phase flip of the transmitted wave. Under forward

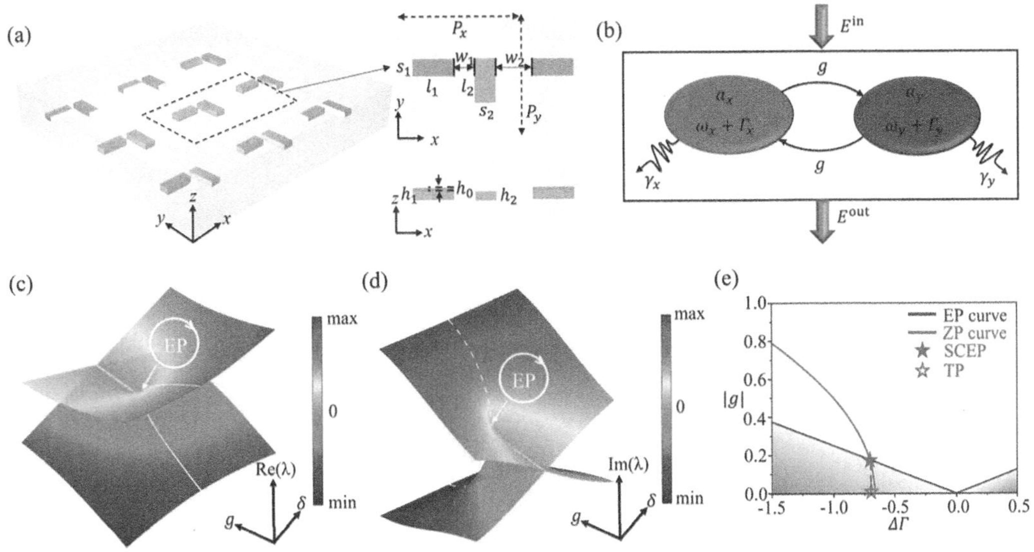

FIGURE 9.5 Design and phase diagrams of the PT-symmetric active metasurface composed of orthogonally oriented nano-resonators [43]. (a) Schematic diagram of the metasurface structure. (b) Theoretical model of two coupled harmonic oscillators. (c, d) Distribution of complex eigenvalues in the parameter space. (e) Evolution of different EPs and phase transition points with parameters.

and backward incidence conditions, light exhibits opposite chiral characteristics, thereby achieving a unidirectional optical invisibility effect. In structures satisfying C_{2y} symmetry, topological phase transition points exhibiting linear polarization characteristics can be observed. Such special points can be utilized to realize circular polarization conversion devices with near-ideal performance. Ultimately, by synergistically utilizing the super chiral EP and topological phase transition points, systematic control over arbitrary polarization states on the Poincaré sphere can be achieved.

9.2.2 Polarization Degree of Freedom Control

In addition to utilizing the geometric structure and material properties of metasurfaces to control the polarization state of a system, the polarization degree of freedom itself can also serve as an actively designable control variable [44,45]. At EPs where specific parameters cause spectral degeneracy in non-Hermitian systems, unique polarization characteristics emerge. Based on the relationship between light polarization and the Poincaré sphere, Qin et al. [46] achieved control over EPs for arbitrary polarization states by manipulating optical states on the Poincaré sphere, as shown in Fig. 9.6. In the general scattering matrix formalism, an EP manifests as a zero-intensity spectral point in one of the off-diagonal elements of the Jones matrix, which can regulate the polarization conversion between two polarization eigenstates. In conventional systems, due to symmetry constraints, the off-diagonal elements of the Jones matrix are equal, restricting the eigenstates at the EP to circular polarization. However, this study introduced extrinsic chirality and asymmetric basis transformations, proposing that if all four elements of the Jones matrix can be

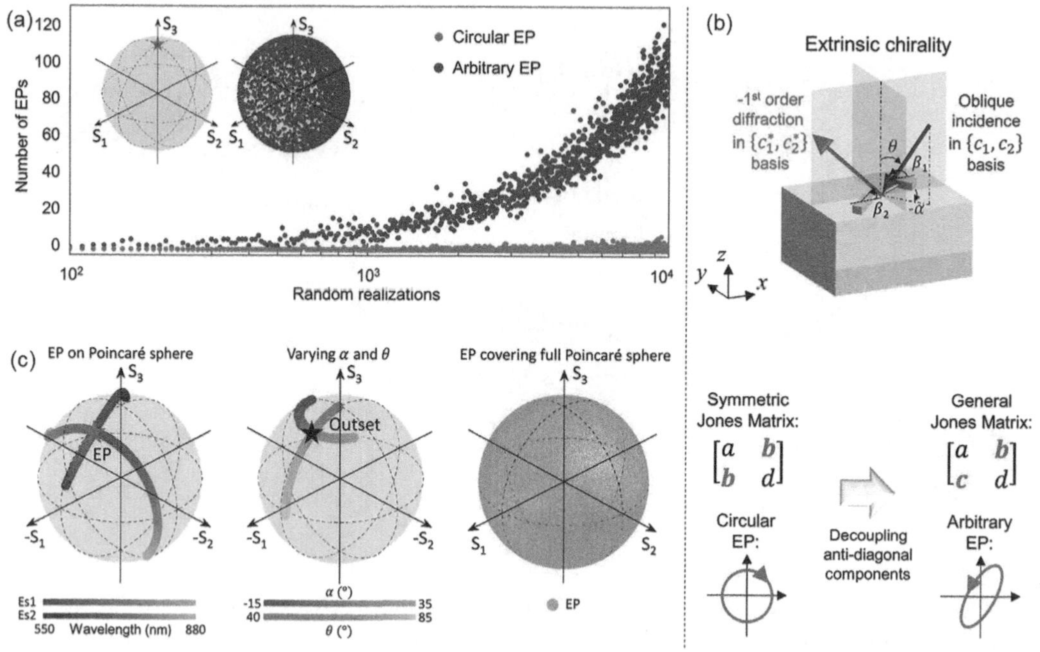

FIGURE 9.6 Achieving full Poincaré sphere EP coverage [46]. (a) Statistical distribution of EPs in random non-Hermitian systems. (b) Evolution of the Jones matrix and transition of EP types; and working mechanism of the extrinsic chiral metasurface, introducing chiral response in asymmetric multilayer structures through oblique incidence combined with first-order diffraction, providing degrees of freedom for arbitrary polarization EPs. (c) EPs evolution under wavelength tuning.

controlled independently when calculating its eigenvalues and eigenvectors, then degenerate eigenstates with arbitrary polarization can be achieved. Through oblique incidence and local basis transformations within the Jones matrix, the off-diagonal elements become decoupled. By tuning the relevant parameters, the distribution of the eigenstate polarization positions on the Poincaré sphere can be precisely controlled. By introducing extrinsic chirality and asymmetric basis transformations, the four elements of the Jones matrix are independently controlled, allowing the eigenstate polarization positions to be arbitrarily distributed across the Poincaré sphere, thus breaking the symmetry constraints of traditional polarization control.

Utilizing the characteristic of eigenstate degeneracy at EPs in non-Hermitian systems enables unique asymmetric control [47,48]. Based on this characteristic of EPs, Yang et al. [49] proposed a plasmonic non-Hermitian metasurface composed of mirror-symmetric micro-nano structural units, as shown in Fig. 9.7. By applying mirror-symmetric transformations to the geometry of the nanostructures, the metasurface can obtain a pair of controllable EPs. This pair of EPs exhibits opposite handedness, making LCP and RCP light two independently controllable degrees of freedom. On the basis of decoupling the two circular polarization channels, and given that any arbitrary polarization state can be decomposed into a superposition of LCP and RCP components, by controlling the amplitude and phase difference of these two circular polarization components, arbitrary polarization

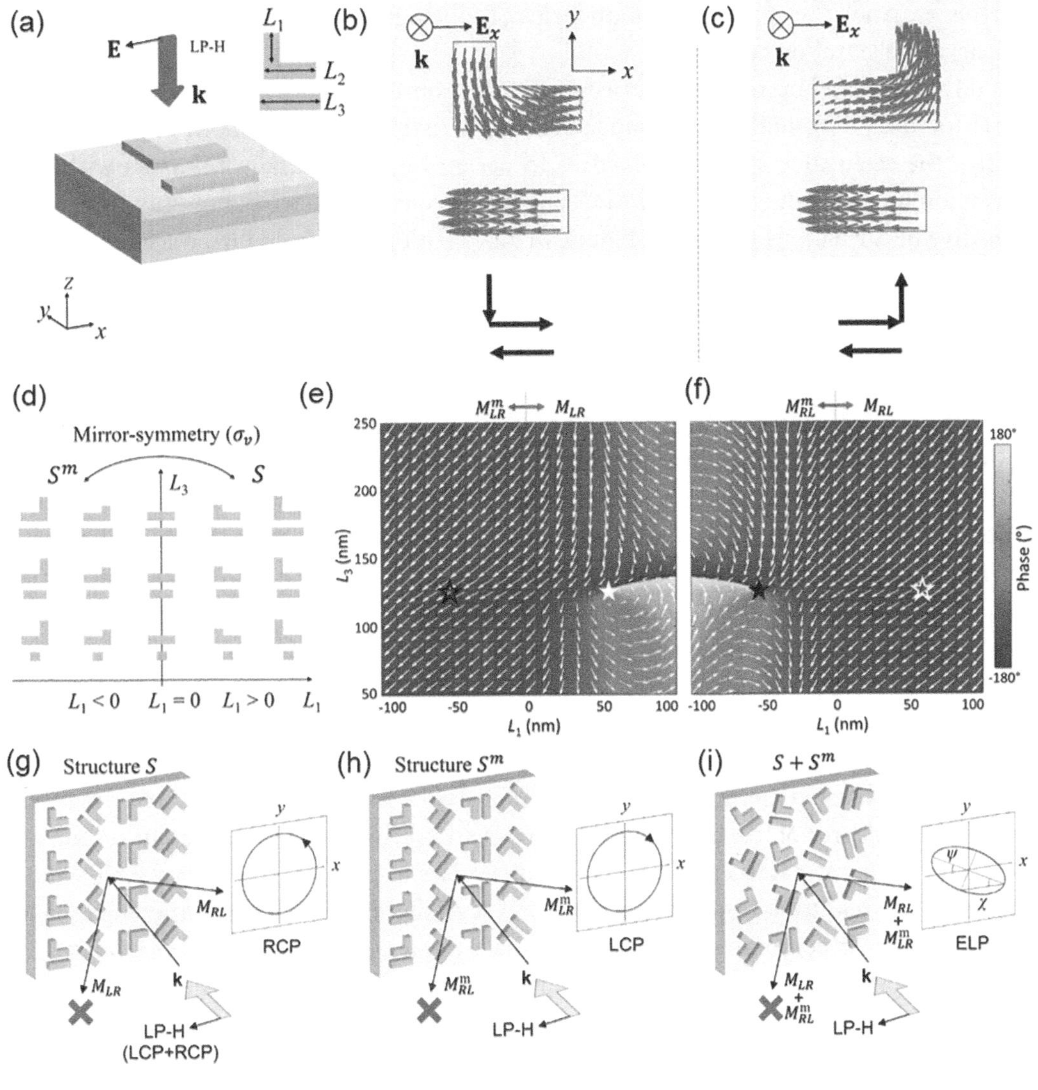

FIGURE 9.7 Polarization control metasurface based on PB phase and mirror-symmetric structure [49]. (a) Three-dimensional structural schematic of the reflective chiral meta-atom. (b, c) Surface current distribution of the meta-atom and its mirror-symmetric structure under x-direction electric field excitation. (d) Schematic diagram of the parameter space for the mirror-symmetric structure. (e, f) Complex amplitude distribution of the circular polarization conversion coefficient in the parameter space. (g, h) Principle of polarization control for a single structure. (i) Metasurface design for full polarization reconstruction.

states on the Poincaré sphere can be achieved. Furthermore, the azimuth and ellipticity angles of the polarization state can be controlled, allowing the distribution of eigenstates on the Poincaré sphere to break free from traditional circular polarization constraints and cover the entire two-dimensional polarization space. This research addresses the issue of limited control dimensions associated with traditional single EPs, providing a universal scheme for asymmetric vector wavefront modulation of arbitrary polarization states, and further expands the application boundaries of topological photonics in the field of high-precision polarization control.

In the field of photonics, the ultra-sensitive response near EPs has become a core principle for designing high-performance sensors. Existing non-Hermitian systems mostly utilize the eigenvalue splitting near EPs to design high-performance sensors with various photonic structures. However, the splitting of eigenvalues arises not only from ultra-sensitive perturbations near the EP but can also result from errors introduced during the fabrication process, which can degrade the overall performance of the system. Due to the perturbation-induced changes in eigenvalues, the overall performance of sensors near EPs cannot be accurately assessed. To address this issue, Shi et al. [50] introduced the polarization degree of freedom to control the position and characteristics of EPs in non-Hermitian systems. When polarized light illuminates the sensor structure, it excites resonances within the structure. If the polarization direction of the light does not align with the resonance direction of the structure, the energy transfer efficiency between the light and the structure changes. Based on this variation in coupling efficiency, a coupled metasurface system can be constructed, consisting of two orthogonal resonators with similar resonant frequencies but different intrinsic losses. By polarization controlling the loss, an effective Hamiltonian is established. By solving the eigenvalues of this resonant structure, the relationship between the incident polarization state and the transmission phase of the coupled system can be obtained, and a phase EP in the polarization space is observed, which can effectively resist fabrication errors.

Unlike traditional methods that directly modify the metasurface structure or material, polarization degree of freedom control employs an approach where the output response is regulated by input parameters. This technique offers important advantages such as non-invasiveness, real-time response, and process reversibility. By adjusting the angle of the incident light or combining a pair of EPs that support mirror circular polarization, an EP with an arbitrary polarization state can be achieved. By matching the incident polarization state with the resonance direction of the metasurface, interference from fabrication errors on sensor performance can be effectively suppressed. Polarization degree of freedom control enables a deep integration of incident light parameters and metasurface characteristics, significantly expanding the functional dimensions and application scope of non-Hermitian metasurfaces, and providing a new technological pathway for designing flexible and controllable polarization devices.

9.2.3 Dynamically Active Control

Above research on non-Hermitian polarization metasurfaces is still mainly focused on static systems. The optical response characteristics of such metasurfaces are determined by their structural parameters; once the device is fabricated, it no longer possesses flexible polarization control functions. The introduction of dynamically active control techniques provides a new avenue for expanding the functionality of non-Hermitian polarization metasurfaces. Dynamically active tunable metasurfaces enable real-time control over the parameter space, thus allowing precise measurement of EPs within a single device and monitoring of light-matter interaction processes in the system. By combining topological control with dynamic tuning, Ding et al. [51] proposed an electrically controlled non-Hermitian metasurface platform, as shown in Fig. 9.8. This platform consists of a two-dimensional symmetry-broken chiral gold meta-atom array on a glass substrate, combined with a thin-film piezoelectric micro-electro-mechanical system (MEMS) reflector. By applying a voltage, the air gap between the meta-atoms and the MEMS reflector can be precisely controlled. The size of this air gap influences the coupling strength of the resonant modes, enabling dynamic control over the parameters of the non-Hermitian system. By adjusting the voltage magnitude, both EPs and diabolic points (DPs) can be observed in the system. At the EP, there exists a unique eigenstate, namely the LCP state. If RCP light is incident, the reflected light is a superposition of both LCP and RCP components. If LCP light is incident, while the polarization state of the

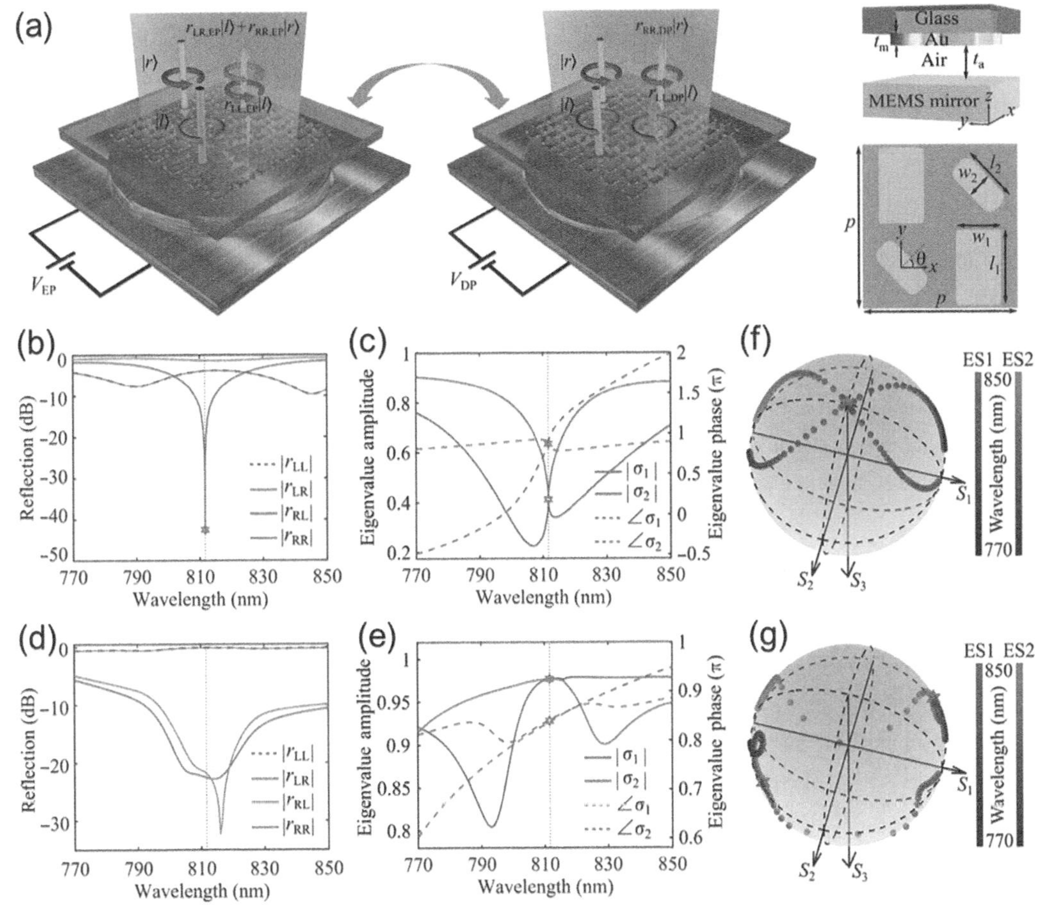

FIGURE 9.8 Structure, topological phase transition, and polarization evolution characteristics of the electrically tunable non-Hermitian MEMS metasurface [51]. (a) Physical schematic of the topological phase transition (EP→DP). The left panel shows the EP state, where the reflection eigenstates degenerate into a single circular polarization state; the right panel shows the DP state, where the reflection eigenstates degenerate into orthogonal polarization states, illustrating the voltage tuning effect on the system's EPs. (b–e) Optical characteristics in the EP state. (f) Eigenstate evolution in the EP state. (g) Eigenstate evolution in the DP state.

reflected light remains unchanged, the polarization conversion process from LCP to RCP is completely suppressed. When the incident polarization state is tuned to specific conditions, the two LCP components in the reflected light undergo destructive interference, resulting in all incident light being converted to the RCP state. Furthermore, when the reflection coefficient for the co-polarized component approaches zero, this non-Hermitian metasurface can function as an ideal absorber for LCP light. Simultaneously, it reflects RCP light and converts it to its orthogonal polarization state, thereby achieving significant circular dichroism. At the DP, the eigenvalues become degenerate, while the corresponding eigenstates remain orthogonal. When LCP and RCP lights are incident, the cross-polarized reflection effects disappear simultaneously. Their investigation into the polarization evolution during continuous variation of the air gap revealed that at specific wavelengths, if the incident light is LCP, the cross-polarization reflection coefficient always approaches zero regardless of adjusting the driving voltage. However, when RCP light is incident, the output polarization state is a superposition of co-polarized and cross-polarized components, and the polarization evolution trajectory on the Poincaré sphere manifests as an irregular closed curve. By adjusting the voltage magnitude, the intensities of both the co-polarized and cross-polarized components can be modulated.

In the research field of circularly polarized thermal radiation control [52], traditional thermal radiation sources typically exhibit broadband characteristics and lack directionality and polarization selectivity during emission. This makes it difficult to meet the practical demands of applications such as infrared sensing and chiral recognition, which require high coherence and spin selectivity. Some research has achieved circularly polarized radiation by designing twisted nanostructures or constructing symmetry-broken metasurfaces. However, these methods often suffer from issues such as structural complexity, insufficient coherence, or weak circular dichroism. Sun et al. [53] designed a planar non-local non-Hermitian metasurface structure based on a Z-shaped pattern arranged in a monoclinic lattice. This structure adopts a gold-alumina-germanium sandwich configuration, as shown in Fig. 9.9. By periodically shifting the straight waveguide array in a direction perpendicular to its length, the mirror symmetry of the structure is effectively broken, forming a monoclinic lattice and thereby achieving a stable chiral optical response. By adjusting the magnitude of the shift, reflected circular dichroism with opposite signs can be achieved, allowing for flexible switching between left-handed and right-handed circularly polarized outputs and realizing spin-selective control. Precise adjustment of the structural shift also enables effective tuning of the Q-factor of the guided-mode resonances. Furthermore, utilizing the critical coupling condition can balance the radiation loss and absorption loss within the system, thereby significantly enhancing the circular dichroism response. This planar structure is simple, maintains a stable chiral response over a wide wavevector range, and does not require complex structures like multi-layer stacking or tilted etching. It can also achieve chiral control based on quasi-bound states in the continuum, offering a practical solution for the development of high-performance, integrated circularly polarized thermal radiation sources.

Constructing non-Hermitian systems typically requires balancing the gain and loss relationships of materials, yet selecting gain materials for metallic metasurfaces presents

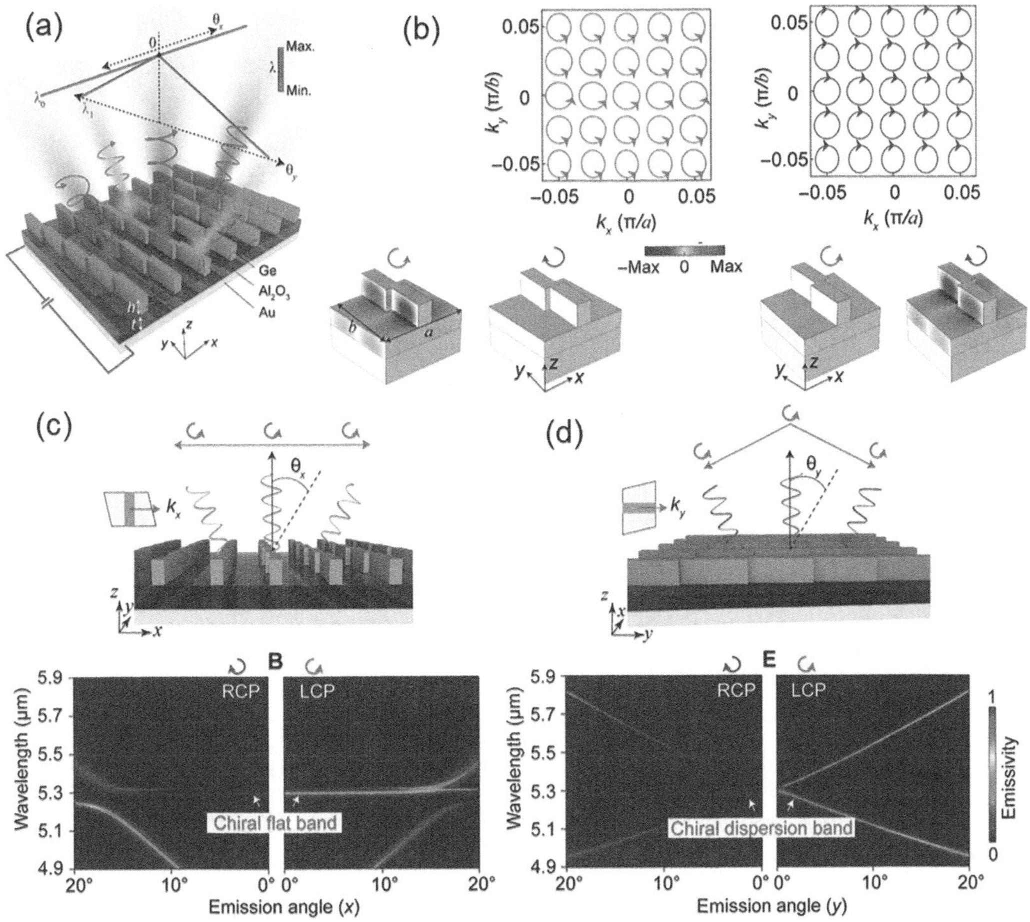

FIGURE 9.9 The monoclinic non-Hermitian polarization metasurface [53]. (a) Metasurface structure and radiation schematic. (b) Real part distribution of the electric field in k-space. (c) Angle-resolved emission spectra for RCP and LCP waves in x-direction. (d) Angle-resolved emission spectra for RCP and LCP waves in y-direction.

significant challenges. By modifying the loss of resonant modes, it is possible to investigate the physical characteristics of PT symmetry breaking and phase transitions at EPs within metasurface systems. Based on coupled-mode theory, He et al. [54] designed a hybrid system composed of an amorphous germanium (Ge) film and a patterned metallic metasurface, as shown in Fig. 9.10. The unit cell of the metasurface consists of a metallic cut wire oriented along the y-direction and a pair of SRRs embedded in the Ge film oriented along the x-direction. This configuration enables the formation of two orthogonally oriented resonant modes. By appropriately adjusting the geometry and materials of this structure, the resonant frequencies and loss rates of the metallic cut wire and the SRRs can be controlled, thereby facilitating the phase transition from the PT-symmetric phase to the PT-broken phase. By tuning the conductivity, the loss of the resonators along the x-direction can also be adjusted, achieving continuously tunable loss. Combined with the support of terahertz

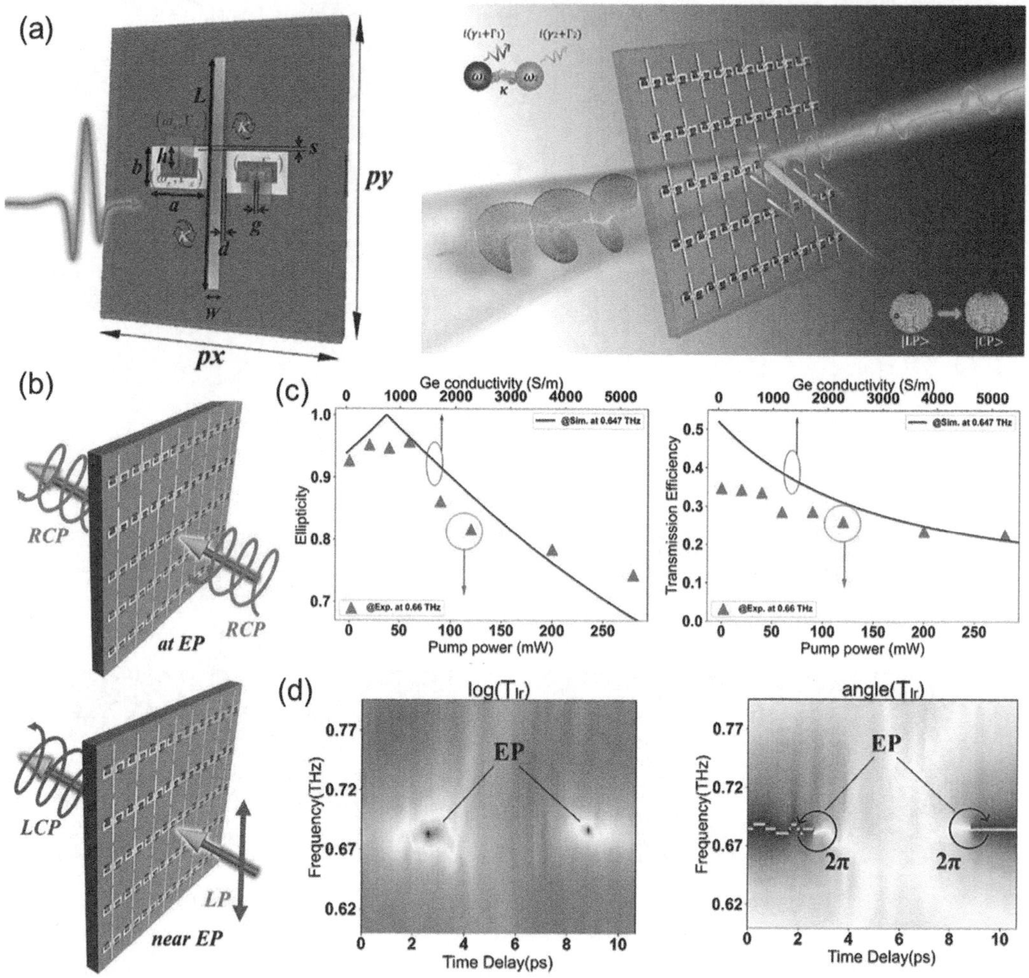

FIGURE 9.10 Optically pumped Ge hybrid non-Hermitian metasurface [54]. (a) Schematic diagram of the device unit cell and function. (b) Schematic diagram for polarization transmission characteristics at the EP. (c) Performance parameter control by pump power. (d) Time evolution of chiral transmission amplitude and phase.

time-domain spectroscopy techniques, the position of the EP can be determined within a two-dimensional parameter space. This control method based on the loss provides an innovative idea for the development of all-optically controlled ultrafast terahertz devices.

Dynamically active control represents an advanced control technique in the field of non-Hermitian polarization metasurfaces. This control method integrates the fundamental principles of structural parameter control, gain-loss control, and polarization degree of freedom control, endowing metasurface designs with reconfigurability, adaptability, and intelligent tuning capabilities. It is a key technology for advancing non-Hermitian metasurfaces from laboratory research towards practical engineering applications. Electrically controlled topological phase transitions based on piezoelectric MEMS reflectors, as well as spectral scanning achieved through the thermo-optic effect, demonstrate that dynamically active control can overcome the functional limitations of traditional passive metasurfaces, providing technical support for the development of high-performance dynamic polarization devices and intelligent photonic devices.

9.3 CHAPTER SUMMARY

This chapter reviews representative research achievements in the field of non-Hermitian polarization metasurfaces. By contextualizing relevant research backgrounds and core issues, it provides a detailed explanation of various polarization manipulation methods based on non-Hermitian metasurfaces. These studies are grounded in non-Hermitian Hamiltonians and TCMT, and focus on PT symmetry as well as the structural and parametric characteristics of the systems. Through the tuning of geometric parameters, material losses, and structural coupling strengths, diverse designs of metasurface structures are realized, ultimately enabling control over polarization states and phase transitions. Covering a broad spectral range from visible light to terahertz, these research outcomes not only break the symmetry constraints and functional limitations of conventional polarization manipulation but also meet the demands for diversified and high-precision polarization control.

REFERENCES

1. Sakurai, J.J. and Napolitano, J. (2020) 'Modern Quantum Mechanics', 3rd edn. Cambridge: Cambridge University Press.
2. Yu, N. and Capasso, F. (2014) 'Flat optics with designer metasurfaces', Nature Materials, 13(2), 139–150.
3. Prasad, C.S., Yang, F., Li, W., et al. (2022) 'Non-Hermitian metasurface with non-trivial topology', Nanophotonics, 11(6), 1159–1165.
4. Fu, H. and Li, J. (2022) 'Research progress of non-Hermitian electromagnetic metasurfaces', Acta Physica Sinica, 71(24), 247802.
5. Xu, H., Mason, D., Jiang, L. and Harris, J.G.E. (2016) 'Topological energy transfer in an optomechanical system with exceptional points', Nature, 537(7618), 80–83.
6. Kato, T. (1995) 'Perturbation theory for linear operators'. Heidelberg: Springer.
7. Mostafazadeh, A. (2002) 'Pseudo-Hermiticity versus PT symmetry: The necessary condition for the reality of the spectrum of a non-Hermitian Hamiltonian', Journal of Mathematical Physics, 43(1), 205–214.
8. Tian, Y., Liu, C. and Liu, Y. (2022) 'Research progress of parity-time symmetry and anti-symmetry', Acta Physica Sinica, 71(17), 171101.
9. Dembowski, C., Gräf, H.-D., Harney, H.L., et al. (2001) 'Experimental observation of the topological structure of exceptional points', Physical Review Letters, 86(5), 787–790.
10. Khan, H.Z., Ahmed, S. and Zhang, Y. (2025) 'Advancements in metasurfaces for polarization control: A comprehensive survey', Next Research, 2(3), 100407.
11. Makris, K.G., El-Ganainy, R., Christodoulides, D.N., et al. (2008) 'Beam dynamics in PT-symmetric optical lattices', Physical Review Letters, 100(10), 103904.
12. Rüter, C.E., Makris, K.G., El-Ganainy, R., et al. (2010) 'Observation of parity-time symmetry in optics', Nature Physics, 6(3), 192–195.
13. Longhi, S. (2018) 'PT symmetry and antisymmetry by anti-Hermitian wave coupling and nonlinear optical interactions', Optics Letters, 43(16), 4025–4028.
14. Bender, C.M. and Boettcher, S. (1998) 'Real spectra in non-Hermitian Hamiltonians having PT symmetry', Physical Review Letters, 80(24), 5243–5246.

15. Kenta Takata, Nathan Roberts, Akihiko Shinya, et al. (2021) 'Non-Hermitian temporal coupled-mode theory: effects of imaginary couplings on exceptional points', In Proceedings: Frontiers in Optics + Laser Science 2021, Washington, DC United States. Optica Publishing Group.
16. Haus, H.A. (1984) 'Coupled-mode theory for guided-wave optics', IEEE Journal of Quantum Electronics, 20(4), 325–332.
17. Fan, S., Suh, W. and Joannopoulos, J.D. (2003) 'Temporal coupled-mode theory for the Fano resonance in optical resonators', Journal of the Optical Society of America A, 20(3), 569–572.
18. Ishimaru, A. (1978) 'Wave propagation and scattering in random media'. New York: Academic Press.
19. Yao, J., Li, J., Zhang, Y., et al. (2023) 'Bound states in continuum in periodic optical systems', Chinese Optics, 16(1), 1–23.
20. Hassan, A.U., Zhen, B., Soljačić, M., et al. (2017) 'Dynamically encircling exceptional points: Exact evolution and polarization state conversion', Physical Review Letters, 118(9), 093002.
21. Li, A., Wei, H., Cotrufo, M., et al. (2023) 'Exceptional points and non-Hermitian photonics at the nanoscale', Nature Nanotechnology, 18(7), 706–720.
22. Nie, R., He, C., Zhang, R., et al. (2023) 'Vanadium dioxide-based terahertz metasurfaces for manipulating wavefronts with switchable polarization', Optics & Laser Technology, 159, 109010.
23. Park, J.-H., Ndao, A., Cai, W., et al. (2020) 'Symmetry-breaking-induced plasmonic exceptional points and nanoscale sensing', Nature Physics, 16(4), 462–468.
24. Oh, D., Baek, S., Lee, S., et al. (2024) 'Complete asymmetric polarization conversion at zero-eigenvalue exceptional points of non-Hermitian metasurfaces', Nanophotonics, 13(24), 4409–4416.
25. Planck, M. (1901) 'Ueber das Gesetz der Energieverteilung im Normalspectrum', Annalen der Physik, 309(3), 553–563.
26. Heisenberg, W. (1925) 'Über quantentheoretische Umdeutung kinematischer und mechanischer Beziehungen', Zeitschrift für Physik, 33, 879–893.
27. Schrödinger, E. (1926) 'Quantisierung als Eigenwertproblem', Annalen der Physik, 384(6), 489–527.
28. Dirac, P.A.M. (1935) 'The principles of quantum mechanics', Nature, 136, 411–412.
29. Schöpf, H. G. (1978) 'Von Kirchhoff bis Planck: Theorie der Wärmestrahlung in historisch-kritischer Darstellung'. Berlin, Boston: De Gruyter. https://doi.org/10.1515/9783112621905
30. Neumann, J. von (2018) 'Mathematical foundations of quantum mechanics'. Princeton: Princeton University Press. (Original work published 1932)
31. Lawrence, M., Zhang, N., Chen, H., et al. (2014) 'Manifestation of symmetry breaking in polarization space with terahertz metasurfaces', Physical Review Letters, 113(9), 093901.
32. Yin, X. and Zhang, X. (2013) 'Unidirectional light propagation at exceptional points', Nature Materials, 12(2), 175–177.
33. Gao, W., Chen, Y., Yang, Y., et al. (2022) 'Experimental demonstration of the non-Hermitian radio frequency metasurface with polarization-dependent unidirectional reflectionless propagation', Advanced Photonics Research, 3(6), 2200019.
34. Yi, C., Li, J., Wang, H. et al. (2025) 'Creating topological exceptional point by on-chip all-dielectric metasurface', Light: Science & Applications, 14(1), 262.
35. Bergholtz, E.J., Budich, J.C. and Kunst, F.K. (2021) 'Exceptional topology of non-Hermitian systems', Reviews of Modern Physics, 93(1), 015005.
36. Song, Q., Odeh, M., Zúñiga-Pérez, J., et al. (2021) 'Plasmonic topological metasurface by encircling an exceptional point', Science, 373(6559), 1133–1137.
37. Cui, T.J., Qi, M.Q., Wan, X., et al. (2014) 'Coding metamaterials, digital metamaterials and programmable metamaterials', Light: Science & Applications, 3(10), e218.
38. Li, Y., Zhang, H., Wang, Y., et al. (2024) 'Independent control of circularly polarized light with exceptional topological phase coding metasurfaces', Photonics Research, 12(3), 534–542.

39. Oh, S.S. and Hess, O. (2015) 'Chiral metamaterials: Enhancement and control of optical activity and circular dichroism', Nano Convergence, 2(1), 24.
40. Barron, L.D. (2004) 'Molecular light scattering and optical activity', 2nd edn. Cambridge: Cambridge University Press.
41. Kolkowski, R., Kovaios, S. and Koenderink, A.F. (2021) 'Pseudochirality at exceptional rings of optical metasurfaces', Physical Review Research, 3(2), 023185.
42. Baek, S., Park, S.H., Oh, D., et al. (2023) 'Non-Hermitian chiral degeneracy of gated graphene metasurfaces', Light: Science & Applications, 12(1), 87.
43. Li, H., Yang, G., Jiang, A., et al. (2024) 'Manifestation of super chiral exceptional points in a plasmonic metasurface', Photonics Research, 12(12), 2863–2872.
44. Li, G., Zhang, S. and Zentgraf, T. (2017) 'Nonlinear photonic metasurfaces', Nature Reviews Materials, 2(5), 17010.
45. Rubin, N.A., Zaidi, A., Juhl, M., et al. (2018) 'Polarization state generation and measurement with a single metasurface', Optics Express, 26(17), 21455–21478.
46. Qin, R., Zhang, Y. and Chen, H. (2025) 'Sphere of arbitrarily polarized exceptional points with a single planar metasurface', Nature Communications, 16(1), 2546.
47. El-Ganainy, R., Makris, K.G., Khajavikhan, M., et al. (2018) 'Non-Hermitian physics and PT symmetry', Nature Physics, 14(1), 11–19.
48. Özdemir, Ş.K., Rotter, S., Nori, F., et al. (2019) 'Parity–time symmetry and exceptional points in photonics', Nature Materials, 18(8), 783–798.
49. Yang, Z., Li, X., Wang, Y., et al. (2024) 'Creating pairs of exceptional points for arbitrary polarization control: Asymmetric vectorial wavefront modulation', Nature Communications, 15(1), 232.
50. Shi, X., Liu, Y., Wang, Z., et al. (2025) 'Polarization-controlled non-Hermitian metasurfaces for ultra-sensitive terahertz sensing', Laser & Photonics Reviews, 19(14), 2500172.
51. Ding, F., Deng, Y., Meng, C., et al. (2024) 'Electrically tunable topological phase transition in non-Hermitian optical MEMS metasurfaces', Science Advances, 10(5), eadl4661.
52. Overvig, A.C., Mann, S.A. and Alù, A. (2021) 'Thermal metasurfaces: Complete emission control by combining local and nonlocal light-matter interactions', Physical Review X, 11(2), 021050.
53. Sun, K., Yang, B., Cai, Y., et al. (2025) 'Circularly polarized thermal emission driven by chiral flatbands in monoclinic metasurfaces', Science Advances, 11(31), eadw0986.
54. He, Y., Zhang, Y., Li, X., et al. (2023) 'Transient loss-induced non-Hermitian degeneracies for ultrafast terahertz metadevices', Advanced Science, 10(12), 2304972.

CHAPTER 10

On-Chip Polarization Manipulation

Abstract

When metasurfaces are integrated with optical waveguides or plasmonic platforms, they give rise to the burgeoning research direction of "on-chip metasurfaces". Such devices not only inherit the powerful light-manipulation capabilities of metasurfaces but also fully leverage the additional design freedom provided by guided-wave modes. They hold great promise as multifunctional interfaces connecting on-chip photonic circuits and free-space light fields. This chapter focuses on on-chip metadevices for polarization control. The core physical processes of on-chip metasurfaces for polarization control involve two key steps: coupling and manipulation. Traditional methods usually separate these two steps, first coupling light via gratings or prisms and then manipulating it with diffractive elements, leading to bulky devices and low efficiency. The advent of metasurfaces makes it possible to directly embed manipulation "instructions" within the coupling process. The core idea is that, by meticulously designing the geometrical parameters of the meta-units, it is possible to simultaneously couple free-space light into surface waves while imprinting these waves with predetermined wavefront and polarization information.

10.1 PHASE-BASED ON-CHIP POLARIZATION CONTROL

PB phase-based polarization control. The key to realizing this functionality lies in the sophisticated utilization of the polarization degree of freedom [1]. Below, starting from the fundamental principles of guided-mode radiation and surface plasmon (SP) coupling [2], this section will provide a categorized introduction based on different manipulation mechanisms. The Pancharatnam-Berry (PB) [3,4] phase is one of the most mature approaches for realizing on-chip polarization manipulation, whose core lies in manipulating the optical

DOI: 10.1201/9781003784821-10

field using geometric phase rather than dynamic phase. By designing the spatial orientation angle θ of anisotropic resonators, a phase delay of $2\sigma\theta$ can be imposed on incident left-handed circularly polarized (LCP) and right-handed circularly polarized (RCP) light ($\sigma = \pm 1$). The advantage of this mechanism is that the imparted phase distribution depends only on the resonator orientation and the spin state of light, and is nearly independent of the operating wavelength and material dispersion, thus exhibiting an intrinsic broadband response.

Within this framework, polarization evolves from a mere transmission carrier into a "switch" and "knob" of the device: by switching the spin state of the incident light, the sign reversal of the optical field phase distribution can be achieved, thereby dynamically controlling the propagation direction of coupled surface waves. Such a manipulation strategy based on the PB phase not only provides a concise and efficient design paradigm for on-chip polarization-multiplexing devices (e.g. polarization splitters, polarization rotators) but also lays a theoretical foundation for realizing reconfigurable metasurfaces, beam deflectors, and spin-dependent photonic functionalities.

Leveraging the principle of phase discontinuity, researchers have designed a metasurface plasmonic coupler based on C-shaped slot resonators (CRs). By controlling the geometric parameters of the CRs (such as radius, width, and opening angle), a controllable phase delay from 0 to 2π can be generated in the cross-polarized channel. Combined with adjustments to the rotation angle β, amplitude modulation of the coupled SPs [5,6] can be achieved using the relation $|E| \propto |\sin 2\beta|$. Simultaneously, a spatial phase discontinuity gradient is constructed based on the generalized refraction formula. When the phase gradient in the x-direction satisfies $d\varphi/dx = k_{SP}$ (the wave number of SPs), the refracted wave from free space propagates along the metal/dielectric interface, with the electric field existing almost entirely as the E_z component and coupling into SPs (Fig. 10.1(a)) [7]. Furthermore, by designing the phase gradient $d\varphi/dy$ in the y-direction, anomalous control over the SP wavefront can be achieved, enabling normal SP emission. This coupler demonstrates a coupling efficiency 4.2 times higher than that of traditional ordinary slot resonators, with significant advantages under large-area excitation. By incorporating amplitude modulation via the rotation angle of the CRs, simultaneous control over both the phase and amplitude of the coupled SPs is realized, achieving dual diffraction modes of the 1st and 3rd orders.

To elevate the application of the geometric phase to a new level, high-efficiency PB meta-atoms connected double-ring resonators in a metal–insulator–metal configuration were designed [8]. By tailoring their reflection coefficients to satisfy $r_{uu} = r_{vv} = 1$ and a phase difference of $\arg(r_{uu}) - \arg(r_{vv}) \approx \pi$, these meta-atoms approximate ideal half-wave plates, significantly enhancing the efficiency of anomalous reflection modes. Peak absolute efficiencies of 0.86 and 0.9 were achieved at 0.4 and 0.63 THz, respectively. Based on these PB meta-atoms, a gradient-phase meta-coupler was constructed and integrated with an "artificial metal" capable of supporting terahertz spoof surface plasmons (SSPs) [9,10]. Leveraging the helicity dependence of the PB phase, the meta-coupler imparts phase gradients of opposite signs upon illumination by LCP and RCP light, enabling chirality-modulated directional excitation of SSPs with an absolute efficiency of 60% at 0.4 THz, far exceeding that of conventional grating couplers shown in Fig. 10.1(b).

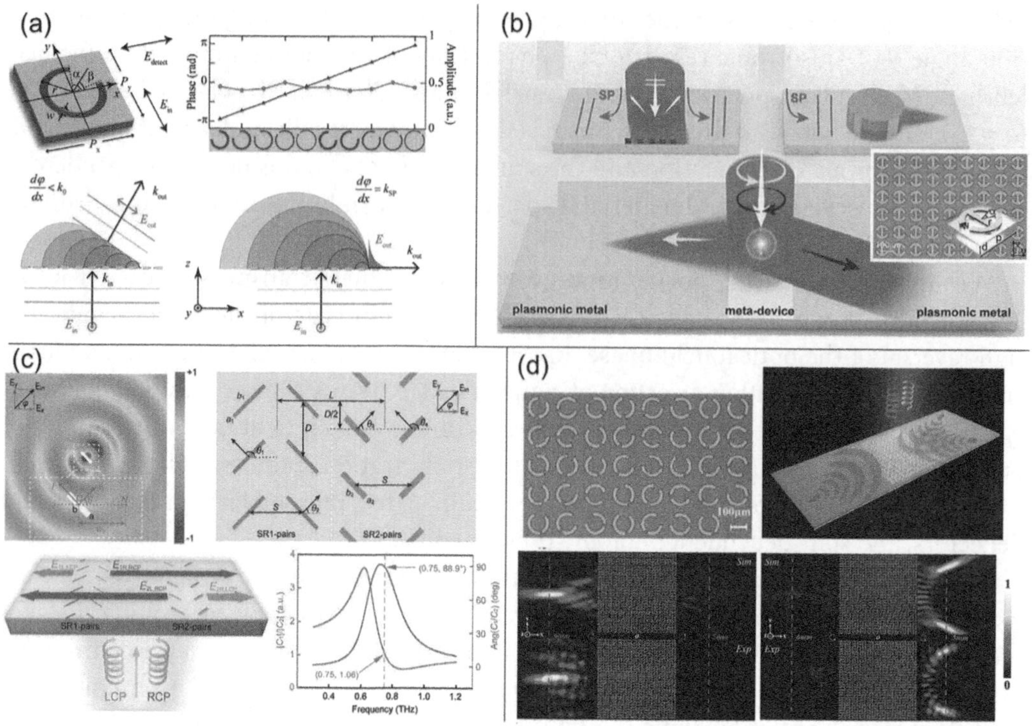

FIGURE 10.1 (a) Schematics of C-shaped SR and phase discontinuity concept. (b) Design and function of a metasurface based on high-efficiency PB meta-atom [8]. (c) Design schematic of the directional SP coupler. (d) Schematics and simulation results of a plasmonic metasurface for circular polarization focusing control [1,13].

Two distinct phase-control strategies were employed to achieve wavefront shaping of SSPs. The first strategy relies solely on the PB phase mechanism: by encoding both gradient and parabolic phase profiles along two orthogonal directions within the meta-coupler, helicity-locked wavefront control was realized. Under LCP incidence, the excited SSPs exhibit a focusing wavefront, whereas RCP incidence leads to a defocusing wavefront, with an efficiency of 40% at 0.4 THz. The second strategy integrates the PB phase with resonance phase mechanisms. By adjusting the opening angle of the meta-atoms, a resonance phase independent of polarization helicity was introduced. When superimposed with the helicity-dependent PB phase, this approach breaks the inherent chirality coupling limitation of the PB phase alone, enabling chirality-decoupled bifunctional wavefront engineering of SSPs. As a result, LCP light excites focusing SSPs, while RCP light generates deflected propagating SSPs, with efficiencies of 30% and 35% at 0.4 THz, respectively. The efficiency reduction is primarily attributed to increased scattering losses and mode-matching issues arising from the integration of multiple functionalities.

The PB phase mechanism provides a core method for the on-chip manipulation of circularly polarized light. By adjusting the orientation angle θ of anisotropic resonators, a geometric phase delay of $2\sigma\theta$ can be imposed on LCP and RCP light. On this basis, the principle of phase discontinuity further achieves momentum matching between free-space light and surface plasmon polaritons (SPPs), laying a theoretical foundation for the design of metasurface couplers. Meanwhile, the research on multi-layer or reflective metasurfaces provides structural optimization ideas for improving coupling efficiency. Through the integration of various phase mechanisms (such as PB phase and dynamic phase), as well as the refined design of meta-atom geometry and materials, researchers have successfully broken through the functional limitations of a single polarization manipulation mechanism, realizing efficient, integrated, and multi-functional excitation of SPs under polarization state manipulation. These advances provide new technical solutions for the application of polarization manipulation in plasmonic integrated photonics and are expected to promote the development of fields such as on-chip optical interconnection, sensing, and quantum information processing.

Polarization control based on interference effects. The core advantage of the geometric phase lies in its spin-locked robustness; i.e. rotating anisotropic resonators can impose a wavelength-independent phase delay on circularly polarized light [11]. However, this advantage also brings a fundamental limitation: the phase distributions of the two spin channels (left-handed and right-handed circular polarization) are conjugate to each other ($\phi = \pm 2\sigma\theta$), which cannot be independently manipulated, resulting in the device's function being locked into symmetric operations when the spin is reversed.

To break this "conjugate locking", researchers have introduced a manipulation strategy based on interference effects. Its core idea is to extend the degree of freedom of polarization from a single geometric phase to the resonant phase (i.e. the dynamic phase determined by the size and shape of the resonator). By designing the relative amplitude and phase difference between the surface wave sources excited by different polarization channels (such as orthogonal linear polarization or left-/right-handed circular polarization), their interference effects are used to synthesize any target wavefront. Under this new framework, the role of polarization transforms from a "switch" to an "interference term": it is no longer merely a knob that triggers a preset phase distribution, but a vector component participating in interference. Its relative weight and phase difference directly determine the result of constructive or destructive interference, thereby achieving precise and decoupled manipulation of the optical field propagation direction. This idea opens up a new path for constructing high-efficiency, reconfigurable on-chip plasmon excitation devices.

A super-unit is constructed using two pairs of slit resonators (SRs) with distinct geometries, introducing new degrees of freedom for manipulation through differences in their resonant responses [12]. By combining this with the manipulation of the relative distance between the SRs, polarization-dependent directional coupling of SPs is achieved. A single high-aspect-ratio SR strongly excites SPs only for the polarization component perpendicular to its long side. Two pairs of geometrically different SRs (SR1, SR2) are arranged with specific angles and spacings to form a super-unit. Due to their structural differences, the two types of SRs exhibit distinct resonant responses for SP excitation, characterized by complex coefficients C_1 and C_2. The excitation direction of SPs within the super-unit is determined by the interference of the resonant responses from the two SR pairs.

By designing the amplitude ratio and phase difference of C_1 and C_2, and combining them with the relative distance L between the two SR pairs within the super-unit, conditions can be satisfied such that $|C_1/C_2| = A_{LCP}/A_{RCP}$, $\arg(C_1/C_2) = \pm\pi/2$, and $L = [\phi_{LCP} - \phi_{RCP} - (2n+1)\pi - \arg(C_1/C_2)]/k_{sp}$. This enables directional SP coupling for arbitrarily polarized incident light. Furthermore, when the incident polarization is orthogonal, the SP coupling direction reverses while the coupling efficiency remains unchanged. For incident light with a fixed polarization ellipticity, different polarization orientation angles can be accommodated simply by adjusting L, without the need to modify the SR geometries, significantly enhancing design flexibility in Fig. 10.1(c). Based on the local polarization-dependent directional coupling and wavefront manipulation capabilities of this strategy, three high-performance polarization-dependent plasmonic devices were constructed and experimentally validated: an SP metalens for uniformly polarized incidence, an SP metalens for non-uniformly polarized vector beam incidence, and a polarization-controlled plasmonic airy beam launcher.

Spin-decoupled control based on resonance and topological phases. Building upon geometric phase and interference-based control, researchers have further explored novel phase manipulation mechanisms to achieve more flexible and robust spin-decoupled regulation. Resonance phase and topological phase: by tuning the resonance response of meta-atoms through their geometric parameters, or by leveraging the exceptional point (EP) characteristics in non-Hermitian systems, polarization-dependent resonance phases or topological phases are introduced as supplements or alternatives to the geometric phase, enabling more flexible spin-decoupled control.

Using a double-split ring resonator with chiral characteristics as the meta-atom [13], a parameter space was constructed by designing two distinct separation angles β. Leveraging the topological properties of non-Hermitian matrices near an EP, the meta-atom forms an EP in the 0.58 THz band. At this point, the cross-polarization reflection coefficient from LCP to RCP drops to zero, and the phase distribution surrounding the EP covers a full 2π range (referred to as the exceptional topological (ET) phase). In contrast, the reflection coefficient from RCP to LCP exhibits no EP, with uniform phase and amplitude distributions. Meanwhile, the rotation angle θ of the meta-atom introduces the PB phase, forming a dual-phase modulation mechanism of "ET phase + PB phase", where SPs excited by LCP incidence are simultaneously modulated by both phases, while SPs excited by RCP incidence are modulated only by the PB phase with opposite sign in Fig. 10.1(d). To address the detection challenge of reflective SP devices, an artificial metal surface with a subwavelength array was designed, allowing SPs to leak electric field energy to the opposite side of the reflector layer during propagation, thereby enabling interference-free detection. This artificial metal surface supports SP propagation at 0.58 THz with a wavevector of $k_{SP} = 1.09k_0$.

Based on the two-dimensional Huygens-Fresnel principle and holographic theory, combined with corrections incorporating surface vortex terms, the phase design equations for the meta-atoms were derived. By selecting points in the parameter space that encircle the EP while maintaining uniform amplitude, the separation angles β and rotation angles θ for each meta-atom were determined, enabling the construction of a topological metasurface device. Experimental and simulation results demonstrate that under LCP incidence, the device focuses SPs at a single focal point on the left side. Under RCP incidence, it focuses

SPs at two focal points on the right side. The wavefront manipulation of SPs under the two polarization states exhibits minimal cross-talk and significant spatial isolation, validating the spin-decoupled polarization control capability. Furthermore, the ET phase demonstrates topological robustness, maintaining stable phase distributions within specific frequency bands and under fabrication tolerances.

Multi-channel polarization multiplexing based on detour phase. The detour phase is a physical mechanism for achieving phase modulation in planar optical devices, such as metasurfaces and diffractive lenses. Its core concept involves introducing controllable phase delays by altering the spatial distribution of subwavelength structures, rather than relying solely on their geometric shapes or material properties. In traditional diffractive optics, phase modulation typically depends on the height, width, or refractive index of the structures. The detour phase takes a different approach: within periodically or quasi-periodically arranged subwavelength units (e.g. nanopillars or holes), the lateral displacement (δx, δy) of these units causes incident light to traverse different optical paths as it passes through units at varying positions, thereby generating the desired phase differences. The detour phase can be easily combined with other phase mechanisms, such as the geometric phase and propagation phase, enabling multi-degree-of-freedom multiplexing.

On-chip metasurfaces offer additional degrees of freedom introduced by guided waves, such as the detour phase. Leveraging this, Ji et al. [14] proposed an on-chip metasurface based on a supercell structure. By integrating the detour phase and geometric phase through a joint modulation mechanism, they constructed an asymmetric, non-unitary equivalent Jones matrix under dual-port guided-wave illumination, achieving full-parameter modulation of the Jones matrix. This design enriches the modulation capabilities of on-chip metasurfaces and significantly enhances multi-channel multiplexing, offering new possibilities for applications in holographic displays, virtual reality, and optical communication. In their work, a silicon metasurface was first fabricated on a lithium niobate waveguide. Under different guided-wave illumination conditions, the on-chip metasurface, jointly modulated by the detour phase and geometric phase, achieved decoupling of the diagonal terms of the Jones matrix. However, at this stage, the Jones matrix remained unitary, meaning the four matrix elements were still interdependent and could not be fully decoupled. To address this, a four-atom combined supercell design strategy was proposed. By coherently superimposing four unitary matrices, a non-unitary, asymmetric equivalent Jones matrix was constructed, enabling each element to be independent. Additionally, thanks to the rich design freedom introduced by the detour phase, geometric phase, and supercell structure (e.g. the rotation angles and in-plane positions of the four atoms within the supercell), each element of the Jones matrix could satisfy arbitrary target complex amplitude distributions. Under guided-wave excitation along the $+x$ and $+y$ directions, and analysed with horizontal and vertical linear polarizers, the guided-mode radiation produced four Greek letter images in the near field. However, when the guided waves propagated in the opposite directions ($-x$, $-y$), the corresponding Jones matrix exhibited a conjugate relationship with that of forward propagation, making multi-directional multiplexing challenging. To overcome this, the study further incorporated the propagation phase into the on-chip metasurface design, as shown in Fig. 10.2. By combining the propagation phase with the

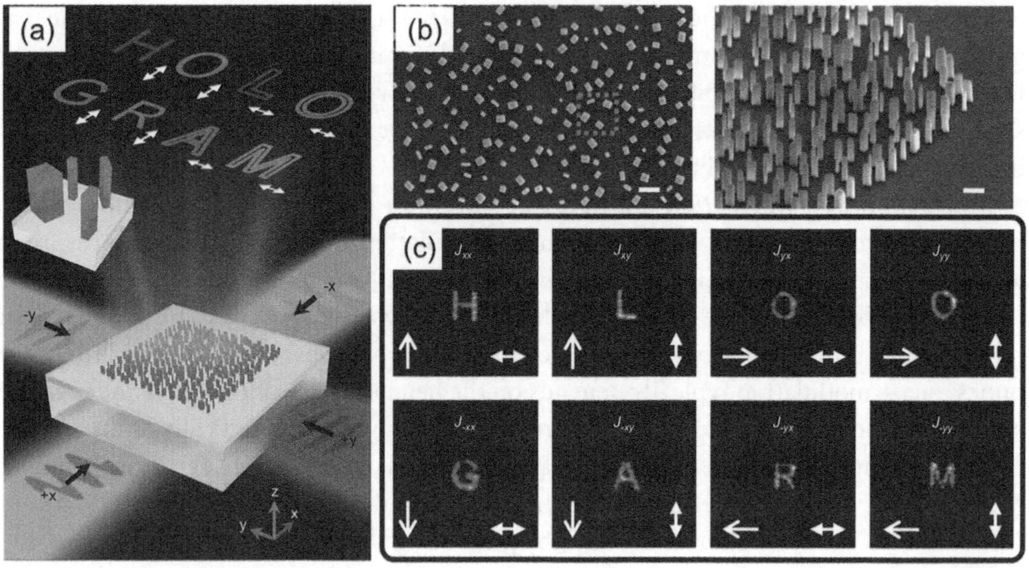

FIGURE 10.2 (a) The schematic device for multi-channel polarization multiplexing based on detour phase. (b) The device SEM pictures. (c) The experimental results of polarization-related multi-channel holographic modulation [14].

detour phase and geometric phase through a joint modulation mechanism, the conjugate relationship under forward and reverse guided-wave excitation was broken, resulting in two sets of mutually independent equivalent Jones matrices. Experimentally, this successfully demonstrated the polarization-related eight-channel direction holographic modulation function.

10.2 ON-CHIP VORTEX DEVICES BASED ON SPIN-ORBIT COUPLING

By combining polarization control with orbital angular momentum (OAM) [15], on-chip plasmonic vortices are generated. The underlying physical mechanism is spin-orbit coupling (SOC): the spin angular momentum (SAM) [16,17] carried by incident circularly polarized light is converted into the OAM of plasmonic vortices through meticulously designed metasurfaces, such as Archimedean spiral slits or slit-pair arrays. This conversion process is inherently polarization-dependent, as the topological charge l of the vortex is determined by both the spin σ of the incident light and the geometric phase factor g of the device, following the relation $l = \sigma(g - 1)$. The geometric phase factor g characterizes the rotational symmetry of the metasurface structure; for example, in an Archimedean spiral design, g represents the number of spiral arms or the rate of azimuthal phase variation. When LCP light $\sigma = +1$ illuminates a metasurface with a given g, the excited plasmonic vortex (PV) acquires a topological charge $l = g - 1$; conversely, RCP light $\sigma = -1$ generates a vortex with opposite topological charge $l = -(g - 1)$. This spin-dependent response enables flexible control over the vortex properties, including its chirality, phase singularity, and OAM density, providing a versatile platform for applications in particle manipulation, information encoding, and on-chip optical communications.

Using subwavelength SRs as the basic unit cell, their dipole response characteristics (excitation only by the polarization component perpendicular to the long side) are exploited [18]. By arranging SRs in pairs and adjusting their spacing and orientation angles, the phase and amplitude of SPs can be synergistically controlled. Based on the two-dimensional Huygens-Fresnel principle, the spin of a SAM of incident circularly polarized light (LCP/RCP) is converted into the OAM of SPs, generating plasmonic vortices with specific topological charges whose electric field distributions follow Bessel functions. By designing inner and outer annular arrays of SR pairs (with a radius difference of $\lambda_{SP}/4$) and leveraging the superposition of spin-dependent vortex generation and spin-independent propagation phases, selective excitation of vortices under different polarization states or multi-vortex interference is achieved. Ultimately, the polarization state, SAM, and OAM information of the incident light are encoded into a single SP interference pattern in Fig. 10.3(a).

However, for plasmonic vortices with the same topological charge, their spatiotemporal evolution behaviours can be endowed with distinct characteristics by varying the combination of the geometric phase factor g and the propagation phase factor p, according to $l=\sigma(g-1)+p$. For instance, when only the geometric phase is employed ($p=0, g=-3$), the SP field reaches its maximum and decays simultaneously along the target orbit. In contrast, when both geometric and propagation phases are utilized ($p=4, g=1$), the SP field appears

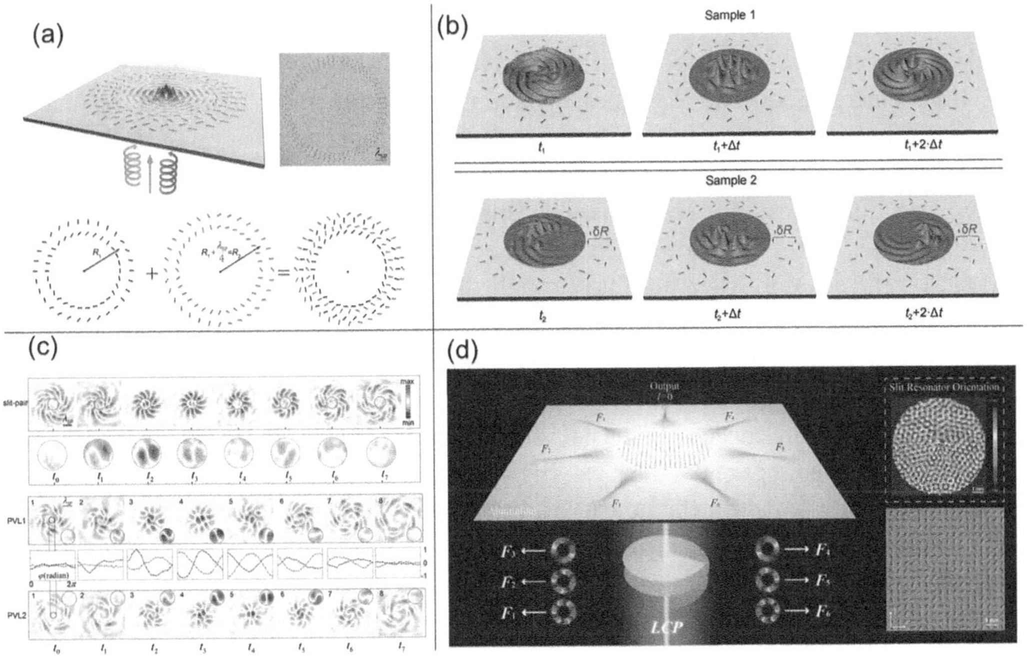

FIGURE 10.3 (a) Schematics and microscopic image of PV interferometers design [1,18]. (b) Schematic diagram of the temporal evolution process of plasmonic vortices of the same topological charge generated by different couplers [19]. (c) Measured time-resolved snapshots of the SP field based on the NSTM system, generated from the slit-pair-based PV lenses [20]. (d) Seven-channel on-chip OAM modes demultiplexing [22].

and evolves sequentially along the target orbit in Fig. 10.3(b) [19]. Yuan et al. [20] further discovered and elucidated the existence of "temporally deuterogenic vortex modes": two vortex modes independently excited by the inner and outer rings undergo destructive interference at the centre frequency due to opposite phases, rendering them invisible in the intensity distribution. However, they can be clearly observed during temporal evolution in Fig. 10.3(c). These works revealed, for the first time, the correlation between phase combinations and spatiotemporal dynamics, identifying and experimentally verifying the existence of time-derived vortices as a frequency-domain hidden mode. By leveraging near-field scanning terahertz microscopy (NSTM) to overcome the limitations of traditional characterization methods, sub-optical-cycle spatiotemporal observations were achieved, providing a new degree of freedom for the temporal manipulation of plasmonic vortices. This approach is particularly suitable for applications such as torque customization in plasmonic tweezers and timing optimization of on-chip light-matter interactions and can be extended to visible and infrared frequency bands.

Within the framework of SOC, LCP and RCP light, as two spin states, can realize diverse optical field manipulations from vortex beams to pure vector beams via geometric phase modulation [21]. The residual OAM in conventional vector beams is eliminated through the vortex phase cancellation mechanism. Arbitrary linearly polarized light can be decomposed into LCP and RCP components with equal amplitude. When passing through a non-axisymmetric half-wave plate whose fast-axis direction varies linearly with the azimuthal angle, the left-handed component is converted to right-handed with a geometric phase $+\theta$, while the right-handed component is converted to left-handed with a geometric phase $-\theta$. Their superposition leads to the cancellation of the geometric phases, and the phase of the output beam becomes uniform. Meanwhile, the vector superposition of the two circularly polarized components forms a radial polarization distribution: at azimuthal angle θ, the polarization direction is exactly radial. This process can be explained by the geometric phase principle on the Poincaré sphere: the incident linearly polarized light corresponds to a point on the equator. After passing through the non-axisymmetric half-wave plate, the polarization state traverses a full cycle along the equator, and the solid angle enclosed by the closed path is zero, resulting in a zero geometric phase. To achieve this function, the authors designed a non-axisymmetric half-wave plate based on internal reflections, fabricated from high-density polyethylene. Four internal reflections generate a π phase delay, which is achromatic within the range of 0.1–1.6 THz. In the experiment, a single-shot polarization measurement method based on the Fourier transform was adopted to reconstruct the spatial distribution of the Stokes parameters of the output beam. The measured average geometric phase was only −0.01 rad, and the residual vortex phase was below 4%, confirming the successful generation of pure vector beams. This work extends SOC from vortex beams to pure vector beams, providing a new optical field modulation tool for applications such as super-resolution microscopy, optical tweezers, and quantum information processing.

By combining the geometric phase with the two-dimensional holographic principle and utilizing subwavelength metallic SRs as meta-atoms, the relationship between the geometric phase and the slit rotation angle enables the coupling of free-space OAM modes into on-chip SPs through holographic design. For LCP incident light carrying a helical phase, each slit is treated as a dipole source. The superimposed field must integrate the propagation phase, in-plane phase, and helical phase. The rotation angle of each slit is then determined by $\theta_m = \arg(H_m^{\sigma})/2$. The complex amplitude of the SP field at any point is obtained by superimposing all slit sources, excited only by the incident polarization component parallel to the slit orientation, ultimately coupling specific OAM modes into focused spots along predetermined directions. Based on this principle, a seven-channel on-chip terahertz OAM demultiplexer was designed [22], achieving seven high-quality focused SP beams within the target frequency band, with an average isolation of −11 dB, excellent focal spot quality, and stable frequency response. For hybrid OAM mode incidence, experimental validation confirmed effective demultiplexing capability, with target channel intensities significantly higher than other channels. Extended designs, including a six-channel demultiplexer and a nine-channel OAM-wavelength joint demultiplexer, demonstrated excellent design scalability. The SP field intensity and phase distributions characterized by near-field terahertz scanning microscopy showed excellent agreement with numerical simulations in Fig. 10.3(d).

This approach breaks through the directional design limitations and functional singleness of traditional schemes, enabling flexible customization of focusing directions, supporting hybrid mode demultiplexing and multidimensional multiplexing expansion. It solves the critical challenge of efficiently coupling free-space OAM modes to on-chip SP modes and routing them, providing a core terminal device for ultra-compact terahertz mode-division multiplexing (MDM) communication systems. Furthermore, it offers a new paradigm for the collaborative control of polarization and OAM in on-chip information processing.

Using subwavelength rectangular slit antenna pairs as fundamental elements, concentric inner and outer PV ring generators were constructed. Leveraging the SOC effect, spin-decoupled PV eigenmodes are excited – by modulating the orientation angles and radial distributions of the slit pairs, geometric phases and propagation phases are introduced, enabling the inner and outer rings to correspond to PV modes with specific topological charges. The radius difference between the two rings creates a sufficient surface wave path difference, allowing the two PV modes excited by a single terahertz pulse to form non-overlapping independent time windows in the time domain, achieving separate control of different modes. In the PV interferometer, liquid crystal (LC) coding units were integrated, with graphene serving as transparent electrodes to divide the LC layer into two independently controllable regions. By applying different bias voltages, the orientation of LC molecules is altered, introducing a tunable phase delay of 0 to π and spin conversion effects, forming four coding levels [23]. Based on a “spatiotemporal correlation coding” mechanism, the first coding level controls the superposition weight of the PV mode from the inner ring within the earlier time window, while the second coding level controls the superposition weight of the PV mode from the outer ring within the later time window. Through 2×2 coding combinations, flexible weighting of the four eigenmodes is achieved, dynamically generating 16 steady-state coherent superposition fields and 8 transient evolution states.

In SOC-based vortex devices, polarization serves as both the “source of angular momentum” and an “information encoding dimension”. The SAM of the incident light acts as the driving force for generating plasmonic vortices, with the two coupled through SOC. The polarization state itself becomes an independent information channel, whose weight and relative phase can be combined with OAM modes for applications such as measurement (e.g. polarimeters, SAM/OAM identifiers), multiplexing (e.g. OAM demultiplexers), or dynamic control (e.g. spatiotemporal coding interferometers). Polarization is not merely an excitation condition but a core element in constructing complex light fields and enabling multidimensional information processing.

10.3 RECONFIGURABILITY AND MULTIFUNCTIONAL INTEGRATION

The functions of static devices are fixed once fabricated. This inherent limitation makes it difficult to meet the urgent demand for dynamically tunable functions in future adaptive and intelligent systems. In practical applications such as communication, imaging, and information encryption, devices often need to adjust their functions in real time according to changes in the external environment, such as dynamically switching channels, adaptively tuning focal lengths, or realizing reconfigurable encryption. Once fabricated, traditional static metasurfaces have locked geometric structures and optical responses that cannot be further modulated, which greatly restricts their application potential in cutting-edge fields including intelligent photonics, adaptive optics, and reconfigurable communications.

Therefore, introducing reconfigurable materials has become an inevitable trend in the development of on-chip metasurface polarization devices. Current research mainly focuses on two material systems: phase-change materials (such as GST and VO_2) undergo reversible phase transitions under thermal, optical, or electrical stimuli, leading to significant changes in refractive index or conductivity, thereby dynamically modulating the electromagnetic responses of meta-atoms. Moreover, the phase-change process is non-volatile – the state can be maintained after removing external excitation, resulting in extremely low power consumption. Liquid crystal materials realize continuously tunable refractive index anisotropy via molecular reorientation under electric fields, producing phase retardation ranging from 0 to π, with continuous, reversible modulation and fast response speed. The introduction of these reconfigurable materials has pushed on-chip metasurface polarization devices from the stage of "static design" to "dynamic modulation", laying a material foundation for future applications such as adaptive optical systems, reconfigurable communication networks, and intelligent photonic computing.

Using the C-shaped SR as the fundamental building block, its geometric parameters are employed to control the amplitude and phase of the cross-polarized output. Combined with the two-dimensional holographic principle, the phase distribution of the CR array is designed by treating the focal points as SP point sources. By superposing the complex electric field amplitudes of these point sources at the positions of the CRs, the parameters of each CR are determined to achieve wavefront focusing of SPs. A $Ge_2Sb_2Te_5$ (GST) [24,25] intermediate layer is introduced between the CRs and the silicon substrate. Leveraging the non-volatile amorphous/crystalline phase transition characteristics of GST (reversibly switchable via optical or thermal stimuli), its conductivity changes significantly with the phase state, thereby dynamically tuning the THz transmission efficiency of the CRs to achieve reconfigurable SPs focusing intensity or focal spot position. When GST is in its amorphous state, the device is in the "ON" state, and the SPs excited by the y-polarized incident light are focused on the target point. In the crystalline state, the transmission efficiency is nearly zero, and the device is in the "OFF" state. By adjusting the pump laser energy, multi-level modulation of the focusing intensity can be achieved, with a maximum extinction ratio of 29.9 dB. The state is non-volatile, remaining stable after the stimulus is removed, and supports multiple reversible switching cycles. As shown in Fig. 10.4(a), a dual-structure array consisting of an active group (GST-integrated CRs) and a passive group (CRs without GST) is designed [26]. When GST is in the amorphous state, the SPs excited by the two groups of structures undergo destructive interference at the f_2 point and ultimately focus at f_1. When GST is in the crystalline state, the active group becomes inactive, and the SPs focus at f_2, achieving focal point switching. The switching contrast ratios in the amorphous and crystalline states are 7.4 dB and −44.4 dB, respectively. The device exhibits stable performance after multiple cycles of optical and thermal stimulation, with a focal spot full width at half maximum of approximately 0.46 to 0.48 mm. The characterization results obtained using near-field terahertz scanning microscopy are consistent with the design.

The metasurface is divided into an active region and a passive region using a block design strategy. The GST layer is integrated only in the active region, leveraging the significant difference in conductivity between its amorphous and crystalline states to achieve non-volatile switching. Both phase states can remain stable for over 10 years without the need for a continuous energy supply. Using the C-shaped SR as the meta-atom, the amplitude and phase of terahertz waves are synergistically controlled by tuning its geometric parameters. Combined with the principles of computer-generated holography and optical path reversibility, the complex amplitude distributions of the active and passive regions are designed [27], where the contributions of the two regions to the far-field function exhibit a π phase difference. In the amorphous state of GST, the far-field contributions from both regions undergo destructive interference, leaving only the active region to achieve amplitude-phase control and functional output of near-field SPPs. In the crystalline state of GST, the active region becomes inactive, and only the passive region operates to achieve

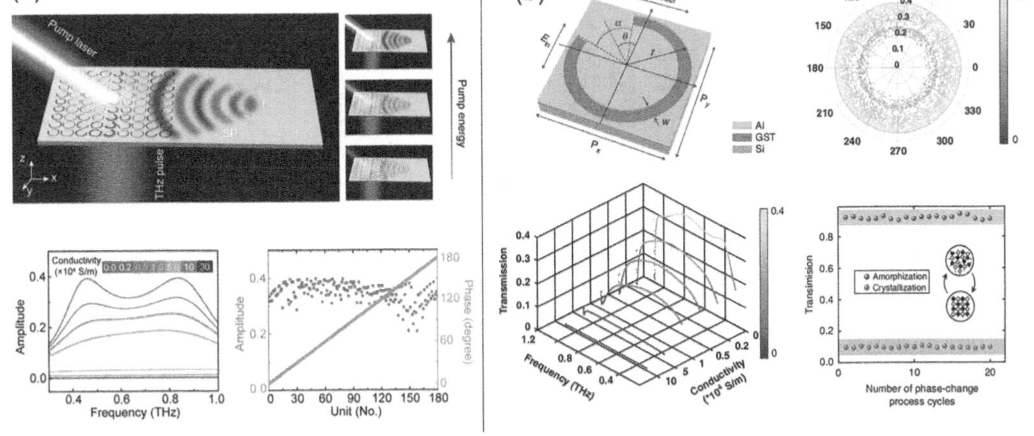

FIGURE 10.4 (a) Non-volatile and reconfigurable THz on-chip metalens working principle. (b) The characterization of GST-integrated CR [27].

wavefront control of far-field propagating waves. This ultimately enables non-volatile switching between near-field and far-field functionalities, while simultaneously achieving amplitude-phase control of both SPPs and free-space waves in Fig. 10.4(b).

Using subwavelength paired slits as the basic unit, multiple sets of mirror-symmetric arc-arrayed pair-slit resonators (APSRs) are constructed [28]. Leveraging the spin properties of circularly polarized light, spin-selective unidirectional excitation of surface plasmonic waves (SPWs) is achieved: LCP incidence excites SPWs that propagate and focus towards the left, while RCP incidence excites SPWs that propagate and focus towards the right. This inherent coupling between spin and propagation direction lays the foundation for polarization manipulation. To achieve broadband operation, slits of varying sizes are designed to correspond to different SPW excitation frequencies. Simultaneously, the radii of each arc array are precisely tuned to satisfy the phase matching condition for constructive interference, thereby eliminating spatial dispersion. This ultimately enables spin-selective achromatic focusing across a broadband range of 0.33–0.60 THz. Experimental results demonstrate that the operating bandwidth of this broadband design is 2.45 times wider than that of narrowband devices, with a maximum extinction ratio of 25 dB, a transverse focal spot resolution of $0.32\lambda_0$, and a horizontal relative offset of only 12%. Stable focusing performance is maintained across the entire broadband range without significant focal spot shift in Fig. 10.5.

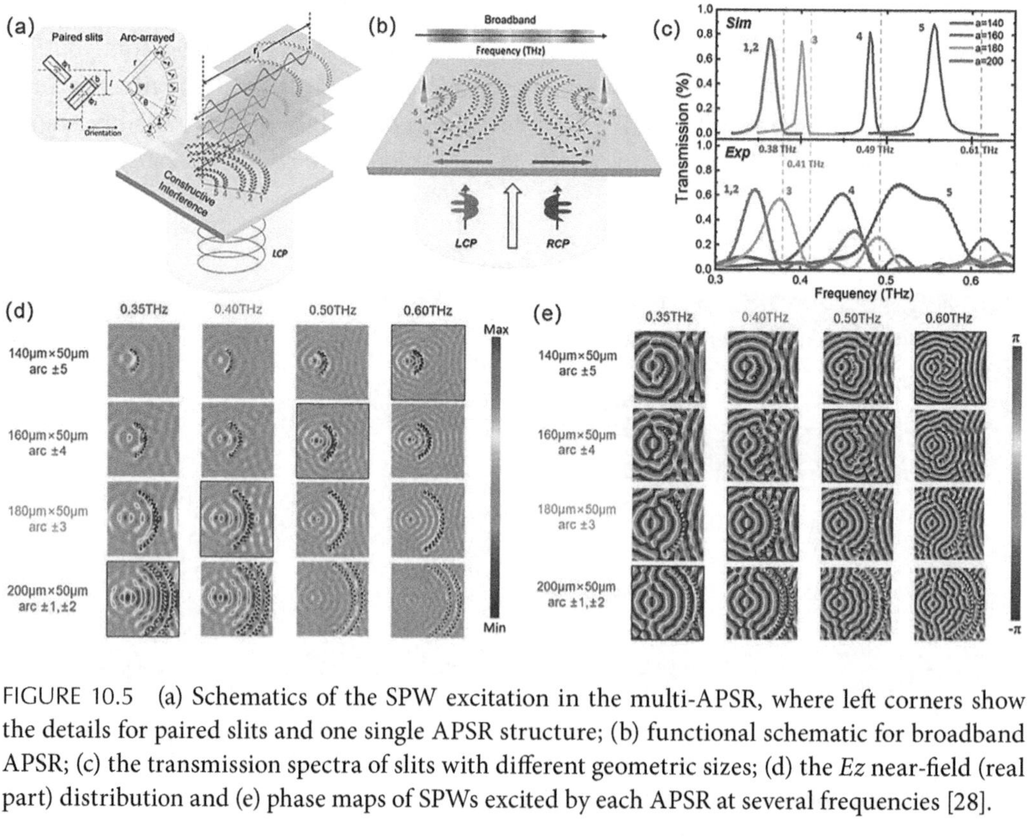

FIGURE 10.5 (a) Schematics of the SPW excitation in the multi-APSR, where left corners show the details for paired slits and one single APSR structure; (b) functional schematic for broadband APSR; (c) the transmission spectra of slits with different geometric sizes; (d) the *Ez* near-field (real part) distribution and (e) phase maps of SPWs excited by each APSR at several frequencies [28].

Building upon the APSR structure, an LC layer is integrated and designed to function as an equivalent broadband quarter-wave plate. By leveraging the electric-field-controlled orientation rotation of LC molecules, polarization conversion is achieved. When a 45° linearly polarized light is incident, an external bias electric field of 0–12 V/mm is applied to rotate the LC molecules from the y-axis to the x-axis. This process converts the incident linearly polarized light into a continuously adjustable mixture of LCP and RCP components. Utilizing the spin-selective excitation property, the dynamic energy distribution of SPWs between the left and right focal spots is realized. Experimental results show that under this polarization control scheme, the device achieves a modulation depth of 73%, with reversible focal spot switching across the entire broadband range of 0.33–0.60 THz, and an extinction ratio of 20 dB. Furthermore, the integration of the LC layer does not compromise the focusing resolution of the SPWs, maintaining the device's control performance.

In addition, the study employed a near-field scanning terahertz time-domain spectroscopy system to detect the transverse near-field E_y and longitudinal near-field E_z components. It was verified that SPWs at different frequencies are excited by arc arrays with corresponding slit sizes. Under polarization state control, the propagation and focusing of SPWs consistently matched the spin characteristics of the incident light, achieving precise manipulation of the SPWs energy distribution through polarization states.

By integrating liquid crystal with APSR metasurfaces, the polarization conversion characteristics of liquid crystals are organically combined with the spin-selective excitation properties of APSRs. This integration not only achieves broadband spin-selective unidirectional focusing of terahertz SPWs but also enables dynamic energy distribution of SPWs focal spots through electrical control of polarization states. This work fills the research gap in active broadband polarization control of terahertz on-chip SPWs and provides a novel device scheme and modulation mechanism for polarization encoding, signal routing, and high-sensitivity sensing in terahertz integrated photonics.

10.4 GENERATION OF ON-CHIP SURFACE WAVES TO FAR-FIELD VECTOR BEAMS

Using PB meta-atoms as the fundamental building blocks, and combining Huygens principle with the interference effect of circularly polarized light, a surface wave (SW) excited metasurface is constructed [9,29]. The generation of vector beams propagating in the far field is achieved in two steps:

i. Directional generation of single-spin circularly polarized far-field beams: Based on the target wavefront phase distribution of a circularly polarized beam, the required phase distribution of the metasurface is derived inversely, and this phase is encoded into the orientation angle of the meta-atoms via the PB phase. An artificial "plasmonic metal" is designed to provide a stable SW excitation source for the metasurface. As the SW propagates along the metasurface, only the circularly polarized light with the target spin can be decoupled to the far field, while the other modes become evanescent waves bound to the metasurface, enabling efficient and directional radiation of a single circularly polarized beam. By tuning the polarization conversion ratio (PCR) of the meta-atoms, a balance between beam generation efficiency and wavefront flatness is achieved.

ii. Synthesis of vectorial beams: Two sets of PB meta-atoms, designed to generate LCP and RCP beams, respectively, are arranged in a checkerboard or odd/even row interleaved pattern to construct a composite metasurface. Upon excitation by the same SW source, the LCP and RCP beams radiated by the two sets of meta-atoms interfere with each other. By controlling the phase difference, amplitude difference, or propagation constant of the two beams, vectorial beams with predefined polarization characteristics are synthesized. This serves as the core control entry point for achieving different polarization modulation effects in the two studies.

As shown in Fig. 10.6(a), in the terahertz band (~0.4 THz), vectorial beams generated by SW-excited metasurfaces are demonstrated, with a focus on the efficient generation of radially polarized Bessel vector beams [30]. This approach addresses the challenges of poor integrability and low efficiency in terahertz vectorial field generation devices. Two foundational metasurface devices were first fabricated, achieving directional radiation of LCP beams (relative efficiency 95%) and fixed-point focusing of RCP beams (relative efficiency 90%) under SW excitation, thereby validating the feasibility and high efficiency of single-spin circularly polarized beam control. By integrating two sets of metasurfaces designed to

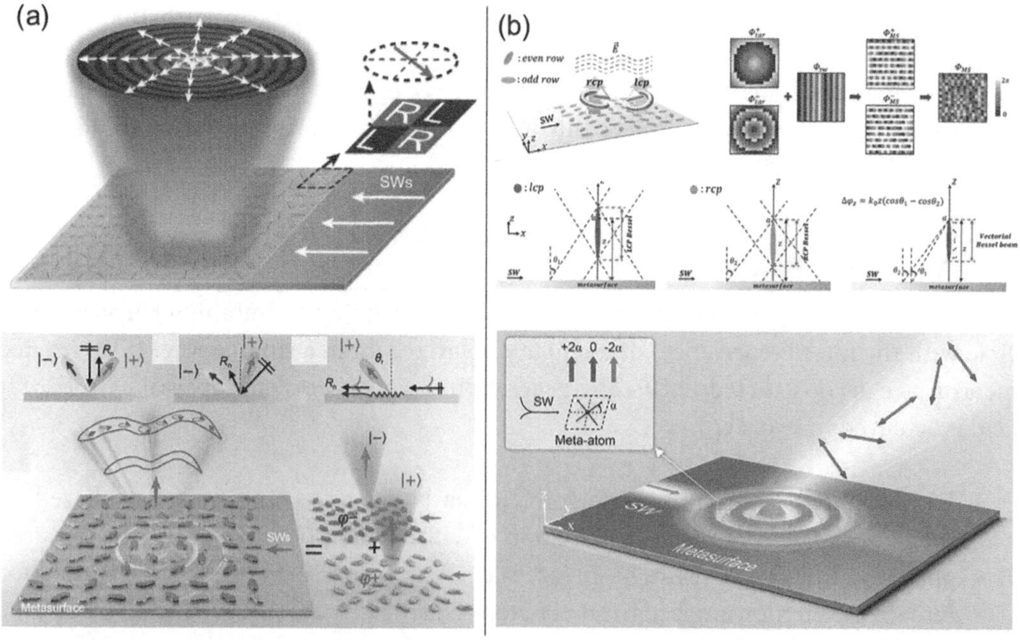

FIGURE 10.6 (a) Schematics of generating a far-field vector beam with arbitrary polarization by the SW metasurface [31]. (b) Schematic illustrations of longitudinal light field generation by the SW metasurface [1,31].

generate LCP and RCP Bessel beams, respectively, the two beams produce opposite OAM and interfere with each other, cancelling the OAM while synthesizing a radially polarized Bessel vector beam. Characterization using terahertz time-domain spectroscopy and near-field scanning techniques shows excellent agreement between experimental results and full-wave simulations, confirming that the generated beam exhibits typical non-diffracting properties and pronounced radially polarization characteristics. This device provides an ultra-compact on-chip solution for applications such as biological sensing, communications, and image multiplexing in the terahertz band.

In the microwave band (12 GHz), longitudinal control of beam polarization was achieved, breaking through the limitation of traditional vector beams that only allow transverse polarization inhomogeneity [31]. For the first time, a Bessel vector beam with longitudinally varying polarization was constructed: two interleaved sets of meta-atoms radiate LCP and RCP Bessel beams with different longitudinal propagation constants, and their interference results in a continuous variation of the beam's polarization state along the propagation direction (longitudinal). Two benchmark devices were first fabricated to generate orthogonally polarized directional plane waves and orthogonally polarized Bessel beams centred at different positions. Subsequently, the core device was constructed to realize the longitudinally varying polarization Bessel vector beam. Near-field scanning measurements revealed that the polarization state of the generated beam continuously transitions from x-polarization to y-polarization along the propagation direction and cycles back, while maintaining the non-diffracting characteristics of the Bessel beam. The experimental results are in perfect agreement with simulations, extending the polarization control dimension of metasurfaces from the transverse to the longitudinal direction in Fig. 10.6(b).

In the future, this design strategy can be extended to higher frequency bands such as infrared and visible light. Meanwhile, the meta-atom materials can be optimized to reduce losses and improve modulation efficiency. Addressing the issue of uneven response of meta-atoms in spatially varying local fields will further enhance device performance and integration density. This technology holds broad application prospects in on-chip photonics fields such as integrated displays, encrypted holography, augmented reality, polarization sensing, and micro-nano particle manipulation.

The development of on-chip terahertz plasmonic metadevices can be summarized as a profound evolution from single-functionality to multifunctional integration and from static modulation to dynamic reconfigurability: early research focused on addressing the challenge of efficient coupling between free-space light and surface waves. Subsequently, researchers began to embed wavefront modulation instructions into the coupling process, realizing complex functions such as anomalous deflection, focusing, and vortex interference through the joint manipulation of phase and amplitude. The role of polarization has also evolved from a mere "excitation condition" to a core "modulation degree of freedom"; through the combined application of geometric phase, resonant phase, and topological phase, independent dual-channel modulation with spin decoupling has been achieved. In recent years, the introduction of phase-change materials and liquid crystals has endowed devices with non-volatile, reconfigurable dynamic tuning capabilities, expanding the modulation dimension from the frequency domain to the time domain and enabling active encoding and modulation of multi-mode superposition states. The latest research has taken a further step to realize the efficient conversion of surface waves into free-space vector light beams, bridging on-chip modulation and spatial light fields, and laying a solid foundation for future applications in communications, on-chip information processing, and high-density data storage.

10.5 CHAPTER SUMMARY

This chapter has provided a comprehensive overview of the rapidly evolving field of on-chip metadevices for polarization control, tracing their development from fundamental principles to advanced multifunctional devices. The core innovation lies in the ability to embed polarization-dependent wavefront manipulation directly into the coupling process between free-space light and surface waves, overcoming the limitations of traditional bulky optical components. The geometric phase mechanism serves as the foundational approach, enabling spin-dependent phase control through simple rotation of anisotropic resonators. This concept has evolved from basic directional coupling to sophisticated wavefront shaping, including focusing, deflection, and vortex generation. However, the inherent conjugate phase relationship between left- and right-handed circular polarizations in pure geometric phase systems motivated the development of interference-based strategies, which break this spin-locking by introducing resonant phase control and achieving fully independent manipulation of orthogonal polarization channels. The chapter further explored advanced spin-decoupled control mechanisms leveraging resonant and topological phases, demonstrating robust and independent wavefront engineering for both spin channels. The integration of SOC principles enabled the generation and manipulation of on-chip plasmonic vortices, where polarization serves as both an angular momentum source and an information encoding dimension. This framework has been extended to vortex interferometry, OAM demultiplexing, and spatiotemporal control of plasmonic fields. The introduction of reconfigurable materials – particularly phase-change materials (GST) and liquid crystals – has transformed static metasurfaces into dynamically tunable platforms, enabling non-volatile switching, continuous modulation, and active spatiotemporal coding. Finally, the decoupling of surface waves from free-space radiation has enabled the on-chip generation of complex vectorial beams with tailored polarization distributions, including radially polarized Bessel beams and beams with longitudinally varying polarization states. Throughout this evolution, polarization has transcended its traditional role as a passive property of light to become an active and versatile control degree of freedom. The convergence of metasurface physics, materials science, and nanofabrication has established on-chip polarization control as a powerful paradigm for next-generation integrated photonic systems, with profound implications for communications, sensing, and quantum information processing.

REFERENCES

1. Lvyi Zhong, Maoqing Guo, Shilong Li, et al. (2025) 'Progress of grating couplers for light exchange between space and on-chip', Advanced Optical Materials, 13(30), e01171.
2. Garcia-Vidal, F. J., Fernández-Domínguez, A. I., Martin-Moreno, L., et al. (2022) 'Spoof surface plasmon photonics', Reviews of Modern Physics, 94(2), 025004.

3. Lin, J., Mueller, J. P. B., Wang, Q., et al. (2013). 'Polarization-controlled tunable directional coupling of surface plasmon polaritons', Science, 340, 331–334.
4. Zhang, X., Xu, Y., Yue, W., et al. (2015) 'Anomalous surface wave launching by handedness phase control', Advanced Materials, 27, 7123–7129.
5. Chen, Y., Zhang, H., Zhang, Z., et al. (2025) 'Dual views of plasmonics: from near-field optics to electron nanoscopy', Photonics Insights, 4, R04.
6. Zhou, H., Li, D., Lv, Q., et al. (2025) 'Integrative plasmonics: Optical multi-effects and acousto-electric-thermal fusion for biosensing, energy conversion, and photonic circuits', Chemical Society Reviews, 54(11), 5342–5432.
7. Xu, Q., Zhang, X., Wei, M., et al. (2018) 'Efficient metacoupler for complex surface plasmon launching', Advanced Optical Materials, 6, 1701117.
8. Wang, Z., Li, S., Zhang, X., et al. (2020) 'Excite spoof surface plasmons with tailored wave-fronts using high-efficiency terahertz metasurfaces', Advanced Science, 7, 2000982.
9. Sun, W., He, Q., Sun, S., et al. (2016) 'High-efficiency surface plasmon meta-couplers: concept and microwave-regime realizations', Light: Science & Applications, 5, e16003.
10. Pors, A., Nielsen, M. G., Bernardin, T., et al. (2014) 'Efficient unidirectional polarization-controlled excitation of surface plasmon polaritons', Light: Science & Applications, 3, e197.
11. McDonnell, C., Deng, J., Sideris, S., et al. (2021) 'Functional THz emitters based on Pancharatnam-Berry phase nonlinear metasurfaces', Nature Communications, 12, 30.
12. Han, J., Xu, Y., Zhang, H., et al. (2022) 'Tailorable polarization-dependent directional coupling of surface plasmons', Advanced Functional Materials, 32, 2111000.
13. Lang, Y., Xu, Q., Xu, G., et al. (2024) 'Topological metasurface for spin-decoupled wavefront manipulation of terahertz surface plasmons', Laser & Photonics Review, 24, 2401281.
14. Ji, J., Li, J., Wang, Z., et al. (2024) 'On-chip multifunctional metasurfaces with full-parametric multiplexed Jones matrix', Nature Communications, 15, 8271.
15. Xie, Z. W., Lei, T., Li, F., et al. (2018) 'Ultra-broadband on-chip twisted light emitter for optical communications', Light: Science & Applications, 7, 18001.
16. Devlin, R.C., Ambrosio A., Rubin N.A., et al. (2017) 'Arbitrary spin-to-orbital angular momentum conversion of light', Science, 358(6364), 896–901.
17. Shen, Y., Wang, X., Xie, Z., et al. (2019) 'Optical vortices 30 years on: OAM manipulation from topological charge to multiple singularities', Light: Science & Applications, 8, 90.
18. Lang, Y., Xu, Q., Chen, X., et al. (2022) 'On-chip plasmonic vortex interferometers', Laser & Photonics Review, 22, 2200242.
19. Yuan, X., Xu, Q., Lang, Y., et al. (2023) 'Tailoring spatiotemporal dynamics of plasmonic vortices', Opto-Electronic Advances, 6, 220133.
20. Yuan, X., Xu, Q., Lang, Y., et al. (2024) 'Temporally deuterogenic plasmonic vortices', Nanophotonics, 13(6), 955–963.
21. Wakayama, T., Higashiguchi, T., Sakaue, K. et al. (2018) 'Demonstration of a terahertz pure vector beam by tailoring geometric phase', Scientific Reports, 8, 8690
22. Jiang, X., Liu, W., Xu, Q., et al. (2024) 'On-chip terahertz orbital angular momentum demultiplexer', Photonics Research, 12(5), 1044–1054.
23. Wang, Y., Zhao, H., Song, C., et al. (2025) 'Spatiotemporal coding terahertz plasmonic vortex on-chip interferometer', Laser & Photonics Review, 25, e01608.
24. Chen, X., Tian, Z., Zhang, S., et al. (2022) 'Reconfigurable and nonvolatile terahertz metadevices based on a phase-change material', ACS Photonics, 9, 1638–1646.
25. Zhang, S., Chen, X., Liu, K., et al. (2022) 'Terahertz multi-level nonvolatile optically rewritable encryption memory based on chalcogenide phase-change materials', iScience, 25, 104866
26. Zhang, S., Chen, X., Liu, K., et al. (2023) 'On-chip non-volatile reconfigurable THz varifocal metalens', Laser & Photonics Review, 17, 2300482.
27. Dong, Y., Chen, X., Zhang, S., et al. (2025) 'Dynamic dual-mode terahertz device with nonvolatile switching for integrated on-chip and free-space applications', Microsystems & Nanoengineering, 11, 169.

28. Wang, Y.-M., Fan, F., Zhao, H.-J., et al. (2024) 'Active broadband unidirectional focusing of terahertz surface plasmons based on a liquid-crystal-integrated on-chip metadevice', Photonics Research, 12(10), 2148–2157.
29. Sun, S., He, Q., Xiao, S., et al. (2012) 'Gradient-index meta-surfaces as a bridge linking propagating waves and surface waves', Nature Materials, 11, 426–431
30. Wang, Z., Pan, W. K., He, Y., et al. (2025) 'Efficient generation of vectorial terahertz beams using surface-wave excited metasurfaces', Opto-Electronic Science, 4(1), 240024.
31. Jin, X., Wang, Z., Liu, M., et al. (2025) 'Vectorial beams with longitudinally varying polarizations generated by surface-wave metasurfaces', Laser Photonics Reviews, 19(12), 2402298.

CHAPTER 11

Inverse Design for Meta-Photonic Polarization Devices

Abstract

Inverse design is a target-oriented design methodology that determines device geometries by first specifying the desired optical performances, such as polarization conversion efficiency, insertion loss, and operational bandwidth, and subsequently retrieving the corresponding structural parameters using optimization algorithms or neural networks. With the rapid evolution of polarization control devices towards higher integration density and superior performance, conventional forward design approaches largely dependent on empirical knowledge and physical intuition have become increasingly inadequate in terms of efficiency and scalability. As a result, inverse design has emerged as a powerful and indispensable paradigm in the development of advanced polarization photonic devices. This chapter introduces two major categories of inverse design strategies for polarization devices: optimization-based inverse design methods and deep-learning-based inverse design methods. Representative polarization control devices realized through these strategies are also discussed.

11.1 OPTIMIZATION ALGORITHMS

Optimization algorithms constitute one of the primary implementation frameworks for inverse design. Their objective is to leverage computational techniques to reduce development time while achieving superior device performance. In the design of polarization control devices, optimization algorithms are generally classified into local optimization algorithms and global optimization algorithms. Different device requirements, such as specific linear mappings or complex functional constraints, necessitate different optimization strategies. Therefore, selecting an appropriate optimization algorithm according to the design objective is crucial for achieving efficient and high-performance inverse design.

DOI: 10.1201/9781003784821-11

11.1.1 Direct Binary Search Method

The Direct Binary Search (DBS) method is one of the most fundamental and straightforward optimization approaches. Strictly speaking, it does not belong to either local or global optimization categories; instead, it can be regarded as a brute-force parameter scanning strategy that replaces manual tuning with automated computation.

As shown in Fig. 11.1, the DBS procedure begins by discretizing the device structure into an M × N array of pixels. Each pixel is assigned a binary state, “0” or “1”, representing unetched and etched regions, respectively. The device performance is evaluated using a figure of merit (FOM). During each iteration, a single pixel is selected and its binary state is flipped. The updated structure is then simulated: if the FOM improves, the new state is retained. Otherwise, the pixel is reverted to its original state. Once all pixels have been

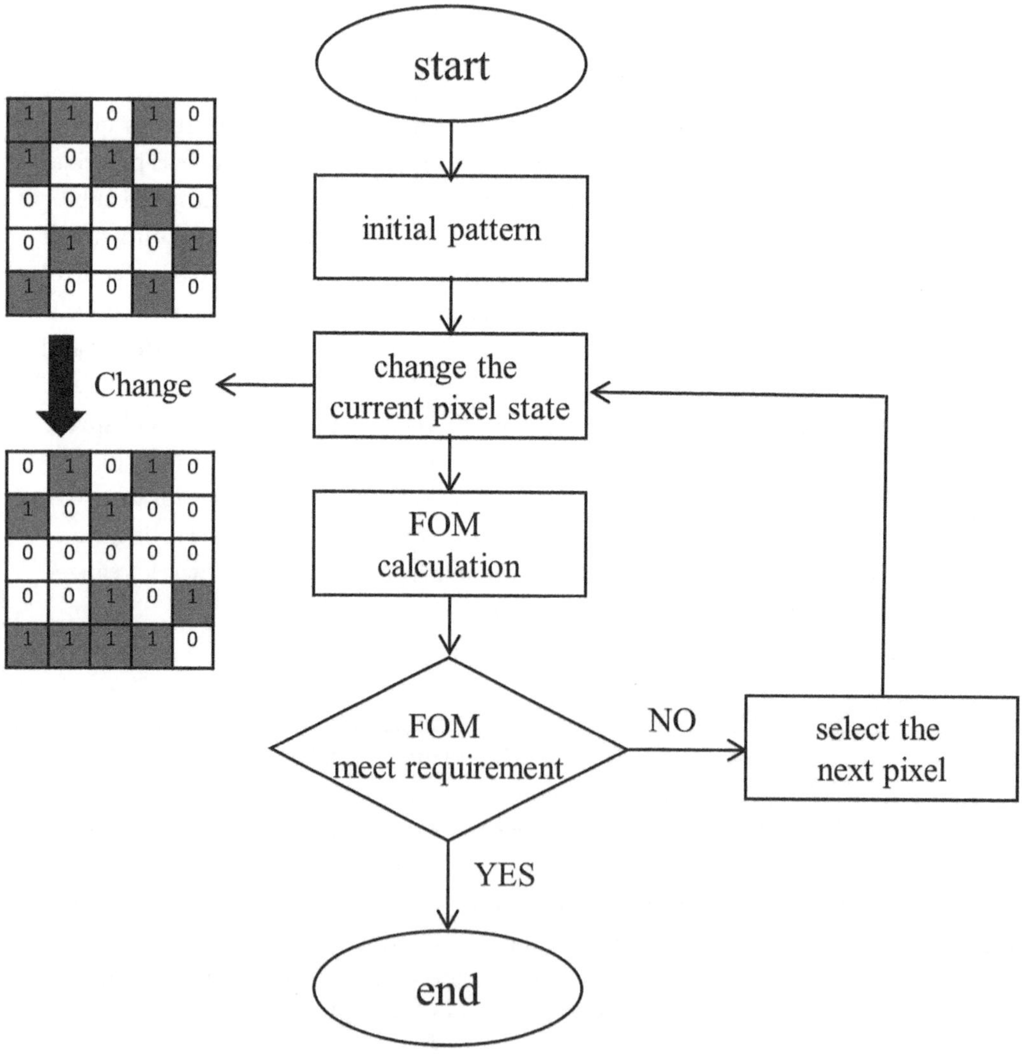

FIGURE 11.1 DBS algorithm optimization flowchart.

scanned, one iteration cycle is completed. The resulting binary pattern serves as the initial configuration for the next iteration. This process continues until either the target FOM is achieved or further improvement becomes negligible.

11.1.2 Local Optimization Algorithms

Local optimization algorithms aim to iteratively minimize an objective function within a neighbouring solution space. Their core principle is gradient descent, which is particularly effective for devices with complex geometries and well-defined parametric mappings [1]. The fundamental idea of gradient descent is to update the design variables in the opposite direction of the gradient to approach a local minimum of a differentiable function. Assume a differentiable objective function $f(x)$, where the goal is to find a point x that minimizes $f(x)$. The gradient of $f(x)$ is defined as $\nabla f(x)$:

$$\nabla f(x)=\left(\frac{\partial f}{\partial x_1},\frac{\partial f}{\partial x_2},\frac{\partial f}{\partial x_3},\ldots,\frac{\partial f}{\partial x_i}\right)^T. \tag{11.1}$$

At each iteration, move from the current point x in the opposite direction of the gradient, with the expectation of reducing the function value, updating the point x: $x_{k+1}=x_k-a_k\nabla f(x_k)$, where a_k is the step size of the kth step, controlling the distance x moves along the gradient. The iteration terminates when the performance criteria are satisfied or when convergence is reached.

Adjoint Method. The adjoint method is an efficient approach for gradient calculation. In the following, a series of equations are used to intuitively explain the design of a photonic polarization controller based on the adjoint method in the context of electromagnetics [2]. For a device to be optimized, the optimization region is denoted as Ω (as shown in Fig. 11.2). Let the electric field intensity at an arbitrary point x_0 inside the region be $E(x_0)$. The influence of the electric field on the FOM is defined as:

$$\text{FOM}=\left|E(x_0)\right|^2. \tag{11.2}$$

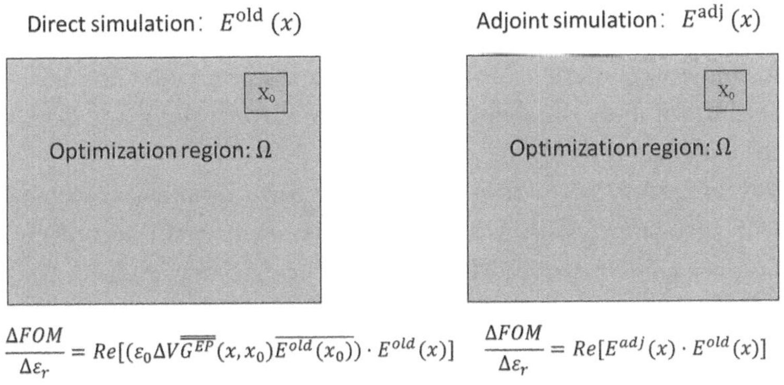

FIGURE 11.2 Schematic diagram of accompaniment method.

Within the optimization region, the relative permittivity ε at each point can be varied. A small perturbation $\Delta\varepsilon$ in permittivity will cause a corresponding change in the FOM at position x_0:

$$\Delta\text{FOM} = Re\left[\overline{E^{old}(x_0)} \cdot \Delta E(x_0)\right], \tag{11.3}$$

where $\overline{E^{old}(x_0)}$ denotes the electric field before the perturbation, and $\Delta E(x_0)$ represents the field variation induced by the change in permittivity. The field variation at the target location x_0, caused by the induced dipole moment, is expressed as:

$$\Delta E(x_0) = \overline{\overline{G^{EP}}}(x_0,x)\,p^{ind} = \varepsilon_0 \Delta\varepsilon_r \Delta V \overline{\overline{G^{EP}}}(x_0,x) E^{new}(x), \tag{11.4}$$

where ΔV is the volume of discrete mesh cells, $\overline{\overline{G^{EP}}}(x_0,x)$ is the Maxwell Green's function relating the electric field at x_0 to the induced polarization density p^{ind}, and $E^{new}(x)$ is the electric field under the updated permittivity. When $\Delta\varepsilon$ is sufficiently small, it can be approximated that: $E^{new}(x) \approx E^{old}(x)$. By rewriting the equation and using the symmetry property of the Green's function, the following expression can be obtained:

$$\frac{\Delta\text{FOM}}{\Delta\varepsilon_r} = Re\left[\varepsilon_0 \Delta V \overline{\overline{G^{EP}}}(x_0,x) \cdot \overline{E^{old}(x_0)} \cdot E^{old}(x)\right]. \tag{11.5}$$

Although each variable $E^{old}(x_0)$, $E^{old}(x)$, and $G^{EP}(x_0,x)$ can be computed individually, separate calculations would be required for every spatial position, which is inefficient. Therefore, the equation is further rewritten using the reciprocity theorem. The adjoint electric field is defined as:

$$E^{adj}(x) = \varepsilon_0 \Delta V \overline{\overline{G^{EP}}}(x_0,x) \cdot \overline{E^{old}(x_0)}. \tag{11.6}$$

The adjoint field can be obtained by placing a dipole source with amplitude $\Delta V \varepsilon_0 \overline{E^{old}(x_0)}$. In summary, the electric field $E^{old}(x)$ in the optimization region can be obtained from a single forward simulation, while the adjoint electric field $E^{adj}(x)$ can be obtained from a single adjoint simulation. Therefore, only one forward simulation and one adjoint simulation are required to compute the shape derivatives for arbitrarily many degrees of freedom throughout the entire design region.

Topology Optimization Algorithm. Topology optimization is a mathematical method that optimizes the structure within a prescribed design domain according to given performance objectives and constraints. It belongs to the category of structural optimization [3,4]. The objective of topology optimization is to minimize the FOM corresponding to the design variable $\varepsilon(r)$:

$$\min_{\varepsilon} \text{FOM}(\varepsilon). \tag{11.7}$$

The material distribution of the device satisfies Maxwell's equation:

$$\nabla \times \left(\frac{1}{u} \nabla \times E(r) \right) - \omega^2 u_0 \varepsilon_0 \varepsilon_r E(r) = -i\omega J. \tag{11.8}$$

The density-based topology optimization expresses the design variable and material interpolation as:

$$\varepsilon_r = \varepsilon_{min} + \rho(\varepsilon_{max} - \varepsilon_{min}),\ \rho(r) \in [0,1], \tag{11.9}$$

where u_0 is the permeability of free space, ε_0 is the permittivity of free space, J is the current source, E is the electric field, ω is the angular frequency, $\rho(r)$ is the continuous density variable, and ε_{min}, ε_{max} represent two materials (typically unetched and etched regions). The structure is updated according to:

$$\rho^{k+1} = \rho^k + \alpha \frac{\partial F}{\partial \rho}, \tag{11.10}$$

where α is the weighting coefficient of the accompanying electric field, and F represents the objective function, a metric used to measure device performance. The level-set method is a topology optimization approach. It uses the same topology optimization framework as density-based topology optimization. The difference lies in the representation of design variables and the update strategy. Density-based topology optimization is essentially parameter optimization, whereas the level-set method describes geometric evolution. The level-set representation is:

$$\varphi(r) = \begin{cases} < 0 & \text{Material} \\ = 0 & \text{Boundary}, \\ > 0 & \text{Air} \end{cases} \tag{11.11}$$

where r is the spatial coordinate and $\varphi(\cdot)$ is the level-set function. The level-set method is naturally binary and does not contain greyscale interpolation. The level-set function evolves according to a governing evolution equation, and the iteration stops when the evolution becomes stable. The update equation is:

$$\frac{\partial \varphi}{\partial t} + \upsilon_n |\nabla \varphi| = 0, \tag{11.12}$$

where υ_n represents the element sensitivity. Topology optimization is essentially a constrained functional optimization problem. In the design of polarization control devices, the ideal objective is to maximize the output power of the target signal at the output port:

$$\max_{\rho(x)} F(E,\rho) = \int_{r_{out}} |E(r) \cdot E^*(r)|^2 \, d\Gamma. \tag{11.13}$$

Then, the gradient E^{adj} is calculated using the gradient descent method and the adjoint equation to solve for $\partial F / \partial p$. In essence, topology optimization is based on Maxwell's equations. By performing variational analysis on both the forward field E and the E^{adj}, the optimal solution can be obtained.

11.1.3 Global Optimization Algorithms

Global optimization algorithms consider the overall device structure and are relatively less dependent on the initial configuration. Moreover, they are less likely to become trapped in local optima. Most global optimization algorithms are inspired by natural phenomena and classical theoretical principles; therefore, they are also referred to as heuristic algorithms. Typical examples include Genetic Algorithms (GA) and Particle Swarm Optimization (PSO).

Genetic Algorithms. Genetic algorithms are metaheuristic algorithms inspired by Darwin's theory of evolution and Mendelian genetics. Their optimization principle follows biological evolutionary rules by simulating inheritance and natural selection processes, forming an adaptive global optimization search strategy. Essentially, GA is a parallel and efficient global optimization method belonging to the class of evolutionary algorithms. GA relies on biologically inspired operators (selection, crossover, and mutation) to generate high-quality solutions for optimizing and searching device structures. Each solution possesses a unique set of characteristics (chromosomes or genotype). Candidate solutions evolve into improved ones through evaluation, ultimately converging towards the global optimum [5]. The main procedure of GA is illustrated in Fig. 11.3.

Initializing the Population: A number of random candidate solutions (e.g. different splitter structures) are first generated. Each solution is represented by a chromosome

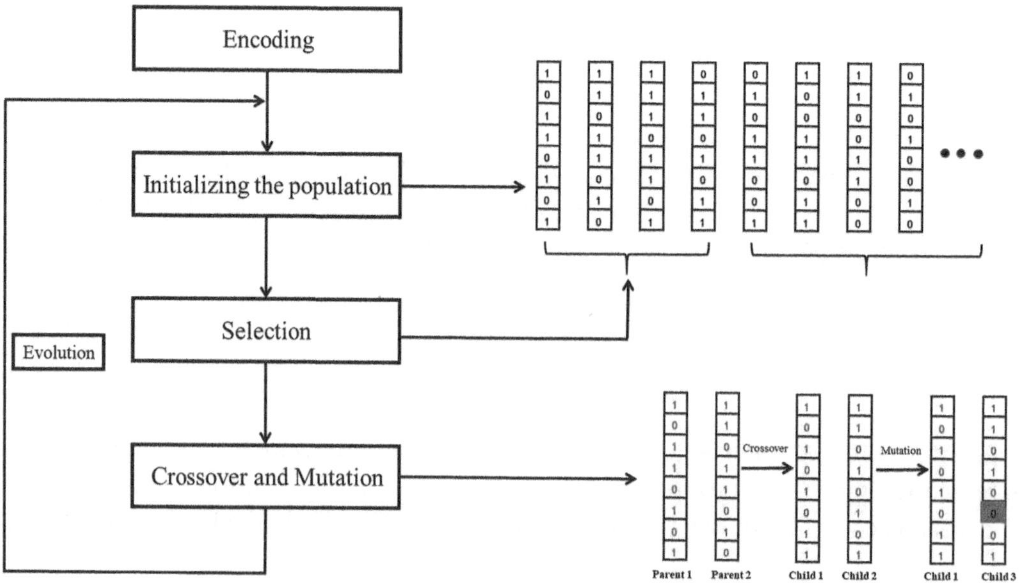

FIGURE 11.3 Genetic algorithm optimization flowchart.

containing multiple parameters encoded as genes, typically in binary form (0 and 1). Each chromosome is associated with a fitness value determined by the fitness function (e.g. insertion loss, extinction ratio). Chromosomes are then ranked according to their fitness. Initial population: $p^{(0)} = \left\{x_1^{(0)}, x_2^{(0)}, x_3^{(0)}, \ldots, x_N^{(0)}\right\}$; Individual: $x_i^{(0)} = \left\{x_{i1}^{(0)}, x_{i2}^{(0)}, x_{i3}^{(0)}, \ldots, x_{iD}^{(0)}\right\}$; Initialization rule: $x_{ij}^{(0)} \sim u\left(x_j^{min}, x_j^{max}\right)$, where $i = 1, \ldots, N$ denotes the individual index, $j = 1, \ldots, D$ denotes the variable index, $u(\cdot)$ represents a uniform distribution, and $p^{(0)}$ denotes the first-generation population. Fitness evaluation: $f_i^{(g)} = f(x_i^{(g)})$, which denotes the fitness of the ith individual in generation g.

Selection: After ranking chromosomes according to their fitness values, the most promising chromosomes are selected to produce the next generation. For example, the top 50% individuals with fitness values closer to the target may be retained, while others are discarded. The selected chromosomes serve as parents for reproduction. Various mating strategies can be applied to randomly generate offspring. The probability that individual i in generation g is selected as a parent is $P_i^{(g)} = f_i^{(g)} / \sum_{k=1}^{N} f_i^{(g)}$. Individuals with higher fitness have a greater probability of being selected.

Crossover and Mutation: Crossover exchanges partial genes between parent chromosomes. For example, offspring 1 may be generated by combining segments of chromosomes from parent 1 and parent 2. Mutation randomly alters certain genes in the offspring (e.g. flipping 0 to 1) to increase genetic diversity. Mutation usually occurs with a low probability, typically between 0.01 and 0.08.

Evolution: The offspring replace the parent generation. Fitness values are recalculated, and the process repeats. After reaching a convergence threshold or a predefined number of iterations, the optimal solution (design structure) is selected.

Particle Swarm Optimization. Particle Swarm Optimization originates from studies of bird flocking behaviour and is also a metaheuristic algorithm. Compared with GA, PSO does not involve crossover and mutation operations, resulting in relatively faster convergence. The PSO procedure is illustrated in Fig. 11.4. Each candidate solution is regarded as a bird, called a "particle". A swarm containing N particles is first initialized. The position and velocity of particle n are denoted as p_n^0 and v_n^0, respectively. The fitness value is calculated as $f(p_n)$. And pb denotes the best position experienced by an individual particle, and gb denotes the best position experienced by the entire swarm. Based on the evaluated fitness, pb and gb are updated. The velocity and position of particle n at iteration $t + 1$ are updated as:

$$\begin{aligned} v_n^t &= w v_n^{t-1} + c_1 r_1 \left(pb_n^{t-1} - x_n^{t-1}\right) + c_2 r_2 \left(gb_n^{t-1} - x_n^{t-1}\right) \\ p_n^t &= p_n^{t-1} + v_n^{t-1} \end{aligned}, \tag{11.14}$$

where w is the inertia weight; c_1 and c_2 are acceleration constants controlling the maximum learning step; and r_1 and r_2 are random numbers. The iteration continues until the stopping condition is satisfied.

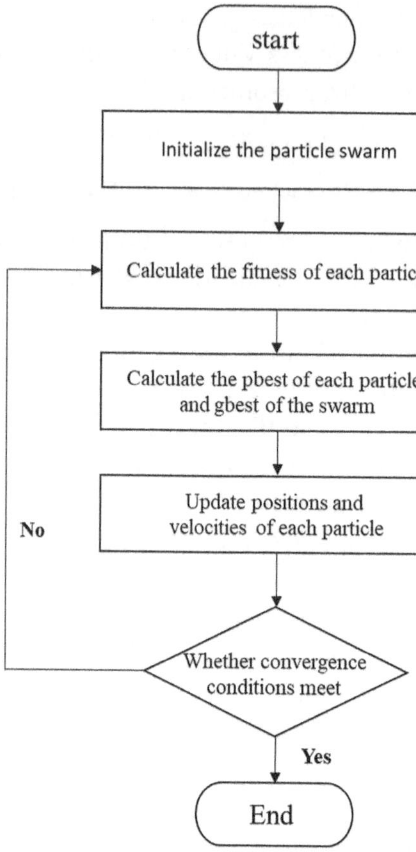

FIGURE 11.4 Optimization process of PSO algorithm.

11.2 POLARIZATION CONTROLLER DESIGN BASED ON OPTIMIZATION ALGORITHMS

11.2.1 Direct Binary Search Method

In 2015, the DBS algorithm was first applied to the design of on-chip polarization controllers [6]. The DBS-designed on-chip polarization beam splitter has a footprint of only $2.4 \times 2.4\ \mu m^2$, which was significantly smaller than any previously reported on-chip polarization beam splitter at that time. However, DBS is a greedy optimization algorithm. To ensure superior device performance, the smallest fabrication-allowed pixel size (typically $120 \times 120\ nm^2$) is usually selected. However, each pixel state flip requires recalculating the entire device performance, resulting in substantial computational cost. To address this issue, a Variable Step Size Binary Search (VBS) algorithm [7] was proposed. It employs varying pixel sizes: first iteration: $480 \times 480\ nm^2$, second iteration: each pixel divided into four $240 \times 240\ nm^2$ pixels, third iteration: further divided into $120 \times 120\ nm^2$ pixels. The optimization process is identical to DBS except for the varying pixel size. The convergence speed was improved by approximately 25%.

11.2.2 Local Optimization Algorithms

In gradient-based local optimization algorithms, the core idea is gradient descent. Since each iteration searches for an optimal solution in the vicinity of the current solution, this method is referred to as local optimization. Traditional gradient methods use numerical differentiation, requiring a full electromagnetic simulation for each variable, which significantly reduces efficiency. In contrast, the adjoint method requires only one forward simulation to obtain $E^{old}(x)$ and one adjoint simulation to obtain $E^{adj}(x)$. The computational cost of the adjoint method is independent of the number of design variables, greatly reducing optimization time. Therefore, the adjoint method is fundamental in gradient-based local optimization.

Shape Optimization. Gradient-based local optimization mainly includes shape optimization and topology optimization. Shape optimization modifies only the device boundaries:

$$\frac{\mathrm{d}F}{\mathrm{d}D} = \int_D g(r)(V \cdot n)\mathrm{d}S, \tag{11.15}$$

where D represents the geometry (material), V is the volume of the integration region, $g(r)$ is the shape gradient density, and n is the normal direction. Shape optimization has relatively low degrees of freedom and strongly depends on the initial structure. In 2019, Lebbe et al. [8] designed a polarization rotator based on multilevel shape optimization. Using gradient algorithms and the adjoint method, they broke vertical symmetry and optimized the bottom and top boundaries separately.

Topology Optimization. Another approach is topology optimization [9], which is mainly implemented using density-based or level-set methods. Compared with shape optimization, topology optimization provides higher degrees of freedom. It allows material addition and removal, enabling changes in topology, such as the creation of new holes or merging and splitting of regions. In 2024, Shen et al. [10] designed a polarization multidimensional photonic integrated circuit based on topology optimization, as shown in Fig. 11.5. The design region was discretized into 20 nm units. The relative permittivity of each unit was mapped linearly to a design variable. The FOM was defined as the transmission combination of four propagation paths corresponding to four modes. Transmission was defined as modal overlap:

$$T = \frac{\left| \int E \times H_m \cdot \mathrm{d}S + \int E_m \times H \cdot \mathrm{d}S \right|^2}{\int Re(E_m \times H_m) \cdot \mathrm{d}S}, \tag{11.16}$$

where E_m and H_m are the electromagnetic field distributions at the cross section of the bus waveguide. The total FOM of the four modal paths is:

$$\mathrm{FOM} = \sum_{mode=1}^{4} \frac{\int_{\lambda_{min}}^{\lambda_{max}} T_{tg}(\lambda) - \int_{\lambda_{min}}^{\lambda_{max}} \left| T_\lambda - T_{tg}(\lambda) \right|}{\lambda_{max} - \lambda_{min}}, \tag{11.17}$$

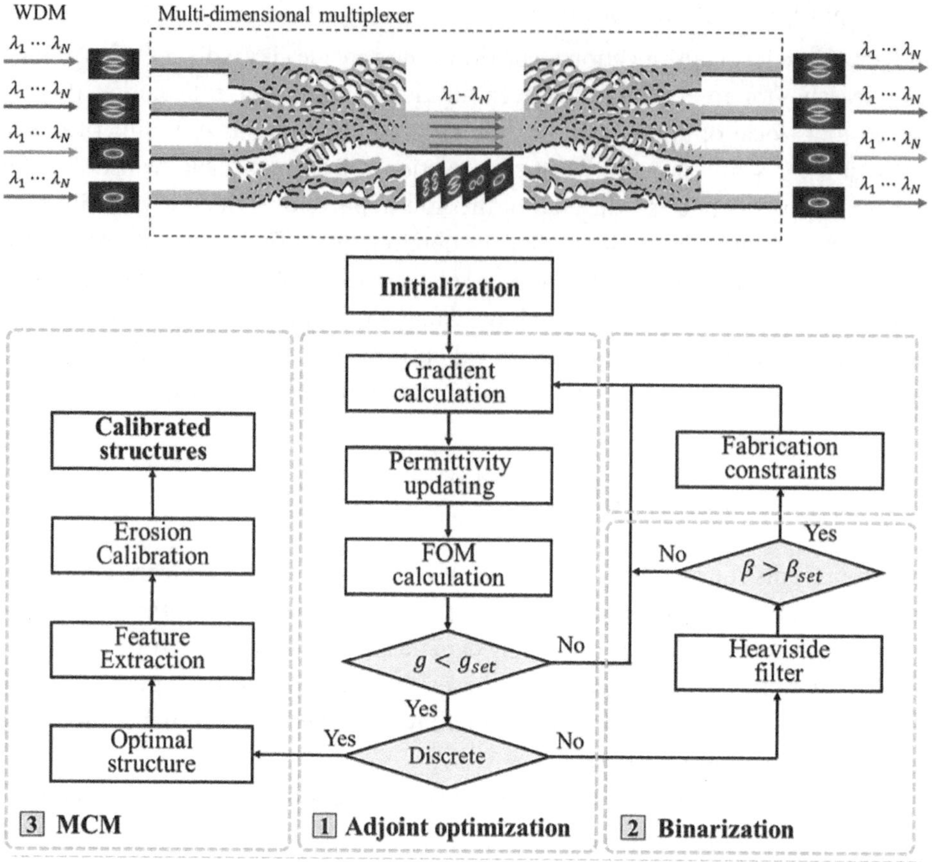

FIGURE 11.5 Polarization multidimensional photonic (de)multiplexer based on topology optimization design [10].

where T_{tg} denotes the target transmission. The adjoint method is also required in this design. Only two simulations are needed per iteration to compute the gradient. The design variables of each unit are updated along the gradient descent direction to improve the FOM. To mitigate fabrication difficulties caused by locally high etching density in topology optimization, a manufacturing calibration method (MCM) with a shrink process was applied. Gradient-based local optimization algorithms achieve fast convergence in high-dimensional design spaces by accurately utilizing derivative information of the objective function. However, they are sensitive to the initial structure of the device.

11.2.3 Global Optimization Algorithms

Heuristic-based global optimization algorithms rely on random or population-based search mechanisms. They possess strong global exploration capability and general applicability. In inverse design problems of polarization control devices, two global optimization algorithms are commonly used: GA [11] and PSO.

In some studies, the objective function is nonlinear; therefore, gradient descent algorithms are not applicable. For example, in 2020, Lu et al. designed a polarization beam splitter based on a GA [12]. The researchers predefined the material, geometry, and operating wavelength, divided the design region into matrix units, and defined design variables accordingly. The harmonic mean of the transmission at the target port was regarded as the signal, while the quadratic mean of the transmission at the non-target port was regarded as the noise. The difference between noise and signal was defined as the fitness function. The global minimum of the fitness function was then obtained using a GA.

The resulting on-chip polarization beam splitter had a footprint of only 970 × 1240 nm^2, which was the smallest silicon-based on-chip polarization beam splitter reported at that time. Similarly, in global optimization-based inverse design, the definition of the FOM is also crucial. The following FOM corresponds to a polarization-independent coupler [13]:

$$\mathrm{FOM} = 1-\eta\cdot\frac{\left(2-T_{\lambda}^{TE}-T_{\lambda}^{TM}\right)}{2}-(1-\eta)\cdot\sum_{i=1}^{N}\frac{\left|T_{\lambda i}^{TE}+T_{\lambda i}^{TM}\right|}{N}, \tag{11.18}$$

where T and λ represent coupling efficiency and wavelength, respectively. The performance of a polarization-independent coupler is mainly characterized by two factors: the peak coupling coefficients of TE and TM polarizations at wavelength λ, and the difference between TE and TM coupling efficiencies over the wavelength range λ_1 to λ_N. The weighting factor η controls their relative contributions to the final FOM.

GA are memoryless evolutionary systems without a unified analytical update equation; instead, they rely on probabilistic jumps and structural recombination. In contrast, PSO is a memory-based continuous dynamical system whose convergence trajectory is continuous. In 2022, Ou et al. designed a tunable polarization-multiplexed achromatic dielectric metalens based on the PSO algorithm [14], as shown in Fig. 11.6. The PSO algorithm was used to optimize the reference radius at each wavelength, thereby constructing a nonlinear

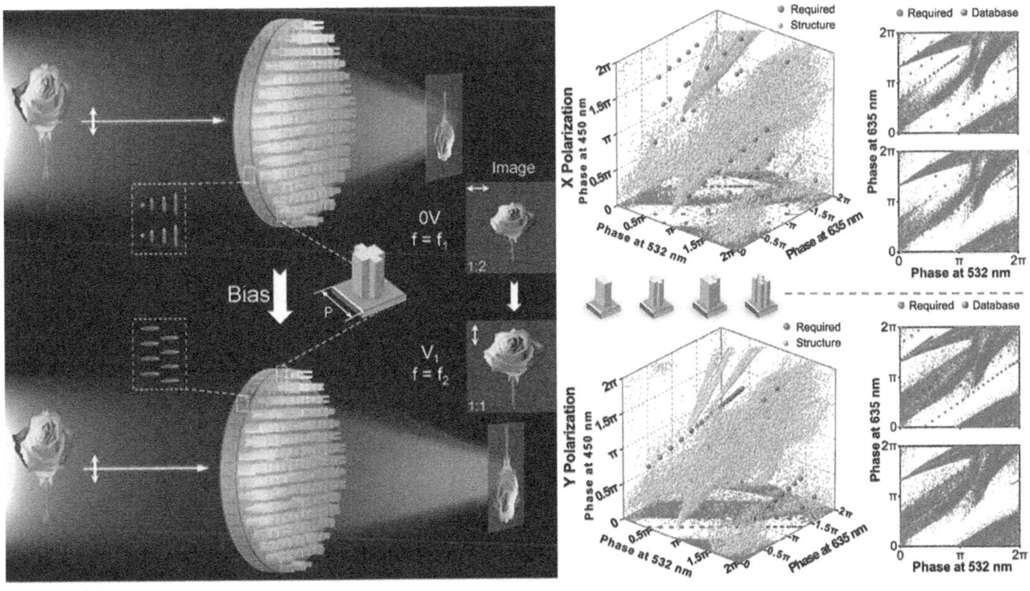

FIGURE 11.6 Adjustable polarization multiplexing apochromatic metamorphic lens designed based on the PSO algorithm [14].

metalens phase distribution. This design enables optimal matching with the dispersion characteristics of a predefined meta-unit database under multiple wavelengths and dual polarizations. The intrinsic dispersion relationship of the meta-unit originates from the waveguide effect:

$$\varphi(\lambda)=\frac{2\pi}{\lambda}n_{eff}(\lambda)H, \tag{11.19}$$

where n_{eff} is the effective refractive index, and H is the height of the meta-unit. The phase $\varphi(\lambda)$ varies monotonically with the wave number $k=2\pi/\lambda$. The researchers proposed a new PSO-optimized phase model:

$$\varphi(r,\lambda)=\frac{2\pi}{\lambda}\left(\sqrt{r^2+f^2}-\sqrt{r_\lambda^2+f^2}\right), \tag{11.20}$$

where r_λ is a wavelength-dependent free parameter, and f is the focal length of the metalens. This allows nonlinear adjustment of the phase-wavelength relationship. The PSO algorithm is used to search for the optimal r_λ, following the optimization procedure described previously.

There are also hybrid optimization strategies that combine local and global optimization algorithms. In 2025, Han et al. proposed a Global Initial Solution-Based Topology Optimization (GISTO) method [5]. First, a two-dimensional encoded non-dominated sorting GA was used for global optimization to generate high-quality initial structures. Then, gradient-based topology optimization was applied to refine the structure. This approach addresses both the low efficiency of global optimization and the dependence of local optimization on initial structures. Optimization algorithms are flexible and diverse. DBS, local optimization, and global optimization algorithms represent different algorithmic strategies. Selecting an appropriate method according to specific device design requirements is essential for achieving efficient design.

11.3 DEEP LEARNING (NEURAL NETWORKS)

In 1943, McCulloch et al. first proposed the MCP neuron model, describing neuron behaviour mathematically [15]. The beginning of deep learning can be traced to 1986, when Rumelhart et al. introduced backpropagation for neural networks [16]. This method repeatedly adjusts connection weights in the network and is significantly faster than earlier learning approaches, marking the emergence of multilayer perceptron (MLP) structures. With the continuous development of computational technology, machine learning has been applied to photonic device design. Deep learning, as an important branch of machine learning, has also been used in the design of polarization control devices.

Traditional machine learning techniques have limited capability in handling raw natural data. In contrast, deep learning is a representation learning method with multiple levels of abstraction. It constructs complex functions by combining simple but nonlinear modules. Each module transforms one level of representation (starting from raw input) into a higher and more abstract representation. By stacking sufficient such transformations, highly complex functions can be learned. As a data-driven inverse design method, deep learning differs from optimization-based approaches that rely on electromagnetic simulations to explore design spaces. Deep learning extracts features from datasets and generalizes within the given design space, enabling rapid acquisition of high-performance device structures without performing numerical simulations one by one.

In deep learning-based polarization device design, two main types of models are used: discriminative models, which map complex relationships between input and output variables. Typical architectures include Fully Connected Networks (FCN) and Convolutional Neural Networks (CNN). Generative models learn from an initial dataset to generate new data. These are particularly suitable when datasets are limited. Typical architectures include Variational Autoencoders (VAE) and Generative Adversarial Networks (GAN).

In deep learning-based design, model training is the most critical stage. The core component of the model is the neural network, which operates through forward propagation and backpropagation. The process from the input layer to the output layer is called forward propagation, while backpropagation updates all connection weights and biases. Datasets are indispensable in deep learning and are typically divided into training, validation, and testing sets. Although the proportions vary depending on the training strategy, the training set usually accounts for more than 60% of the total dataset.

11.3.1 Discriminative Models

The neuron is the fundamental unit of a neural network. As shown in Fig. 11.7, the forward propagation process of a single-layer neural network is equivalent to linear regression: each input x is assigned a weight, producing an output vector z. To enable the network to handle complex problems, activation functions [17] are introduced to process the linearly transformed values and provide nonlinearity. Linear transformation (weighted sum and bias): $z = wx + b$; nonlinear transformation (three type of activation functions): $\delta(x) = max(0, z)$ (ReLu type), $\delta(x) = tanh(z)$ (Tanh type), and $\delta(x) = 1/(1 + e^{-z})$ (Sigmoid type). Here w is the weight, b is the bias, x is the input value, z is the linear output, and δ is the nonlinear output.

Moreover, it is required for the model to define a loss function to optimize weights w and bias b. The choice of loss function depends on the specific problem. Two common loss functions are Mean squared loss and Logarithmic loss. Mean squared loss is expressed as $L(\hat{y}, y) = (\hat{y} - y)^2/2$, and Logarithmic loss is expressed as $L(\hat{y}, y) = -[y \log \hat{y} + (1-y) \log(1-\hat{y})]$, where $\hat{y}$ is the predicted value and y is the true value. The iterative update of

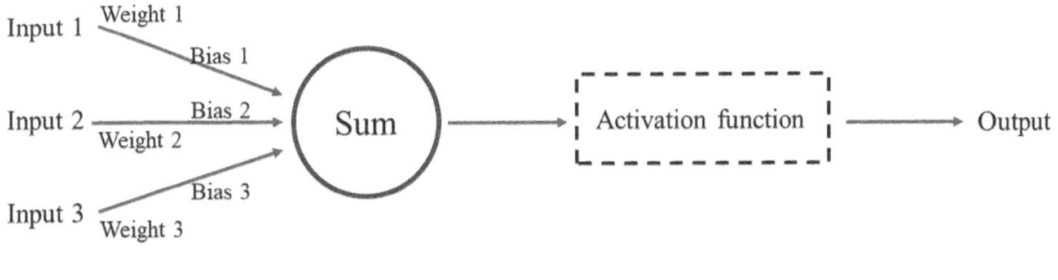

FIGURE 11.7 Forward propagation process of a single-layer neural network.

neural networks is performed using gradient descent, which is an idea of correcting errors with errors. An iterative update is expressed as:

$$w = w - \alpha \frac{\mathrm{d}L(w)}{\mathrm{d}w}, \tag{11.21}$$

where α and w denote learning rate and weights, respectively. The core principle is the chain rule:

$$\frac{\mathrm{d}L(a,y)}{\mathrm{d}w} = \frac{\mathrm{d}L(a,y)}{\mathrm{d}a} \cdot \frac{\mathrm{d}a}{\mathrm{d}z} \cdot \frac{\mathrm{d}z}{\mathrm{d}w}. \tag{11.22}$$

***Fully Connected Neural Network* (*FCN*).** As the most fundamental discriminative neural network architecture in deep learning, a FCN with L layers consists of one input layer, one output layer, and L-2 hidden layers. Each node represents a neuron. Neurons in adjacent layers are fully connected, while neurons within the same layer are independent of each other. The propagation principle is similar to that of a single-layer neural network. As the number of layers increases, the capability to handle complex problems improves, enabling multi-input and multi-output tasks. As shown in Fig. 11.8, the forward propagation process of an FCN can be expressed as follows:

$$a_j^m = \sigma\left(\sum_k w_{jk}^m a_k^{m-1} + b_j^m\right), \tag{11.23}$$

where σ is the activation function, a_j^m denotes the value of the jth neuron in the mth layer, b_j^m denotes the bias term, and w_{jk}^m denotes the connection weight between neurons in adjacent layers. When handling complex problems, activation functions must be introduced. Stacking linear transformations alone cannot represent complex nonlinear mappings and may lead to dead regions in the network. The backpropagation process uses the previously

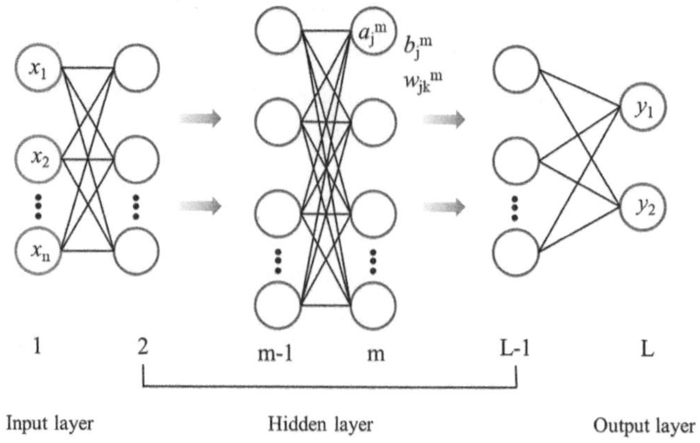

FIGURE 11.8 Forward propagation process of a fully connected neural network.

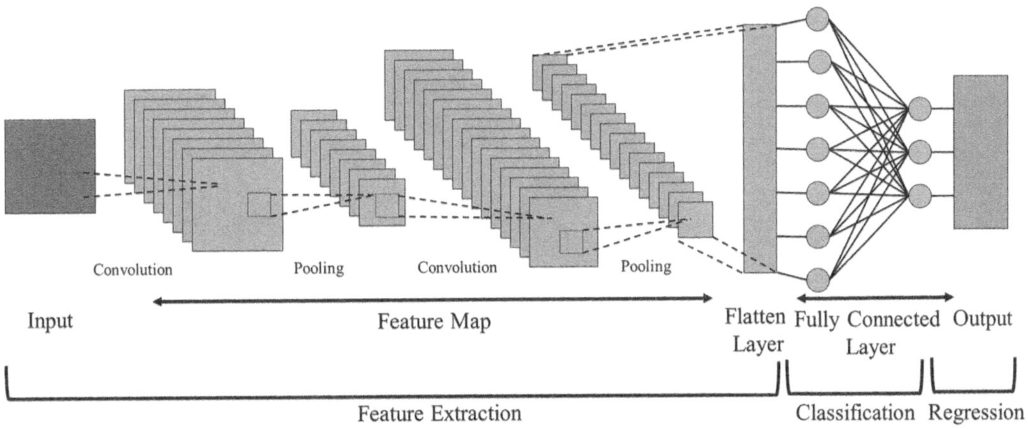

FIGURE 11.9 Schematic diagram of CNN model.

defined loss function and gradient descent method to update weights and biases iteratively until the model converges.

***Convolutional Neural Network* (CNN).** In inverse design of polarization control devices, the design region is typically represented as a two-dimensional pixel matrix (e.g. an $N \times N$ binary structure). Although FCN provides an effective design approach, high-degree-of-freedom devices require a large number of weights and biases: $z = wx + b$. This leads to excessive parameters, low training efficiency, high computational cost, and potential overfitting. Moreover, two-dimensional or three-dimensional device structures must be flattened into one-dimensional vectors to be compatible with FCN, which destroys spatial correlations, especially along the vertical direction.

CNN architectures effectively address this issue. As shown in Fig. 11.9, a CNN also consists of an input layer, one or more hidden layers, and an output layer. However, unlike conventional neural networks, CNN data are arranged in three dimensions: width, height, and depth. CNN performs propagation using local convolution kernels:

$$Y_{i,j} = \sum_{m,n} K_{m,n} X_{i+m,j+n}, \tag{11.24}$$

where K is the convolution kernel, X is the input feature map, and Y is the output feature map. In FCN, each spatial position has independent weights. In CNN, the same convolution kernel K is shared across all spatial locations. CNN also incorporates pooling layers, which progressively reduce the spatial size of the data. This reduces the number of parameters in the network and significantly decreases computational cost.

11.3.2 Generative Models

In deep learning-based inverse design of polarization control devices, a pretrained forward model is typically required as a surrogate solver. The accuracy of the forward model directly affects the final device performance. For discriminative models to accurately predict structural parameters, large datasets are required to construct a forward model. In

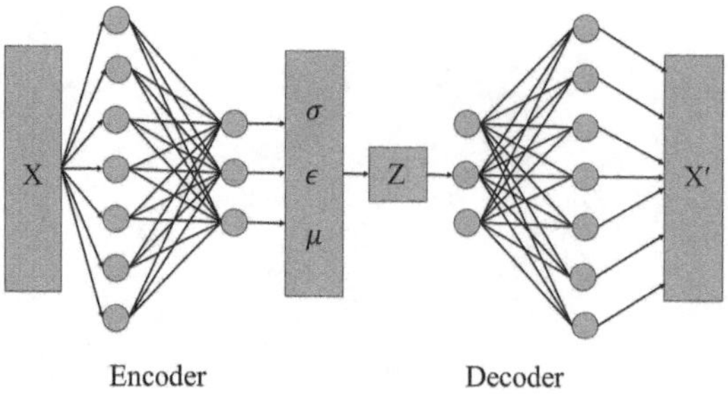

FIGURE 11.10 Schematic diagram of VAE model.

complex polarization device design, the large data volume and extensive search space significantly limit the applicability of discriminative models. Generative models, however, learn the latent distribution of input data and generate new samples similar to the training data, thereby alleviating the burden of dataset collection. Moreover, generative models effectively address the one-to-many mapping problem in inverse design.

***Variational Autoencoder* (*VAE*).** VAE is a generative model derived from the autoencoder framework. The basic idea of an autoencoder is to encode real samples into a compressed representation and then reconstruct them through a decoder. When reconstructed samples closely match real samples, the autoencoder is successfully trained. VAE introduces variational inference into the autoencoder framework. Its architecture is shown in Fig. 11.10. Given input sample X, the encoder transforms high-dimensional data into a low-dimensional latent vector capturing the main features of the dataset. Through regularization, the latent space outputs a mean μ and standard deviation σ, describing a probability distribution. Using the reparameterization technique, the latent vector z is generated as $z = \mu + \sigma \odot \varsigma$, where ς follows a Gaussian distribution. The decoder then maps the latent vector back to the original space to generate new data. To maximize the Evidence Lower Bound (ELBO) while ensuring structured and continuous latent space representation, the VAE loss consists of reconstruction loss and Kullback-Leibler (KL) divergence:

$$L = E_{q(z|x)}\left[\log(x|z)\right] - \beta \cdot KL\left(q(z|x) \| p(z)\right). \tag{11.25}$$

The reconstruction term ensures that the decoder reconstructs data close to the original input. The KL divergence forces the latent variable distribution to approach a normal distribution. The parameter β controls the weight of the KL divergence.

***Generative Adversarial Network* (*GAN*).** Another generative model is the GAN, which consists of two components: a generator and a discriminator. The generator attempts to produce fake data that closely resemble real data. The discriminator distinguishes whether the input data are real or generated. The two networks compete adversarially. Through this competition, highly realistic samples are generated. The architecture is shown in Fig. 11.11. The discriminator performs binary classification training to optimize its ability to distinguish real and generated data. The generator updates its parameters based on feedback from the discriminator, attempting to generate more realistic samples to deceive the discriminator.

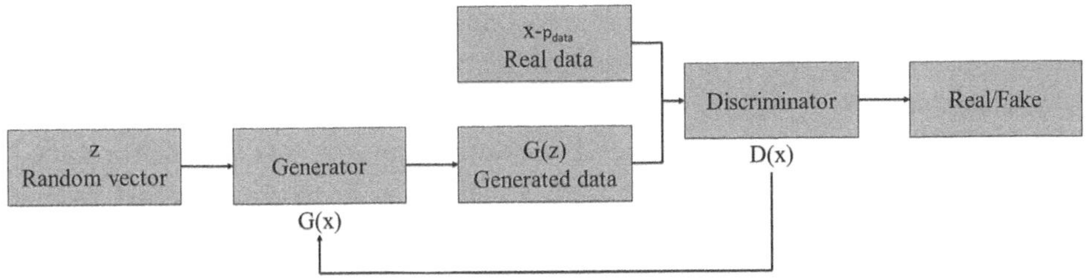

FIGURE 11.11 Schematic diagram of GAN model.

Training alternates between optimizing the discriminator and the generator until the discriminator can no longer distinguish real and generated data, or a predefined number of iterations is reached. The discriminator loss function is:

$$L_D = -E_{x \sim P_{data}}\left[\log D(x)\right] - E_{z \sim P_z}\left[\log\left(1 - D\left(G(z)\right)\right)\right], \tag{11.26}$$

where $E_{x \sim P_{data}}$ is the expected value of the real data sample x, $E_{z \sim P_z}$ is the expected value of the random vector sample z, and $D(\cdot)$ is the discriminator function. The discriminator aims to maximize the correct classification probability. The generator loss is:

$$L_G = -E_{z \sim P_z}\left[\log D\left(G(z)\right)\right]. \tag{11.27}$$

The generator attempts to minimize the probability that generated samples are correctly classified as fake.

11.4 POLARIZATION CONTROL DEVICE DESIGN BASED ON DEEP LEARNING

11.4.1 Discriminative Models

The most fundamental neural network architecture is the fully connected neural network (also known as the multilayer perceptron, MLP). In 2020, Ren et al. implemented pixel-level dynamic control of three-dimensional vector polarization states based on a multi-layer artificial neural network (MANN) [18]. The MANN established a mapping between a "2D vector field on the hologram plane" and a "3D vector field on the imaging plane". The model architecture consisted of: an input layer with a six-element array containing the amplitudes and phases of three orthogonal components of the target 3D vector field; four hidden layers, each containing 1000 neurons; an output layer with a six-element array containing the amplitudes and phases of the azimuthal and radial spatial components on the hologram plane.

In inverse design of polarization control devices, inverse models often encounter a one-to-many mapping problem. As mentioned previously, for discriminative models, one target parameter ideally corresponds to one structural parameter; otherwise, the inverse design result may fail to converge. Currently, two common solutions are adopted: dimensionality reduction methods, which transform data from high-dimensional space into low-dimensional space to establish a one-to-one mapping [19,20]. Cascaded networks, which address mapping non-uniqueness through network concatenation [21].

In cascaded networks, optimization algorithms are often integrated to accelerate the generation of structural parameters [22]. For example, an inverse network combined with the GS optimization algorithm [23] generates meta-unit structural parameters based on target intensity and phase distributions.

Another approach is a bidirectional dual-network stacked architecture composed of a primary network and an auxiliary network. Unlike cascaded networks, the two networks operate in parallel and independently. In 2018, Ma et al. optimized chiral metamaterials using such a bidirectional dual-network stacked neural architecture [24], as shown in Fig. 11.12. The forward combiner corrected the reflection spectrum predicted by the auxiliary network using the circular dichroism (CD) information from the primary network.

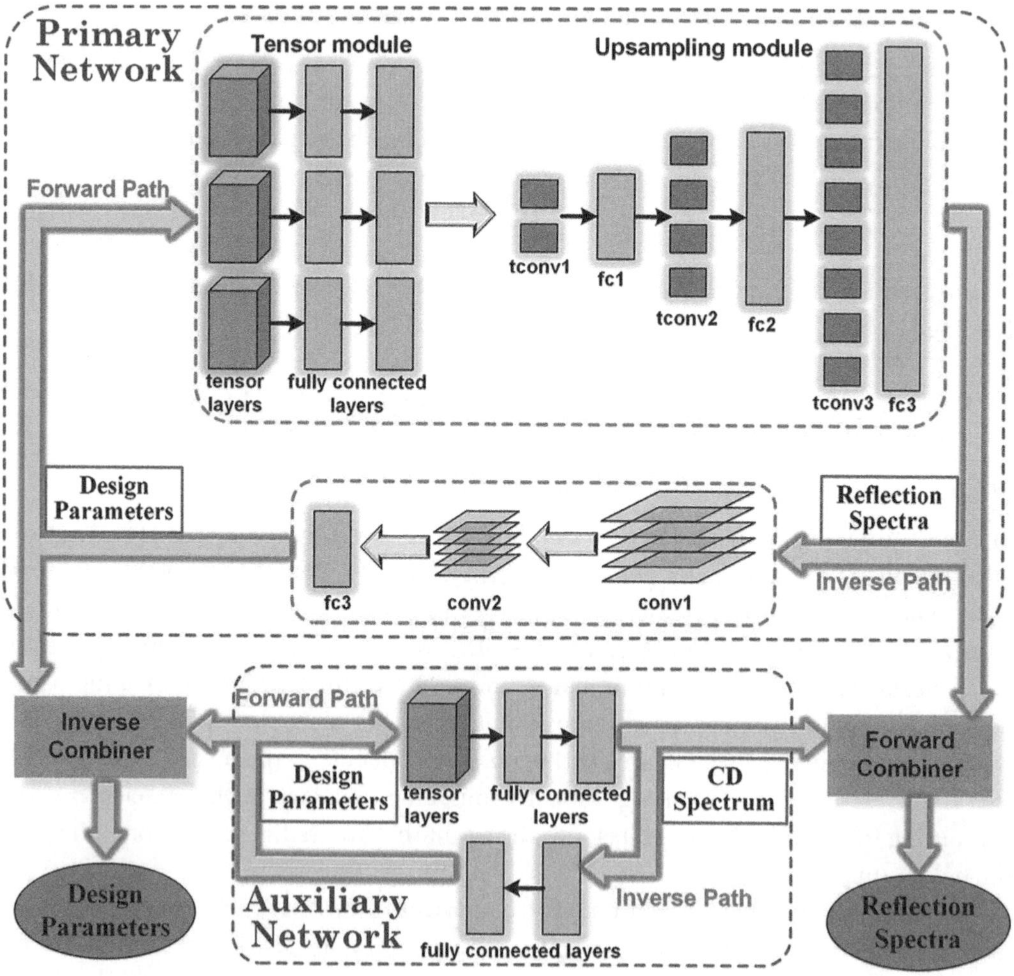

FIGURE 11.12 Schematic diagram of the bidirectional dual-network stacked neural architecture [24].

The inverse combiner performed a weighted summation of design parameters retrieved by both networks to output the final design parameters.

In 2023, Guo et al. proposed a deep learning architecture [25] consisting of two core components: a hologram optimization module and a crosstalk-correction fully convolutional neural network (FCN). The FCN was based on a CNN architecture with a symmetric "downsampling-upsampling" structure. The input and output were both 1024 × 1024 tensors, matching hologram image resolution. This architecture addressed non-ideal crosstalk caused by multi-channel multiplexing. It extracted and corrected features from noisy optical system outputs, improving decryption accuracy. CNNs can also assist in efficient and accurate polarization state detection [26].

11.4.2 Generative Models

The design principles of polarization control devices involve complex polarization state manipulation, resulting in complicated mapping relationships. Consequently, obtaining high-quality design datasets is challenging. Therefore, generative models can effectively reduce dependence on large datasets in deep learning-based polarization device design. In 2019, Ma et al. constructed a VAE-based deep generative probabilistic graphical model [27]. By introducing probabilistic representation in the latent space, they achieved one-to-many mapping solutions. Combined with a semi-supervised learning strategy, data utilization efficiency was improved.

The deep generative model architecture consists of three models: recognition model, prediction model, and generative model. The core inverse design module is the generative model. It takes the preset polarization optical response y and a randomly sampled latent variable z as dual inputs to generate geometric patterns of polarization metamaterials that satisfy the target requirements. In 2022, the same group proposed an end-to-end design architecture integrating statistical machine learning [28]. The framework consists of a deterministic prediction model (forward model) and a stochastic generative inverse model. The deterministic prediction model adopted a hybrid CNN-RNN architecture. The stochastic generative inverse model was based on a VAE architecture, randomly generating diversified nine-dimensional geometric design parameters of meta-atoms to form a meta-atom candidate pool.

Based on VAE, variants such as CAVE [29] and A-CAVE [30] were proposed. These addressed the issue that conventional VAE models tend to directly encode target optical performance into latent variables during training, leading to overfitting of conditional information. VAE is fundamentally based on probabilistic Bayesian learning and aims to fit the probability density function of the data distribution. However, its fitting capability is weak for nanostructures without shared common features.

In 2021, Zhu et al. proposed a hybrid deep model architecture [31], as shown in Fig. 11.13. The framework consists of a neural network generator, a neural network simulator, and an evolutionary strategy (ES) optimizer. The neural generator is based on a GAN-driven compositional pattern-producing network (CPPN). CPPN does not require common nanostructural features and can generate arbitrarily complex two-dimensional nanostructures suitable for metasurfaces, enabling high-performance polarization control device design.

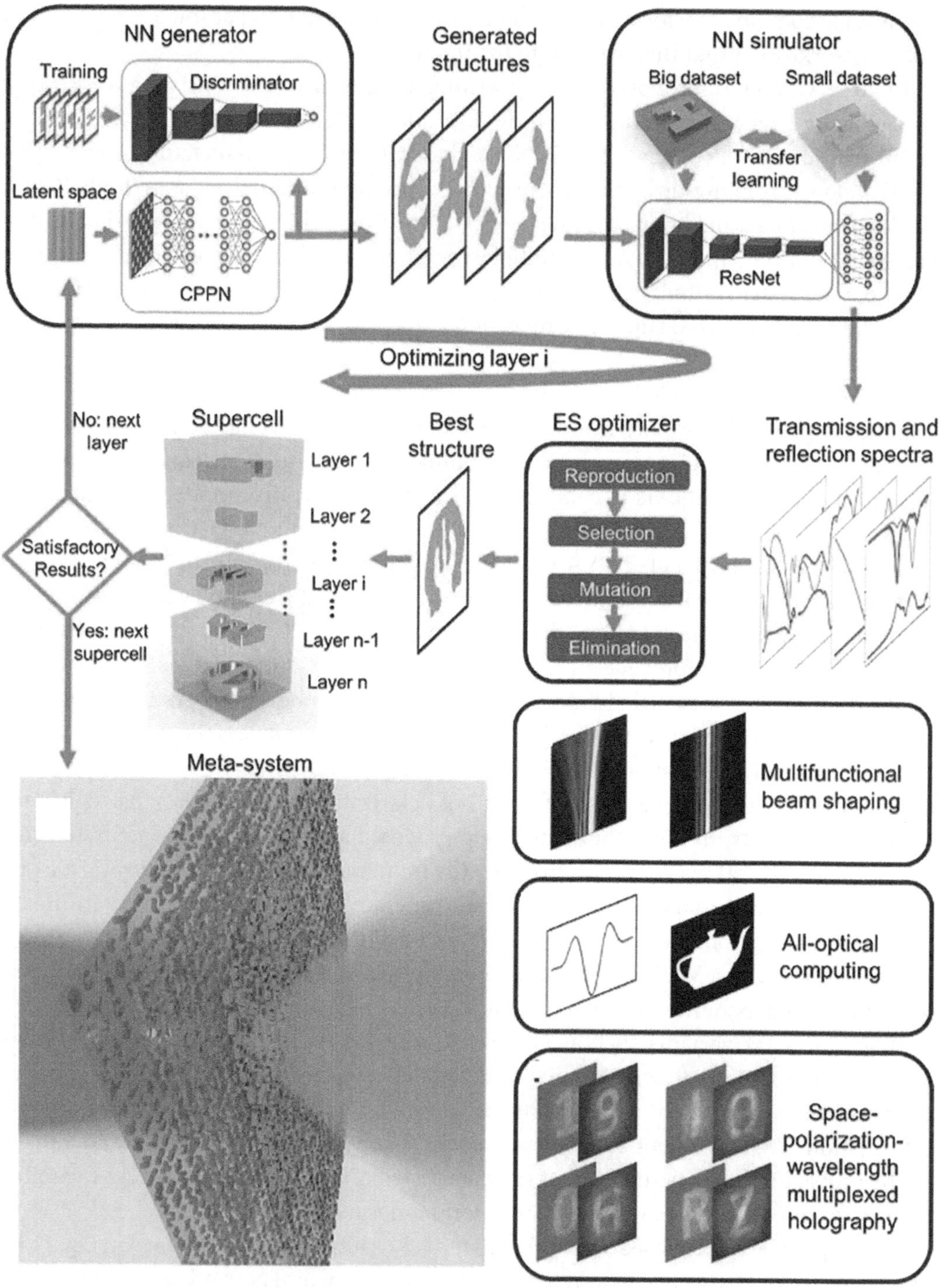

FIGURE 11.13 GAN-based hybrid deep model architecture [31].

11.5 CHAPTER SUMMARY

Inverse design methods have gradually become an important category of design approaches for polarization control devices. They enable the development of higher-performance devices within shorter time frames and demonstrate significant potential. This chapter introduced two main categories of inverse design methods: optimization algorithms and deep learning approaches. Representative design examples were also presented. The choice of inverse design method depends on the structural complexity and performance requirements of specific devices: local optimization algorithms are suitable for high-complexity devices with a given initial structure. Global optimization algorithms are suitable for devices without a clear initial structure and with strong sensitivity to local optima. Deep learning approaches are more appropriate for standardized devices requiring rapid iteration and large-scale production. It should be emphasized that this chapter only analyses inverse design methods from selected perspectives. In practical design processes, many additional factors must be considered. Readers are encouraged to consult relevant literature for more comprehensive discussions.

REFERENCES

1. Chen, C. T., and Gu, G. X. (2020) 'Generative deep neural networks for inverse materials design using backpropagation and active learning', Advanced Science, 7(5), 1902607.
2. Lalau-Keraly, C. M., Bhargava, S., Miller, O. D., et al. (2013) 'Adjoint shape optimization applied to electromagnetic design', Optics express, 21(18), 21693–21701.
3. Lin, X., Wei, M., Lei, K., et al. (2023) 'Compact mid-infrared chalcogenide glass photonic devices based on robust-inverse design', Laser & Photonics Reviews, 17(2), 2200445.
4. Phan, T., Sell, D., Wang, E. W., et al. (2019) 'High-efficiency, large-area, topology-optimized metasurfaces', Light: Science & Applications, 8(1), 48.
5. Han, Z., Zheng, Y., Xu, M., et al. (2025) 'High-efficiency polarization beam merging metagrating design via global initial solution-based topology optimization', Optics Letters, 50(5), 1637–1640.
6. Shen, B., Wang, P., Polson, R., et al. (2015) 'An integrated-nanophotonics polarization beamsplitter with 2.4 × 2.4 μm^2 footprint', Nature Photonics, 9(6), 378–382.
7. Yuan, Z., Wang, Y., Fan, H., et al. (2025) 'Multichannel parallel mode order converter for on-chip reconfigurable PDM-MDM transmission', Laser & Photonics Reviews, 19(9), 2401689.
8. Lebbe, N., Glière, A., and Hassan, K. (2019) 'High-efficiency and broadband photonic polarization rotator based on multilevel shape optimization', Optics Letters, 44(8), 1960–1963.
9. Xu, M., He, Q., Pu, M., et al. (2022) 'Emerging long-range order from a freeform disordered metasurface', Advanced Materials, 34(12), 2108709.
10. Shen, R., Hu, F., Hong, B., et al. (2024) '100 nm broadband and ultra-compact multi-dimensional multiplexed photonic integrated circuit for high-capacity optical interconnects', Photonics Research, 12(12), 2891–2900.
11. Wang, J., Yermakov, O., Gu, Y., et al. (2025) 'Fiber tip empowered by axial-asymmetric four-sector nanostructure for broad-angle light coupling', Laser & Photonics Reviews, 19(5), 2400688.
12. Lu, C., Liu, Z., Wu, Y., et al. (2020) 'Nanophotonic polarization routers based on an intelligent algorithm', Advanced Optical Materials, 8(10), 1902018.
13. Wang, X., Yu, H., Huang, Q., et al. (2021) 'Polarization-independent fiber-chip grating couplers optimized by the adaptive genetic algorithm', Optics Letters, 46(2), 314–317.
14. Ou, X., Zeng, T., Zhang, Y., et al. (2022). 'Tunable polarization-multiplexed achromatic dielectric metalens', Nano Letters, 22(24), 10049–10056.

15. McCulloch, W. S., Pitts W. (1943) 'A logical calculus of the ideas immanent in nervous activity', Bulletin of Mathematical Biology, 5, 115–133.
16. Rumelhart, D. E., Hinton, G. E., and Williams, R. J. (1986) 'Learning representations by back-propagating errors', Nature, 323(6088), 533–536.
17. Han, Z., Yu, S., Lin, S. B., et al. (2020) 'Depth selection for deep ReLU nets in feature extraction and generalization', IEEE Transactions on Pattern Analysis and Machine Intelligence, 44(4), 1853–1868.
18. Ren, H., Shao, W., Li, Y., et al. (2020) 'Three-dimensional vectorial holography based on machine learning inverse design', Science advances, 6(16), eaaz4261.
19. Kiarashinejad, Y., Abdollahramezani, S., and Adibi, A. (2020) 'Deep learning approach based on dimensionality reduction for designing electromagnetic nanostructures', npj Computational Materials, 6(1), 12.
20. Melati, D., Grinberg, Y., Kamandar Dezfouli, M., et al. (2019) 'Mapping the global design space of nanophotonic components using machine learning pattern recognition', Nature communications, 10(1), 4775.
21. Tanriover, I., Lee, D., Chen, W., et al. (2022) 'Deep generative modeling and inverse design of manufacturable free-form dielectric metasurfaces', ACS Photonics, 10(4), 875–883.
22. Wang, C., Cheng, X., Wang, R., et al. (2024) 'Flexibly designable 2D chiral metasurfaces with pixelated topological structure based on machine learning', Laser & Photonics Reviews, 18(7), 2300958.
23. Lei, Z. D., Xu, Y. D., Lei, C., et al. (2023) 'Dynamic multifunctional metasurfaces: an inverse design deep learning approach', Photonics Research, 12(1), 123–133.
24. Ma, W., Cheng, F., and Liu, Y. (2018) 'Deep-learning-enabled on-demand design of chiral metamaterials', ACS Nano, 12(6), 6326–6334.
25. Guo, J., Zhang, Y., Ye, H., et al. (2023) 'Spatially structured-mode multiplexing holography for high-capacity security encryption', ACS Photonics, 10(3), 757–763.
26. Chen, C., Xiao, X., Ye, X., et al. (2023) 'Neural network assisted high-spatial-resolution polarimetry with non-interleaved chiral metasurfaces', Light: Science & Applications, 12(1), 288.
27. Ma, W., Cheng, F., Xu, Y., et al. (2019) 'Probabilistic representation and inverse design of metamaterials based on a deep generative model with semi-supervised learning strategy', Advanced Materials, 31(35), 1901111.
28. Ma, W., Xu, Y., Xiong, B., et al. (2022) 'Pushing the limits of functionality-multiplexing capability in metasurface design based on statistical machine learning', Advanced Materials, 34(16), 2110022.
29. Yang, Y., Zhang, X., Liu, K., et al. (2023) 'Exploring the limits of metasurface polarization multiplexing capability based on deep learning', Optics Express, 31(10), 17065–17075.
30. Tang, Y., Kojima, K., Koike-Akino, T., et al. (2020) 'Generative deep learning model for inverse design of integrated nanophotonic devices', Laser & Photonics Reviews, 14(12), 2000287.
31. Zhu, D., Liu, Z., Raju, L., et al. (2021) 'Building multifunctional metasystems via algorithmic construction', ACS Nano, 15(2), 2318–2326.

Index

Note: Page numbers followed by f, and t indicate figures, and tables, respectively.

For Product Safety Concerns and Information please contact our EU representative GPSR@taylorandfrancis.com Taylor & Francis Verlag GmbH, Kaufingerstraße 24, 80331 München, Germany

Batch number: 10784201

Printed by Printforce, the Netherlands